Handbook on Mechanical Properties of Rocks

Other volumes in the
Series on Rock and Soil Mechanics

W. Reisner, M. v. Eisenhart Rothe:
Bins and Bunkers for Handling Bulk Materials
– Practical Design and Techniques –
1971

W. Dreyer:
The Science of Rock Mechanics
Part I Strength Properties of Rocks
1972

T. H. Hanna:
Foundation Instrumentation
1973

C. E. Gregory:
Explosives for North American Engineers
1973

M. & A. Reimbert:
Retaining Walls Vol. I
– Anchorages and Sheet Piling –
1974

Vutukuri, Lama, Saluja:
Handbook on Mechanical Properties
of Rocks Vol. I
1974

M. & A. Reimbert:
Retaining Walls Vol. II
– Study of Passive Resistance in Foundation Structures –
1976

H. R. Hardy, Jr. & F. W. Leighton:
First Conference on Acoustic Emission/Microseismic Activity
in Geologic Structures and Materials
1977

L. L. Karafiath & E. A. Nowatzki:
Soil Mechanics for Off-Road Vehicle Engineering
1978

Baguelin, Jézéquel, Shields:
The Pressuremeter and Foundation
Engineering
1978

Editor-in-Chief
Professor Dr. H. Wöhlbier

Serieson Rock and Soil Mechanics
Vol. 3 (1978) No. 1

HANDBOOK
ON
MECHANICAL PROPERTIES
OF ROCKS
– Testing Techniques and Results –
Volume II

by

R. D. Lama

CSIRO
Division of Applied Geomechanics
Australia

V. S. Vutukuri

Department of Mining Engineering
Broken Hill Division
University of New South Wales
Australia

First Printing
1978

TRANS TECH PUBLICATIONS

Distributed by
TRANS TECH S. A.
CH-4711 Aedermannsdorf, Switzerland

Copyright © 1978 by
Trans Tech Publications
Clausthal, Germany

International Standard Book Number

ISBN 0-87849-021-3

Printed in Germany
by Druckerei E. Jungfer, Herzberg

D
624.1513
LAM

FOREWORD

Rock mechanics is devoted not only to the prevention of rock failure, thus ensuring the stability of mining and civil engineering excavations, but also to the promotion of rock failure in applications involving caving in mines and rock comminution processes such as drilling, blasting, crushing and grinding. It is obvious that as a result of these applications, rock mechanics is concerned with both dynamic and static phenomena.

While the static phenomena in the field of rock mechanics have always received considerable attention, the dynamic aspects have been given particular prominence recently due to underground and surface structures designed in earthquake-prone areas. In addition, ground vibrations due to heavy machinery are of particular importance, in large plants as environmental controls become more stringent. Damage caused by blasting in deep open pit mines has to be given careful consideration due to the problems of stability of high slopes and associated environmental factors.

Encompassing static and dynamic phenomena in rock mechanics requires that the rock engineer must be familiar with both the static and dynamic properties of rock materials. In this respect, the authors of this volume have undertaken a very useful task by bringing together information which will be pertinent when embarking upon a research project or a design undertaking. This second volume, which is a part of the four-volume "Handbook on Mechanical Properties of Rocks", is an important contribution in the field of rock mechanics. Devoted to the static and dynamic constants of rock, the volume compiles data scattered in many papers and books published around the world. The test methods described in this volume and the results listed constitute a valuable reference source saving the user much time and effort.

Prof. Z. T. BIENIAWSKI, D.Sc (Eng)
Professor of Mineral Engineering
Pennsylvania State University
University Park, Pennsylvania
February, 1978   U.S.A.

PREFACE to
Volumes II, III and IV

The knowledge of mechanical properties of rocks is essential in any rock mechanics investigations connected either with mining, tunnelling, drilling, blasting, cutting or crushing. After predicting the state of stress, strain or stored energy from the analysis of loads or forces being applied to the rocks, the behaviour, i.e., fracture, flow, or simply deformation of the rock can be estimated from these mechanical properties.

The mechanical properties of a rock depend primarily on its mineral composition and constitution, i.e., its structural and textural features. They also depend upon the condition it is in when tested (e.g., temperature, water content).

In reviewing the factors which influence the mechanical properties, it is helpful to work progressively from the scale of the single mineral to that of the rock mass. The properties of a single mineral are a function of its chemical composition, lattice structure (which determines glide systems), and lattice defects such as vacancies and dislocations. They also depend on its orientation relative to the applied stress field and on the mode of load application.

In bulk specimens of intact rock the mechanical properties depend not only on the properties of the individual minerals, but also upon the way in which the minerals are assembled. The relevant information is given by a full petrographic description, which includes the mineral composition of crystals, grains, cementing materials and alteration products and also the structure and texture, including size, shape, distribution and orientation of crystals, grains, pores and cracks. The degree of isotropy or anisotropy is also important and varies with the size of the body of rock under consideration. For example, in schist, gneiss, and other foliated rocks, the constitutive properties vary with direction even at the microscopic scale, and to the extent that the mechanical properties even of a small specimen are affected. However, in sedimentary rocks, which are generally laminated, the rock within a lamina may be relatively isotropic, whereas, at a scale that includes the separation between lamina, the same rock may be relatively anisotropic. On the other hand, other rocks may be strongly anisotropic even within very thin sheets. Primary anisotropy, brought about by preferential orientation during crystallisation, or by recrystallisation during sedimentation or metamorphic processes, may be distinguished from secondary anisotropy, brought about by geologic deformation of the rock. The rock mass contains planes of weakness that affect its mechanical properties, making it mechanically anisotropic.* These planes of weakness may be joints, faults, fractures, partings between beds, or in bedded or laminated rock, layers of lower strength rock. The mechanical properties of a rock mass depend upon the following factors:

1. The mechanical properties of the individual elements constituting the system.
2. The sliding characteristics of the planes of weakness.
3. The configuration of the system with respect to the directions of loading.
4. The operating stress field.

The size or scale of the rock body that is being considered is an important factor in deciding the testing program for the determination of the mechanical properties of rock. From an engineering standpoint, if bodies of rock are being considered at a macroscopic scale, planes of weakness generally are not a point of concern. Operations such as crushing, grinding, and drilling occur at this scale. However, at a megascopic scale, e.g., in the excavation of large underground openings, in large open pit mines, planes of weakness generally dominate the failure process. This conclusion is borne out by the observation that the larger part of fracture surfaces created by underground or slope failures occurs on planes of weakness rather than through fresh rock.

Virtually all data on the mechanical properties of rock have been obtained from tests at a macroscopic scale, i.e., on specimens of rock, hand size or larger, but specifically below a size that would include planes of weakness of geological origin, e.g., joints, bedding planes, etc.

As yet, data are meager on the mechanical properties of large bodies of rock, i.e., from tests at a megascopic scale. The requirement for this information has been realised for some time, but the problems of preparing in situ test specimens and developing equipment for applying loads of the required magnitude have been slow in materialising. Several investigators have studied the effects of mechanical defects such as joints and bedding by testing specimens containing either real or simulated planes of weakness in the laboratory.

Various testing techniques (both laboratory and in situ) have been developed for the determination of mechanical properties and the results obtained are often dependent on many factors.

---

* This anisotropy is in addition to that due to mineral fabric (schistosity, etc.) that may be present both on macroscopic and megascopic scales.

Probably because of the relative importance of this subject, a large amount
of literature is available from a great number of papers scattered over many
engineering and scientific periodicals and texts, and the proceedings of
several conferences and symposia on the subject of rock mechanics, rock
pressure, mining, etc.

The purpose of this Handbook is to present a detailed treatment of the
subject from the widely scattered literature in a simple, clear and logical
form.

The authors apologise for the delay in bringing out the Volumes II, III and
IV of the Handbook. In the light of valuable reviews published on Volume I
of this Handbook and the suggestions of several colleagues, the subject
matter of the remaining chapters has been greatly expanded, which shall
be clear when compared with the Table of Contents for Volume II as
published in Volume I. A number of appendices have been included which
were not intended earlier. Appendix II dealing with Laboratory Mechanical
Properties of Rocks contains results of tests on over 2000 rock types from
different locations around the world. Appendix III contains results of in
situ tests on over 300 different projects.

As a result of greatly extended subject matter and still keeping the size of
each volume handy, it was essential to divide the remaining material into
three volumes. The distribution of the material in four volumes is as follows:

Volume I – Chapter 1 – Specimen Preparation for Laboratory Tests;
Chapter 2 – Compressive Strength of Rock; Chapter 3 – Tensile Strength
of Rock; Chapter 4 – Shear Strength of Rock; Chapter 5 – Strength of Rock
under Triaxial and Biaxial Stresses; and Appendix I – Stiff Testing
Machines.

Volume II – Chapter 6 – Static Elastic Constants of Rock; Chapter 7 –
Dynamic Elastic Constants of Rock; and Appendix II – Laboratory Mechanical Properties of Rocks.

Volume III – Chapter 8 – Large Scale Testing of Rock; Chapter 9 – Time-
Dependent Properties of Rock; Appendix III – In Situ Mechanical Properties of Rocks; and Appendix IV – Crack Propagation Velocity in Rocks.

Volume IV – Chapter 10 – Mechanical Behaviour of Jointed Rock; Chapter
11 – Classification of Rock; Chapter 12 – Miscellaneous Properties of
Rock; Appendix V – Stereographic Projections; Appendix VI – Definitions
of Some Rock Mechanics Terms; and Appendix VII – Imperial, Metric and
S.I. Units.

The authors apologise for this change.

It is hoped that this Handbook will serve students, designers, practising
engineers, members of teaching staff and research workers in equal measure
in the fields of mining engineering, civil engineering, geological engineering
and petroleum engineering.

ACKNOWLEDGEMENTS

The authors gratefully acknowledge the inspiration given by the many authors from whose publications materials have been liberally drawn for the preparation of this Handbook.

Permission to reproduce figures from their publications was kindly given by various publishers and authors. The authors express their thanks to them. The authors offer their most grateful thanks to Professor L. *Müller* – Salzburg, formerly Head of the Rock Mechanics Division, University of Karlsruhe, for his continued inspiration; Dr. Z. T. *Bieniawski,* Professor of Mineral Engineering, Pennsylvania State University (formerly Head of the Geomechanics Division, National Mechanical Engineering Research Institute, S.A.C.S.I.R.); Professor L. A. *Endersbee,* Dean of the Faculty of Engineering, Monash University, and Dr. E. *Hoek,* Principal, Golder Associates Ltd. (formerly Professor of Rock Mechanics, Imperial College of University of London) for going through Volumes II, III and IV once again and writing forewords respectively.

The authors wish to express their sincere gratitude to Dr. G. D. *Aitchison,* and Dr. C. M. *Gerrard,* Chief and Acting Chief respectively, of the Division of Applied Geomechanics, C.S.I.R.O., and Professor J. E. *Anderson,* Director of the Broken Hill Division of University of New South Wales, for continuous support in the preparation of these volumes.

The authors are deeply indebted to the following specialists for reviewing the manuscript (either in whole or in part) and offering many valuable comments, suggestions and positive criticisms:

1. Dr. Z. T. *Bieniawski,* Professor of Mineral Engineering, Pennsylvania State University, U.S.A. (formerly Head, Geomechanics Division, National Mechanical Engineering Research Institute, South African Council for Scientific and Industrial Research, Pretoria, South Africa).
2. Professor L. A. *Endersbee,* Dean of Faculty of Engineering, Monash University, Melbourne, Australia.
3. Dr. E. *Hoek,* Principal, Golder Associates Ltd., Canada (formerly Professor of Rock Mechanics, Imperial College, London).
4. Professor S. *Budavari,* Professor of Rock Mechanics, University of Witwatersrand, South Africa.

5. Mr. P. G. *Chamberlain*, Head, Property Determination Research Support, Fundamentals of Rock Failure, Twin Cities Mining Research Centre, U.S. Bureau of Mines, U.S.A.
6. Mr. L. G. *Alexander,* Principal Research Scientist, Mr. K. G. *Grant,* Principal Research Scientist, Dr. C. M. *Barton,* Principal Research Scientist, Dr. G. P. *Price,* Research Scientist and Dr. P. G. *Fuller,* Senior Research Scientist, all of the Division of Applied Geomechanics, C.S.I.R.O., Australia.

The authors also wish to thank Professor A. H. *Willis,* Pro-Vice-Chancellor, Professor I. K. *Lee,* Head of School of Civil Engineering and Professor F. C. *Beavis,* Head of School of Applied Geology of University of New South Wales, Professor E. L. J. *Potts,* Head of Department of Mining Engineering of University of Newcastle-upon-Tyne and Professor H. R. *Hardy,* College of Earth and Mineral Sciences of the Pennsylvania State University for their encouragement in this work.

The authors are indebted to many colleagues in the preparation of the manuscript. Among the many others, the authors wish to mention especially the following:

Mr. J. D. *Dover,* Miss M. *Pert,* Miss J. *Elliot,* Mrs. J. *Gross* and Mrs. H. *Topp* of the Division of Applied Geomechanics, C.S.I.R.O., Mr. P. *Hern,* Mr. P. *Longrigg,* Mr. K. *Murray,* Mr. *Ray Ellice,* Mr. J. *Haddon,* Mesdames S. A. *Nejaim,* L. R. *Tweedie,* B. M. *Lean* and A. F. *Grossole* of the Broken Hill Division of the University of New South Wales and Mrs. F. *Bell,* Mrs. L. *Howe,* Mrs. Y. *Rouse,* Mr. L. *Separovich,* Mr. R. *Devarajan,* Mr. V. R. *Sambandam* and Mrs. M. *Browne.*

The authors wish to express their deep appreciation and their profound gratitude to their wives, Mrs. *Barbara Lama* and Mrs. *Venkata K. Vutukuri* for their patience and understanding during the preparation of the manuscript.

CONTENTS

Volume II

CHAPTER 6

Static Elastic Constants of Rock

6.1. Introduction

The deformation constants of a material are the most important parameters in any design and their determination involves the use of measuring techniques both for load and deformation. The amount of deformation that most of the rocks undergo is extremely small and its measurement requires special techniques. This chapter deals with the various instruments that have been successfully used for the measurement of deformation of rock specimens. Methods for determination of various constants under different stress conditions are described followed by a detailed discussion of the influence of various environmental and inherent factors on these parameters especially the Young's modulus and Poisson's ratio. Anisotropy in the modulus values and rock dilatancy in compression have been discussed separately. The elastic properties of different rocks are given at appropriate places with a comprehensive compilation of worldwide data in Appendix II.

6.2. Definitions of Terms

A material is called to be elastic when it recovers its original state after being subjected to a loading-unloading cycle. The relationships between stresses and strains are represented by constants called the *elastic constants* of the material.

The important elastic constants are:
1. Modulus of elasticity
2. Poisson's ratio
3. Modulus of rigidity
4. Bulk modulus
5. Lamé's constant

The total load acting on an area divided by the area is called the stress. The displacement between two measurement points on application of force is called deformation. This deformation divided by the original length is called the strain.

When a test specimen is loaded in either compression or tension within the range of elastic strain, the ratio of the stress to the strain in the direction of the applied stress is called the modulus of elasticity or YOUNG's modulus and is usually designated by the letter E. If the specimen is tested in shear, the ratio of the shear stress to the shear strain is called the shear modulus or modulus of rigidity and is designated by the letter G.

The axial strain in a loaded specimen is accompanied by transverse or lateral strain. In compressive tests, lateral strain is positive (expansion) and in tensile tests, lateral strain is negative (contraction). The ratio of lateral strain to axial strain is known as POISSON's ratio and is designated by the symbol ν.

Bulk modulus is the ratio of hydrostatic pressure to the volumetric strain and is designated by K. Compressibility is reciprocal of bulk modulus.

All these constants are related with each other if assumed that rock under consideration is isotropic, homogeneous and perfectly elastic. Accordingly, the following relationships hold good.

$$
\begin{aligned}
E &= 2\,G\,(1+\nu) \\
&= 3\,K\,(1-2\nu) \\
&= \frac{9\,KG}{(3\,K+G)} \\[4pt]
G &= \frac{E}{2\,(1+\nu)} \\
&= \frac{3\,EK}{(9\,K-E)} \\
&= \frac{3\,K\,(1-2\nu)}{2\,(1+\nu)} \\[4pt]
K &= \frac{E}{3\,(1-2\nu)} \\
&= \frac{EG}{3\,(3\,G-E)} \\
&= \frac{2\,G\,(1+\nu)}{3\,(1-2\nu)} \\[4pt]
\nu &= \frac{(E-2\,G)}{2\,G} \\
&= \frac{(3\,K-E)}{6\,K} \\
&= \frac{(3\,K-2\,G)}{2\,(3\,K+G)}
\end{aligned}
\tag{6.1}
$$

It is seen that for an ideal, isotropic, homogeneous, elastic material, only two constants need to be known, and the others can be easily calculated. The determination of these involves measurement of strains for a known applied stress. The stress may be tensile, compressive or shear or any combination of the two. However, it is always advantageous to subject the specimen to a simple stress field, usually uniaxial compression or uniaxial tension. Due to the complex nature of the rocks, the values of the various constants so determined are different depending upon the stress field and the test conditions.

6.3. Test Requirements

The three requirements of any test are:

1. A properly prepared specimen.
2. A suitable load application arrangement.
3. A suitable deformation measurement device.

The shape of the specimen and the method of preparation depend upon the type of test (compression, simple tension, Brazilian, torsion, etc.) to which it is to be subjected. A detailed description of the methods of preparation of specimens, specifications and the precautions are given in Chapter 1 and at other appropriate places in Volume I of this book where these tests are described.

The load application arrangement should have sufficient capacity and be able to apply load at rates required under specific tests. In general, except when required otherwise for certain special tests, the loading device should be capable of applying load continuously and without shock to produce approximately constant rate of load or deformation such that fracture point is reached after about 5–15 minutes. The loading system should permit taking of at least 20–50 readings at suitable intervals from its initial value to the maximum value attained before the failure of a specimen.

The deformation measuring device should be such that it has a range at least equal to the maximum deformation to be measured and an accuracy of at least $\pm 2\%$ of the value of the reading above 250 μm (0.01 in) and within ± 5 μm (0.0002 in) for readings lower than 250 μm (0.01 in) (A.S.T.M., 1974). Most of the deformation measuring devices available conform to this accuracy and many surpass this limit by 100 or even 1000 times.

The deformation measuring device should preferably be mounted on the specimen between points marked on the specimen and be capable of measuring deformation independent of the movement of the loading platens. The points of measurement should be so selected that the stress gradients between these points are small and that stress value in the area is known quite precisely. For example, in the uniaxial compression and direct tension tests, high stress gradients prevail near the specimen ends and the deformation measuring devices should be so mounted that these are fairly away from the loading platens.

OBERT and DUVALL (1967) showed that, for hardened ground steel platens and aluminium test specimen, radial deformation at the ends is only 20% less than that at the mid-plane. If the specimen h/d ratio is two or more, stress tends to be more uniform at the centre and results will be more reliable if strain measurements are limited to the central one-third the height of the specimen. The error is substantially less than 5% and can be disregarded for practical engineering purposes. Similarly, in the Brazilian test, the lengths should be limited to $1/4$–$1/8$ the diameter of the disc and gauges located at the centre of the specimen where stress is more uniform. Similar considerations should be given to other tests. In specimens of special shape, it is always advisable to study the stress distribution at different points on the specimen surface before selecting the measuring points.

The reliability of any measurement is also dependent upon the length of the measuring base. GUSTKIEWICZ (1975b) has shown that for medium-grained granite, the total base length required for an accuracy of ± 0.03 mm (0.001 in) should be 50 mm (2 in) and if two gauges are used, the minimum length of the gauge should be 30 mm (1.2 in). The base length is dependent upon the average grain size. In general it may be said that base length should be at least 10 times the average grain size. In fine-grained massive rocks the base length can be limited without any serious error, but in bedded specimens the longer the base length, the more average the results.

Deformation of the specimen should be measured at more than one point with the measuring devices placed symmetrical to the geometric axis or the axis of load application. Readings obtained from only one point could be misleading particularly where there is misalignment. It is suggested that minimum of two measuring bases be laid for each type (lateral or longitudinal) of deformation to be measured. It is preferable to have three such bases placed at suitable points dividing the whole specimen into three equiangular segments.

The number of specimens to be tested is dependent upon the rock type. Chapter I, Section 1.4. gives the method to be adopted in determining the number of specimens to be tested.

For weaker rocks, it may not be possible to mount heavy gauges on to the specimen for measurement of deformation, or use other techniques such as electrical resistance strain gauges. (The high strength cement used can unduly strengthen the surface giving lower deformation.) In such cases there is no alternative except to measure displacement of the end platens. Proper preparation of the specimen surface and alignment of the platens becomes extremely important in such cases and special care should be taken. This method, however, has the following disadvantages:

1. The end surfaces of the specimens may not be completely flat resulting in excessive deformation at low stress.
2. The press platens may move in early stages of loading if they are not properly locked.
3. Unhardened press platens (or even hardened press platens at extremely high loads) may distort effecting the deformation values measured.

In this respect, it should be noted that for a 10 cm (4 in) prism of strength 6.89 MPa (1000 lbf/in²) (68.03 kgf/cm²) with a modulus of elasticity of 6890 MPa (10 $\times$ 10⁵ lbf/in²) (68030 kgf/cm²) total deformation will be about 0.1 mm (0.004 in) before failure. Hence an error of the order of 0.01 mm (0.0004 in) in the setting of the platens would cause an error in the measurement of axial deformation of 10%. The errors are greatly accentuated if capping materials are used or the end surfaces are treated by some other way (placing of softer materials between platens and specimen surface). The errors can be minimised by taking the tangent value of modulus (see Section 6.6.) or by loading the specimen to a small initial load say 100 kgf (200 lbf).

When mounting gauges on the specimen (other than electrical resistance strain gauges) it is sufficient to measure the length of the base with an accuracy up to 0.3 mm (0.01 in).

6.4. Measurement of Deformation

The measurement of small deformation of rock specimen which is usually far below the magnitude which can be detected by senses, requires magnification to a varied degree.

A number of deformation measurement techniques are available and have been successfully used. Most of these can be grouped into one of the following groups:

1. Mechanical devices
2. Optical devices
3. Electrical gauges

One of the simplest techniques is the use of lever principle. If a lever gives a magnification of 10:1, another lever could be added to enlarge it to give a total magnification of 100:1. The problem, however, seems more difficult because of the errors involved due to friction, slip, weight and inertia of the flexible parts.

These difficulties led to the development of optical gauges and electrical gauges. In the optical gauge an advantage is taken of a beam of light which is weightless, inertia-less and sufficiently rigid. In the electrical gauge, displacement is transformed into an electrical signal which can be accurately and remotely measured. The advances in electronics have made the use of electrical gauges almost universal and particularly in dynamic work where inertia of mechanical parts is a great hindrance and where points of measurements are inaccessible.

A good deformation or strain measuring gauge should have the following characteristics:

1. The gauge should have the capacity of measuring strain with an accuracy of 1 μ.
2. The response of the gauge should be high (low inertia) to permit the measurement of dynamic strains.
3. The gauge calibration should be sufficiently stable for the temperature and time ranges of the test and the humidity fluctuations during the test.
4. The gauge size should be small to allow its mounting on small specimens and to measure point strains.
5. The response of the gauge should be linear to the deformation it is measuring. It should possess remote read out possibility.
6. The installation of the gauge and auxiliary equipment should be simple and economically feasible.

Before discussing further any group of equipment, some general points are discussed here which refer to mainly mechanical and optical gauges.

1. The usual method of fixing the measuring instruments to any specimen is to provide a rounded point forced by a spring which acts upon the surface of the specimen whose displacement is measured. This point then measures the relative movement of the contact point to the body of the instrument. It is important that the relation between this motion and that of the body is definite and analysable. For example in Fig. 6-1 the direction of motion of the gauge is not in line with the direction of motion of the body. As a result, the gauge is not measuring the actual motion of the body, but some resultant value depending upon the angle φ.
2. In cases where the gauge points are sharp (e.g. diamond pivots in mirror extensometers) these dig into the specimen making indentations. As long as these indentations are small compared to the specimen size, the normal and tangential forces exerted by the clamps and gauge may not effect the specimen behaviour. But for weak materials, the forces acted

upon the clamps may be important. For brittle materials tested to failure, the indentations may form points of higher stress concentration and sources of crack propagation. In such cases, it is important to keep both these indentations and forces as small as possible.

3. Forces that vary between wide limits during the test may cause flexure or other type of deformation resulting in inaccuracies. Take, for example, the case of a diamond pivot used in the extensometer for measurement of the lateral deformation of the specimens. If an arrangement for keeping the clamping force constant is not provided, the expansion of the specimen will increase the indentation (Fig. 6-2). Consequently, the

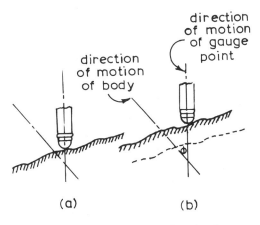

Fig. 6–1: Factors affecting measurement of motion by a contact point
(a) initial position
(b) after displacement
(after HETENYI, 1960).

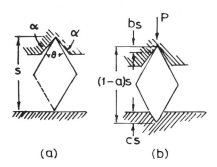

Fig. 6–2: Diamond pivot.
(after HETENYI, 1960).

length s will decrease by a factor a due to the compression of the diamond. Besides the diamond digs into the specimen (cs) and into the groove (bs) in which it operates decreasing the lever arm further by factors b and c. These values are related to each other as follows:

$$a = \frac{2P}{\theta\ Est} \log_e \left(\frac{\theta s}{2\ w} \right)$$

$$b = \frac{P}{4\ \alpha\ Est}$$

$$c = \frac{P}{\theta\ S_y\ st} \tag{6.2}$$

where P = force at any point

t = thickness of the diamond at right angle to the plane of the figure

E = modulus of elasticity of the diamond

S_y = yield limit of the specimen

θ, α = angles measured in radians (Fig. 6-2) and

w = width of the diamond points

Taking P = 10

θ = 1

α = 0.03

E = 30,000,000

S_y = 30,000

s = 0.16

t = 0.06

w = 0.0001

we get

a = 0.0005, b = 0.0003, c = 0.035

It is clear that the value of c (digging into the specimen) is a critical factor and it may change the magnification by several percent.

4. Temperature may influence the magnification ratios due to dimensional changes in parts particularly when the gauge material and specimen material have different coefficients of expansion. When taking zero readings, it is important that sufficient time elapses when the specimen or gauge is removed from one temperature condition to the other. It may be desirable to use instruments made out of invar metal or use temperature compensated gauges.

The sensitivity of any instrument is either determined by the minimum value (l_{min}) that can be read or by the difference (e) between the true deformation (t) and the indicated value (i), ($t - i = e$) whichever is greater. $l_{\min}$ can usually be taken as $1/5$ of the smallest scale division and is represented as the percentage of the actual reading.

The forces involved in the setting and operation of the instrument such as clamping, weight, inertia, frictional resistance may influence the $l_{\min}$ values. The frictional forces resisting the movement of the various mechanisms acting perpendicular to the direction of operating motion of the instrument are serious in dynamic monitoring. Similarly, inertia forces are very serious in dynamic tests.

6.4.1. Mechanical Gauges

The most important instrument used in measuring static deformation is the dial gauge which uses gears for measuring small motions. The instrument consists of a small encased gear train actuated by a rack cut in the spindle which follows the motion to be measured. The spindle is spring loaded (30–100 g) (1–3 oz) to provide a positive contact and overcome lost motion. A lightweight pointer mounted at the end of the gear train moves on a graduated scale.

The dial gauges are available with different dial diameters and different least counts. Most common are least counts of 0.01 and 0.001 mm (0.001, 0.0005, and 0.0001 in) and the usual range is 2.5 to 10 mm (0.100 to 0.250 in) though some special gauges are available with a range up to 10–15 cm (6 in).

The weight of most of these dial gauges is comparatively large (150–200 g) (6–8 oz) but small gauges of only about 50 g (1–$1/2$ oz) are also available with a smaller range 0.5 mm (0.02 in) with graduations up to 0.001 mm (0.0001 in).

Dial gauges can be mounted between the press platens, or on the cylindrical or prismatic specimens using special fixtures. A general arrangement for measuring lateral and longitudinal deformation is given in Fig. 6-3. Fig. 6-4 shows the mounting of dial gauges using knife edges fixed on two independent holders joined together with two strips of spring steel. The top holder is provided with a bore through which the dial indicator spindle can be inserted. The spindle ends act against a flat surface of the bottom holder of the second knife edge. Line indentations are made with sharp edges on the specimen surfaces at points placed at distances equal to the distances between the knife edges. Two sets of duralumin frames hold the knife edges with the deforming steel-strip springs. Straight knife edges are used for prisms and concave knife edges for circular cylinders.

Measurement of lateral deformation is done by using special holders to keep a set of dial indicators with their spindles pressing against the specimen surfaces. In such cases, the width of the specimen forms the measuring base and hence cannot be varied for a given specimen dimension.

The method has been found to be very reliable for rocks of medium and high strength. The method is not suitable for soft rocks since the knife edges may cut deep into the specimen and the specimen may flake off under shear stresses induced by the spring steel strips.

To avoid damage, the gauges have to be invariably removed before the failure of the specimen.

6.4.2. Optical Gauges

The optical gauges consist of a short mechanical arm with a small mirror mounted on it which reflects a beam of light falling upon it. Following the law of reflection, the reflected beam moves twice as fast as the short arm and the magnified angular movement of the reflected beam is either measured at some distance from the pivot point of the system or by using an optical system thereby magnifying the displacement of the end of the short mechanical arm.

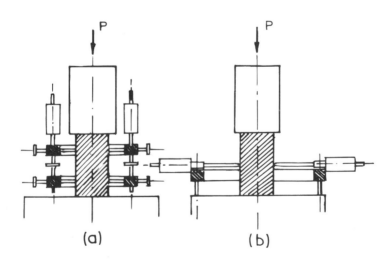

Fig. 6–3: Schematic diagram for measuring deformation by dial indicators
(a) Measuring longitudinal deformation
(b) Measuring transverse deformation
(after Protodyakonov, 1961).

Fig. 6–4: Specimen with dial gauge mountings for horizontal and vertical displacements (after LAMA, 1966).

A large variety of optical instruments is available with magnifications from 10–10,000 times and range from 5 mm to 0.002 mm (¹/₄ in to 0.0005 in) using one, two, or three mirror systems. One of the best known is the single mirror MARTENS extensometer which has been quite frequently used in static rock testing and is described below.

The system consists of a fixed knife edge mounted on one end of a fixed arm and a small mirror mounted on a diamond which has one diamond edge resting in the groove of the arm forming the pivot point and the second diamond edge resting on the specimen (Fig. 6-5b). A fixed scale facing the mirror is mounted near the telescope at a distance H from the mirror and initial reading is taken by reading the scale reflected through the mirror. As the specimen deforms, the displacement (ΔL) causes rotation of the diamond which in turn rotates the fixed mirror by a certain angle α (Fig. 6-5b) giving a reading S on the scale, reflection of which is read through the telescope. The magnification of the system is given by

$$M = \frac{S}{\Delta L} = \frac{H \tan 2\alpha}{r \sin \alpha} \qquad (6.3)$$

For small angles magnification is given by

$$M = \frac{2H}{r}$$

where H = distance between scale and mirror
         r = distance between specimen and reflecting surface of the mirror.

The displacement ΔL is given by

$$\Delta L = \frac{S}{M} = \frac{r \cdot S}{2H}$$

and strain

$$\varepsilon = \frac{\Delta L}{L} = \frac{S}{M \cdot L} \qquad (6.4)$$

The following points should be taken into account while using this equipment:

1. The accuracy is dependent upon no relative movement between the specimen, scale and the telescope. Increase in distance (H) increases magnification.
2. If the displacement is small and difficulties arise in reading the scale, it is advisable to increase the distance (H) rather than to use finely divided scale.
3. The scale must be in the plane of travel of the reflector scan and perpendicular to the line of sight of the telescope.

MARTENS extensometer is provided with a double telescope, two mirror extensometers and two scales so that measurement can be taken at two points placed at $180°$ on the specimen.

The system is quite cumbersome to set up and is time consuming in taking of readings. The telescope many times requires refocussing during the test particularly when a high-power telescope is used. Because of the possibility of damage to the mirror, the extensometer has to be removed before the fracture point of specimens.

The system is capable of giving a magnification up to 10,000 times and is quite useful in rock specimen testing in the pre-failure range.

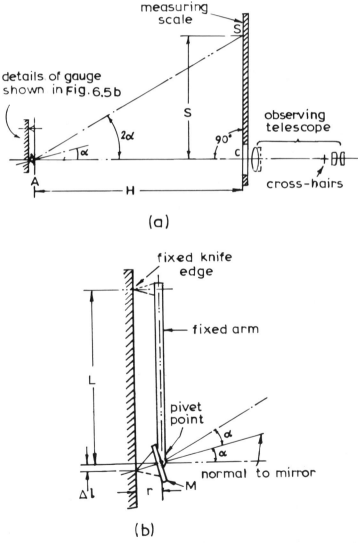

(a)

(b)

Fig 6–5: Martens single mirror extensometer
(after Hetenyi, 1960).

The error due to penetration of the diamond into soft rocks has been overcome in roller extensometers (LAMB, 1922) where the mirror is mounted
on a roller placed between the top of the knife edge and the fixed arm such
that the roller is lying between two metal surfaces (Fig. 6-6) thus eliminating
penetration error. This has been used successfully for creep measurements
by a number of investigators (PRICE, 1964; HOBBS, 1970).

Fig. 6–6: LAMB's roller extensometer mounted on a rock specimen
(after PRICE, 1964).

6.4.3. Electrical Gauges

Electrical measurement of deformation and many other mechanical parameters such as force, acceleration, angular movement, has been made
possible using such effects as change in electrical resistance of the wires
when stretched, or a movable part as in the rheostat, or a change in self-
inductance or capacitance of a movable magnet element. The two types of
electrical gauges very commonly used in the measurement of rock deformation are: (1) linear variable differential transformer (LVDT) and (2)
resistance strain gauge. These are described in detail here.

6.4.3.1. Linear variable differential transformers

The linear variable differential transformer (LVDT) is an electromechanical transformer which produces an electrical output proportional to the displacement of a movable core. It consists of 3 coils equally spaced on a cylindrical coil form with the central coil forming the primary coil and the outer two forming the secondary coils (Fig. 6-7a). A cylindrical movable magnetic core inside this coil forms the magnetic flux linking the coils. The secondary coils are connected in series opposition so that the two

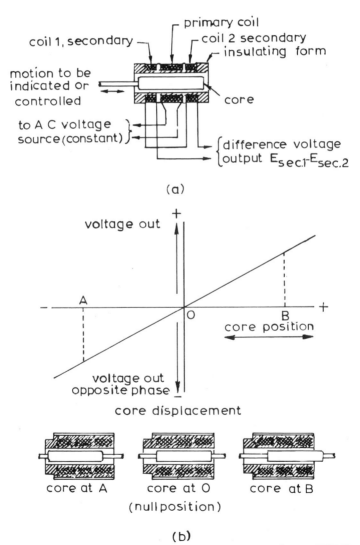

(a)

(b)

Fig. 6–7 (a): Operation of the linear variable differential transformer (LVDT).
(b): LVDT output voltage and phase as function of core position, linear graph

voltages in the secondary circuit are opposite in phase and the net output of the transformer is the difference of these voltages. When the core is in central position, the net output is zero and this point is called the balance point or null point. As the core moves on either side the voltage induced in the coil towards which it moves increases while in the other decreases giving a differential output with a proper sign. The output varies linearly with the displacement of the core (Fig. 6-7b).

In reality the voltage at null position does not become zero (Fig. 6-8) but in most of LVDTs this is about 0.25 % of the maximum output.

The linearity is generally 0.25 % of full scale though special LVDTs are available with linearity of 0.1 % and even lower. The sensitivity is dependent upon the frequency and the input voltage. The output can be easily resolved up to 0.1 % by any reasonable null balance indicating equipment. Most of the LVDTs are designed for input voltage range from 6 to 24 volt and frequency ranges from 50 to 20,000 Hz. At full displacement most of them give an excitation voltage of $\pm$ 3 to $\pm$ 16 volt, the usual being $\pm$ 10 volt. The linear range varies in wide limits from $\pm$ 0.5 to $\pm$ 250 mm (0.02 to 10 in).

The LVDT being an AC device, while read out and recording equipment mostly operate on DC input, and often have higher power level, as such a transducer signal conditioner is an essential unit required with any LVDT use. The signal conditioner converts the line voltage of a given frequency

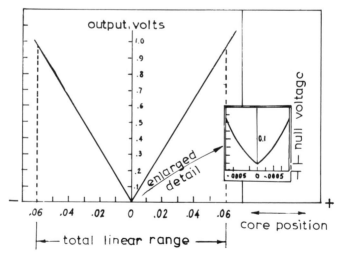

Fig. 6–8: „V" graph, absolute magnitude of LVDT output voltage as function of core position (insert shows magnified view of null region).

into excitation voltage of suitable frequency. It provides carrier signal demodulation and filtration to remove carrier ripple as well as amplify the signal for feeding to the recorder.

Many units are provided with a built-in electronic package which includes a DC/AC converter, a demodulater and a filter. These units have DC excitation of $\pm$ 6 to $\pm$ 24 volt and output of $\pm$ 5 to $\pm$ 15 volt. They are very often called DCDTs.

The LVDTs require very low driving force (fraction of a gram) which varies linearly from zero as the core is moved from null point. It is simple and rugged. LVDTs are limited in use because of the mechanical attachment problem which is common to all other gauges except the strain gauges. The dynamic response is limited by the mass of the core and supporting attachment assembly as well as the carrier frequency of the applied AC excitation voltage. The carrier frequency should be at least 10 times the maximum frequency to be measured.

LVDT output may be affected while working at high temperatures and in the presence of stray magnetic fields of considerable magnitude. The temperature rise during the test could result in zero shift and change in sensitivity. The errors can be overcome by using special circuits and by optimising excitation frequency. The presence of stray magnetic field is to add an unwanted voltage component and this can be measured by moving the core to the extreme position while the LVDT is disconnected. If the voltage is high, and unacceptable, the problem can be solved as follows: Disassemble the transformer and the core, wedge the core inside the transformer to the limit of rated displacement range and measure the output by moving the LVDT through different angles and at different locations to obtain the position with a minimum induced voltage. Special magnetically shielded LVDTs are available for locations subjected to stray magnetic currents.

LVDTs can be mounted between the platens or onto the specimen. The common method of mounting is by means of a non-magnetic two piece split block or any other suitable holder bored to fit the outside diameter of the LVDT. The block can be clamped to the specimen using knife edges or with screws placed at 120° angles (Fig. 6-9) or between the machine platens.

A linear variable differential transformer has been used by LEEMAN and GROBBELAAR (1957b) for measurement of lateral deformation. The extensometer is shown in Fig. 6-10. The extensometer consists of a ring with two legs (F & G) which "hinge" about a thin section of the extensometer ring (E). The extensometer is attached at the mid-section of the specimen (B)

using two diametrical clamping screws (C) one for each leg of the extensometer. The clamping pressure is provided by the spring clip (D). A small LVDT (linear range 0.25 mm or 0.01 in) with a coil assembly connected to one leg and the core (K) to the other is used to detect the magnified movement. The magnification of $(a + b)/a$ times the displacement of the clamping screws is obtained at the end of the measuring arms of the extensometer. This ratio of 10–15 is quite sufficient for measurement on rocks. The arrangement of LVDT gauges for measuring axial and radial deformation in uniaxial tension is given in Fig. 6-11.

Fig. 6–9: Mounting of LVDT for a rock specimen
(after Lama, 1976).

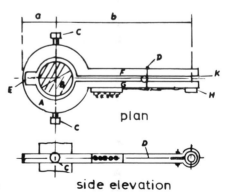

Fig. 6–10: LVDT lateral extensometer
(after Leeman and Grobbelaar, 1957b).

Fig. 6–11: Arrangement of LVDT gauges for measuring axial and radial deformations in uniaxial tension
(after HAWKES and MELLOR, 1970).

6.4.3.2. Electrical resistance strain gauges

(1) Principle

Electrical resistance strain gauges are based upon the principle discovered by Lord KELVIN in 1856 that a wire when stretched increases its electrical resistance and that the increase in electrical resistance is different for different materials. In the last 20 years, because of intensive developments in this field, these gauges today dominate the field of strain measurements.

They very closely meet all the requirements of a good strain measuring method.

Four classes of resistance strain gauges are available on the market; these are:

Unbonded wire gauges
Bonded wire gauges
Bonded foil gauges
Piezoresistive gauges.

The first three classes are based upon the principle found by Lord KELVIN. Piezoresistive gauges are based upon the use of semiconductors as the sensing element which changes its electrical resistivity with applied stress. These piezoresistive gauges have high sensitivity (gauge factor almost 60–70 times that of wire or foil gauges) and are useful for dynamic stress measurements or ultra-low static analysis where high precision is required. For most rock mechanics applications, bonded wire and foil gauges suffice and only these are discussed here in more detail.

(2) Construction of resistance strain gauges

A resistance strain gauge consists of several inches of 0.001 in diameter wire (wire strain gauge) or an extremely thin (0.0001 in) foil (metallic foil strain gauge) in the form of a grid mounted between two layers of paper or some other backing material to protect it from damage and facilitate handling and application. In wire resistance strain gauges with gauge lengths greater than 0.5 in (10 mm) the grid is flat while those with smaller gauge length have bobbin type construction with two layers of grid or a helical core flattened and cemented between two layers of paper.

Since the strain gauge filament is very small in cross-section it is considerably weaker than the cement, it allows the cement to transmit both compressive and tensile strains from the specimen to which it is cemented to the gauge conductor.

The foil strain gauges offer several advantages over the wire strain gauges. These have high heat-dissipation properties, and permit higher voltage to be applied which in turn gives better resolution. Because of the larger width, the foil gauges provide larger ratio of bonding area to the contact area than the wire grid element; this reduces stress in the adhesive and thereby stress relaxation.

Two qualities are of particular importance in strain gauges; the change in gauge resistance and the strain. The dimensional relationship between these is called the gauge factor of the gauge. Its value can be calculated from the properties of the gauge material and is given by

$$F = \frac{\Delta R/R}{\Delta L/L} \tag{6.5}$$

where
F = gauge factor
R = initial resistance
ΔR = change in resistance
L = original length
ΔL = change in length

The gauge factor is a measure of the strain sensitivity of the gauge; the higher it is, the more sensitive the gauge.

An ideal strain gauge material should have high resistance, a larger change in resistance, high elastic limit, insensitive to temperature in both its physical and electrical properties, and constant gauge factor. Table 1 gives the characteristics of different strain gauge materials.

Most gauges use advance or iso-elastic alloys. Advance alloy gauges have very low temperature sensitivity and are useful for long term work. Iso-elastic alloy is useful only for dynamic work because of high gauge factor. High temperature sensitivity of this material is not important in these measurements because of the fast changes in strain and temperature changes in this short period shall be insignificant except in such cases where high heat is being generated or absorbed during the test.

A large number of different gauges for use under a variety of conditions and on a variety of test materials is available. Special temperature compensated gauges are available for a number of materials such as quartz, titanium, mild steel, stainless steel, aluminium, magnesium, plastics, etc. It need be pointed out that these gauges are not compensated over the comlete range of temperature but only for a definite temperature range.

The base lengths of the gauges vary from as small as 0.5 to 25 mm ($1/64$ to 1 in) with resistance from 60 to 1000 ohms for foil gauges, and from 1 to 200 mm ($1/16$ to 8 in) with resistance from 60–2000 ohms for wire gauges.

TABLE 1
Properties of strain gauge conductors[1])
(after PERRY and LISSNER, 1962)

Common name	Composition	Gauge factor	Temp coef. of resistance[2])	Resistance, ohms per ft, in 1-mil diam	Stress equivalent to 10°C on steel, lbf/in^2
Nichrome	Ni–0.80, Cr–0.20	+2.0	+300	638	+2,000
Manganin	Ni–0.04, Mn–0.12, Cu–0.84	+0.47	Nil	260	−400
Advance Copel Constantan[3]) }	Ni–0.45, Cu−0.55	+2.1 +2.4 +2.1	±2	290	−66 −200 −60
Chromel-C	Ni–0.64, Fe–0.25, Cr–0.11	+2.5		640	+980
Iso-elastic	Ni−0.36, Cr-0.08, Fe-0.52, Mo-0.005	+3.5	+175	680	+5,000
Nickel		−12.1	+6,000	70	−13,500[4])
Platinum		+4.8	+3,000	80	
Soft iron		+4.2	+5,000	68	
Carbon		+20.0	−500	45,000	

[1]) These data are not to be taken too literally, since most of the characteristics vary markedly with small changes in composition, with degree of cold-working, etc.

[2]) Ohms per ohm per degree centigrade x 10^6

[3]) Constantan is the name also applied to a 60–40 alloy with somewhat different properties

[4]) Unstable

For special applications such as measurement of principal strains at a point, a combination of strain gauges in the form of rosettes are available with 2, 3 or 4 gauges mounted in rectangular forms, 45° orientations, delta or T – delta orientations. It is not possible to discuss these in the limited space here. A comprehensive treatment on the subject is given by PERRY and LISSNER (1962).

(3) Selection of resistance strain gauges

There are three most important factors influencing the selection of a gauge:

(i)   Operating temperature.
(ii)   The state of strain including the gradient magnitude and its time dependence.
(iii)   The stability requirements.

(i) Operating temperature: The operating temperature determines the backing material and the gauge alloy. Paper backed gauges using nitrocellulose cement are useful only up to 55 °C. Epoxy based gauges can be used from —170° to +100 °C whilst phenol resin and glass-fibre carriers may be used for temperatures from —200° to +230 °C. At temperatures beyond 450 °C it is advisable to use strippable foil gauges, without the carrier. These have been successfully used in static conditions up to 650 °C and 930 °C under dynamic conditions. The working range of some of the gauge alloys is given in Table 2.

TABLE 2
Effective temperature ranges for strain-sensitive alloys

Material	Static condition	Dynamic condition	Gauge factor
Advance	−170° to +200°C	−170° to +260°C	2.1
Iso-elastic	not recommended	room temp. only	3.6
Nichrome V	not recommended	1000°C	2.0
Karma	315°C	830°C	2.0
Platinum-iridium	not recommended	1300°C	6.0

TABLE 3
Strain gauge cement summary
(after Baldwin-Lima-Hamilton Corporation)
(reproduced from PERRY and LISSNER, 1962)

Item	Organic				
	Solvent release, room-temperature curing		Thermo-plastic, melting	Chemical setting	
				Room-temperature curing	
Base	Nitro-cellulose		Shellac	Acrylic	Epoxy
Cements and properties	SR-4 and post yield	Duco	DeKhotinsky hard	Eastman 910, F-88	EPY-150
Cure temperature, °F	Room to 150	Room to 150	Melts at 275	Room	Room to 150
Cure time	2−10 hr	12−48 hr	When cool	1−5 min	10−70 hr
Cure pressure lbf/in^2	1−5	1−5	1−15	1−15	5−15
Maximum operating temperature, °F	150	150	100	150	150
Strain gauge compatibility	Use with quick-drying thin paper-backed gauges	All paper-backed gauges	All	All except paper wrap-around construction	All
Specimen material compatibility	All except plastics soluble in MEK and acetone and unbondable plastics	All except plastics soluble in MEK and acetone and unbondable plastics	All except some plastics	All except some plastics	All except some plastics
Strain limit	> 10 % at room, 1/2 % at −320 °F	> 10 % 1/2 % at −320	2–3 %	> 2 % > 1/2 % at −320	> 2 % 1/2 % at −320
Electrical properties	Excellent over operating temperature range	Excellent over operating temperature range	Excellent	Excellent	Excellent
Humidity resistance	Fair, absorbs up to 2 % water	Fair, absorbs up to 2 % water	Good	Fair, absorbs up to 0.3 % water	Good, absorbs up to 0.1 % water

TABLE 3 (continued)

Organic		Ceramic			
Chemical setting		Drying, heat curing	Chemical setting, heat curing	Fusing	
Heat curing				Molten spray	Heat curing
Epoxy	Phenolic	Silicate	Phosphate	Refractory oxide	Glass
EPY-400	Bakelite	RX-1	AL-P1, PBX, Brimor	Rokide A, Rokide C	L6AC
250−500	250−350	220	600	None	1,800
2−10 hr	5−6 hr	1 hr at temperature	1 hr	None	$^1/_2$ hr
5−50	50−100	None	None	None	None
400 normal, 500 after proper cure	300 continuous to 500 for short time	500	> 1,000	> 1,000	Approx. 1,500
All that will stand cure temperature. Exception: all paper-backed	Phenolic-backed only	Strippable, transferable, and ceramic-insulated	Strippable, transferable, and ceramic-insulated	Strippable, transferable, and ceramic-insulated	Strippable, transferable, and ceramic-insulated
All except some plastics and reactive metals	All except some plastics and reactive metals	All except some plastics and reactive metals	All except some plastics and reactive metals	All except some plastics	All metals that will stand cure temperature
2 % 1 % at −320	2−3 % $^1/_2$ % at −320	$^1/_2$ %	$^1/_2$ %	2 %	$^1/_2$ %
Excellent	Excellent to 300 °F. Poor in 400−500 °F range	Deteriorates with increase in temperature. Limits useful temperature	Deteriorates with increase in temperature. Limits useful temperature	Excellent but deteriorates at high temperature	Excellent
Good, absorbs up to 0.1 % water	Fair, absorbs up to 0.2 % water	Poor, is hygroscopic, soluble in water	Fair, is hygroscopic	Good, is porous and somewhat hygroscopic	Excellent

(ii) State of strain: In static strain fields, advance alloy gauges are more useful while gauges made from iso-elastic alloy are useful only at low dynamic strains requiring high resolution. Strains up to 5–10% can be measured with advance alloy gauges while copper-nickel alloy gauges can give range of 10–15%. At places of very high stress gradients it is advisable to use gauges with smaller gauge length. Under uniform stress fields, gauge length should be as long as possible. The longer gauges are easier to apply than shorter gauges and are more stable. Foil gauges have longer fatigue life than wire gauges. The fatigue life is dependent upon the strain induced. Advance alloy foil gauges show fatigue life in excess of one million cycles at strain levels of 1500 μ in/in. Advance alloy wire strain gauges are quite suitable for most laboratory or field work in rock mechanics testing.

(iii) Stability: For tests running over longer times, it is advisable to use temperature compensated gauges. The gauge length should be as large as possible to minimise the effect of stress relaxation in the adhesive. Epoxy carriers with water proofing is essential to minimise the effect of changes in humidity.

(4) Factors influencing gauge behaviour

Besides the 3 factors influencing the selection of the gauge and hence the behaviour also, there are several other important factors, which an experimentor must take into account while using electrical resistance strain gauges. These are discussed here in detail.

(i) Choice of cement

The selection of the cement for bonding the gauge to the specimen depends upon the gauge backing material, the environmental conditions and the time available. The strain gauge manufacturers in general supply the instructions and recommendations for the appropriate cement to be used with the gauges. Table 3 gives a summary of the different cements, their principal characteristics and the type of strain gauges with which they are compatible. In general, for gauges with paper backing, cellulose cement or epoxy plastic; cellulose, epoxy or acrylic cement should be used for room temperature work. Epoxy and acrylic cements can be cured at room temperature. These have short curing times and superior mechanical properties. At high temperatures, creep of cement does not transmit all the strain from the specimen to the gauge. High temperature work requires bakelite or epoxy backed strain gauges with bakelite or high temperature epoxy cements. Bakelite back strain gauges and bakelite cement are recommended for long term stability.

(ii) Strain cycling

The strain gauges show zero shift, as they are cold worked during the first few cycles of loading, as well as show hysteresis. The zero shift may be of the order of 35–50 μ strain for a total strain of 2000 μ (Fig. 6-12). Wire strain gauges show longer shift than foil strain gauges. To avoid this it is advisable to subject specimens to a load cycle at strains at least equal to the maximum strain to be measured. In many cases, however, it may not be possible and approximate correction may have to be applied.

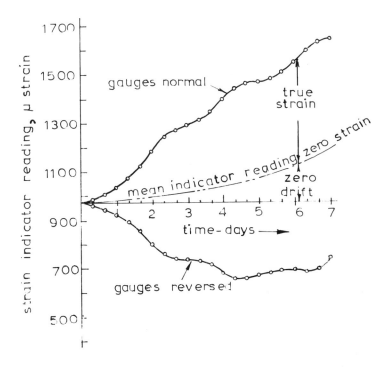

Fig. 6–12: Plot of hypothetical strain readings illustrating the determination of zero drift by electrically reversing active and dummy gauges
(after PERRY and LISSNER, 1962).

Larger cycles may damage the gauge leads and in such cases it is advisable to use double intermediate leads or mount gauge leads at such an angle that strain in the leads is zero (Fig. 6-13).

(iii) Gauge current

While the ambient temperature fluctuations occurring in the test conditions influence the test results, the gauge current strongly controls the gauge temperature and its influence is more marked than a few degree changes in ambient temperature. The most notable influence of gauge current is the

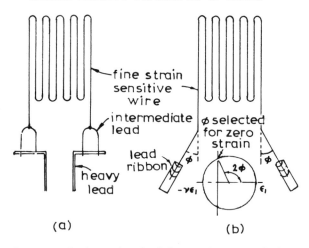

Fig. 6–13: Lead concepts for improving the fatigue performance of wire strain gauges
(a) Dual lead concept
(b) Zero strain concept
(after DALLY and RILEY, 1965).

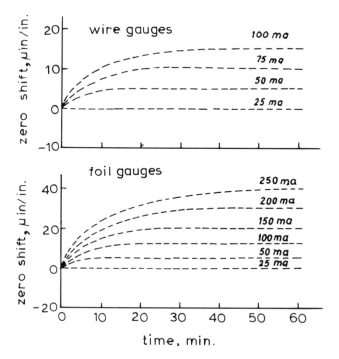

Fig. 6–14: Zero shift as a function of gauge current and time
(after DALLY and RILEY, 1965).

zero shift. Fig. 6-14 shows the influence of various current values and time on the zero shift in temperature compensated wire and foil gauges mounted on a good heat sink. The foil type gauges are much better in this regard. The error due to gauge current can be compensated using dummy gauges provided that sufficient time is allowed for the temperature equilibrium to attain. When a number of active gauges use one dummy gauge, the dummy gauge is constant in circuit while the other active gauges are switched on in sequence. This results in higher temperature for the dummy gauge. This system will inevitably result in errors in measurements. The error can be eliminated by using suitable correction factors.

(iv) Time

Where a test continues for a long time (e.g. creep tests) many of the individual factors may become significant. In all such cases, it is important that each factor be analysed individually. As already indicated, advance foil type temperature compensated gauges with either bakelite or epoxy base should be used for all long term tests. These should be cemented with phenolic cement for the bakelite carrier or the epoxy cement for the epoxy carrier. The creep resistance of epoxy cements can be increased using 20 % chopped fibre glass in the resin. The gauge should be properly water-proofed. A low input voltage should be used with three wire lead system to eliminate temperature induced resistance changes in the lead wires.

The drift in the measuring system can be eliminated by reversing the active and dummy gauge positions while taking the readings (Fig. 6-15); the mean value of the two gives the actual reading.

When readings are taken intermittently, say once or twice a day, it is important that both the active and dummy gauges are switched on sufficiently in advance to attain equilibrium position. Under conditions of extreme variations in temperature, the strain measuring equipment should be placed in a constant temperature chamber to avoid change in resistance of the measuring bridge.

(v) Hydrostatic pressure

Pressure acting on the sensing elements increases the resistance of the gauge due to its pressure sensitivity. The errors may also result due to change in the geometry of the gauge. In later case, bobbin type gauges are not recommended and foil gauges have a distinct advantage. The pressure sensitivity of various materials is constant and the error can be corrected. For advance alloy the error introduced is $+15$ μ strain per 66.7 bars (1000 lbf/in²). BRACE (1964) found that for constant foil gauges the pressure effect is $+3$ μ strain for a pressure of 66.7 bars (1000 lbf/in²). It is important that thickness of the strain gauge installation be minimum to avoid deformation of the gauge due to pressure. Thus surface should be smooth and all air bubbles be carefully avoided. In porous rocks, the strain gauge can be damaged due to bonding of the strain gauge with the pores giving erroneous results.

(5) Surface preparation of specimens, and checking the adequacy of bonding

Specimen surface needs to be carefully prepared for any strain gauge application. Rock specimen having rough surface should be smoothed out in an area sufficiently larger than that occupied by the strain gauge. While no sharp points underneath the strain gauge can be tolerated, the surface should not be polished for this decreases adhesion. Successive applications of 2–3 grades of abrasive paper finishing with the finer paper (say 180 grit) is usually sufficient for all purposes. The surface should be cleaned with some volatile solvent such as acetone, trichloroethylene, toluene, etc. The surface may then be marked to show the direction and orientation of the gauge by either drawing pencil lines showing the gauge axis or the peri-

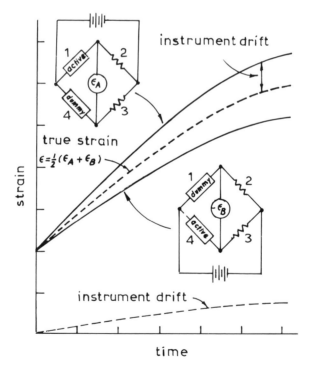

Fig. 6–15: Reversing the positions of the active and dummy gauges to eliminate strain-indicator drift
(after DALLY and RILEY, 1965).

meter of the gauge. These markings should not run under the gauge, since a burr raised by the marker may pierce the gauge resulting in its damage. It is essential to follow the application and curing instructions required for each cement as prescribed by the manufacturer. The cement and strain gauge should be applied immediately after cleaning the surface. There are several indications which can be used to determine if the bonding has been right depending upon the type of cement used. When cellulose nitrate cement is used, the influence of minute traces of solvent or moisture in the adhesive lowers the resistance of the adhesive layer. In such a case a properly dried installation will exhibit a resistance of 1000 megaohms. Epoxy cements develop residual stresses on curing and if the gauge shows strain with time, it is an indication that curing is not complete. Normally, resistance between 100–1000 megaohms means that the cement is not fully cured.

The presence of voids and air or gas bubbles between gauge and the specimen greatly influences the readings. Their presence can be checked by tapping the gauge with a soft erasing rubber and noting the change in the strain indicator. If the indicator records a change in reading, the bonding is not satisfactory.

Moisture proofing of strain gauges is important in many field and laboratory investigations. To protect the gauge from the specimen moisture, a piece of thin aluminium foil can be cemented to the specimen surface on which the gauge can be later cemented. The gauge can be further protected by giving successive coats of auxiliary materials such as petrosene wax, certain sealing compounds, synthetic rubbers, etc. The most successful method of insulation against moisture under extreme conditions consists of multi-layer treatment using wax, synthetic rubber, stainless steel shim stock and more rubber. The complete method recommended by DEAN (1957) is given below:

"Carefully clean the gauging area to ensure that all the surfaces to be water-proofed are free from grease, oil, and fingerprints. Using absorbent cotton dampened with acetone, swab around the gauge and up and down the lead wires until a fresh piece of cotton shows no discolouration. For paper-base gauges, apply a wax buffer precoat of Zophar C-276 or Di Jell 171[1] over each gauge. This prevents direct contact of the 3-M[2] compounds with gauges. Bakelite gauges ordinarily do not require this precoat. If 3-M, EC-864 synthetic rubber is to

[1] L. Sonneborn & Sons, Inc., Building Products Division, New York.

[2] Minnesota Mining & Manufacturing Company, Detroit, Mich.

be used as the principal coating, a single thin coating of a 3-M metal primer should be applied next. No primer is needed for the number EC-801 synthetic rubber as it contains added bonding resins. EC-853 primer thinned 50 per cent with methyl isobutyl keytone may be used for all steels, including stainless, and for S-T types of aluminium. Brush the primer over the area of freshly cleaned metal around the gauge. The primer may not be applied over the wax buffer coating nor over bakelite gauges, but should cover all adjacent bare metal surfaces thoroughly. Allow the primer to dry for at least 1 hour at room temperature or longer if humidity is high. Mild heat up to 130 °F will speed the drying. No adverse effects have been noted if the primer is allowed to dry for an extended time period prior to the application of 3-M compound provided the installation has been kept free of oils, fingerprints, dust, etc. Clean the plastic insulation on connecting wires with acetone and coat thinly with EC-1217 thinned 50 per cent with methyl isobutyl keytone. Clean rubber insulation with acetone and naphtha solvent and then coat thinly with EC-853. Primer-application brushes should be washed out with acetone. A partial coating of either EC-864 or EC-801 mixed with EC-807 accelerator is applied next. Mix 10 parts by weight of the base to 1 part of the accelerator in absolutely clean mixing vessels and do all mixing thoroughly. If the accelerator has settled out in storage, stir or shake jar vigorously until any top fluid is completely blended. EC-864 may be mixed in clean cans or bowls. EC-801 requires more thorough mixing on a flat surface such as a slab of safety glass. Stir and fold in the accelerator with a stiff spatula. Do not permit the accelerator to dry out around the edges and flake into the fresh mix. Do not mix more material than can be used in the next 30 min. EC-864 and EC-801 are available in 1-pt cans, and the proper bonding accelerator is furnished in separate glass jars. Apply the mixed compound over the gauge area with a putty knife or spatula to the desired thickness. A cap of stainless-steel shim stock 0.002 in. thick is rolled on and pressed down into the fresh 3-M waterproofing compound, and then the coating of synthetic rubber is applied over the entire placement. When using EC-864, the stainless-steel shim cap requires a primer such as EC-853. With the use of EC-801, however, no primer coat is required for the stainless steel shim stock. The total thickness of the build-up over the gauge will be approximately $1/8$ in. After the 3-M compound has cured, apply several coats of Herecrol RC-9 primer as a surface sealer. This is quick-drying and may be applied with a brush. A cross-section of a completed waterproofing installation is shown in Fig. 6-16. This type of protection has worked satisfactorily at the David Taylor Model Basin for gauges subjected to extremely severe conditions of submersion, erosion, and long-time applications.''

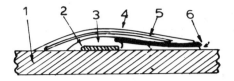

Fig. 6–16: The DEAN shim-cap method of waterproofing strain gauges:
(1) metal specimen under test;
(2) mounted strain gauge;
(3) soft wax, Di Jell 171 or Zophar Mills wax C–276;
(4) self-vulcanised rubber cover;
(5) stainless-steel shim cap 0.002 in thick;
(6) connecting cable.

(6) Principle of Wheatstone bridge

The change in resistance of a strain gauge is measured using a strain indicator. The most widely used strain indicators make use of the principle of Wheatstone bridge both for dynamic and static measurements. There are two ways of using the bridge:

1. As a direct read-out system where the output voltage is measured and related to the strain.
2. As a null balance device system where the output voltage is adjusted by adjusting the resistive balance of the bridge.

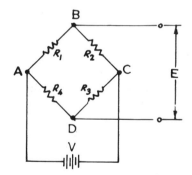

Fig. 6–17: The Wheatstone bridge circuit.

The principle of operation is given in Fig. 6-17 which indicates four resistances R_1, R_2, R_3 and R_4. A voltage V is applied between points A and

C and the output is measured between points B and D. Using KIRCHHOFF's laws, it can be shown that

$$E = \left(\frac{R_1 R_3 - R_2 R_4}{(R_1 + R_2)(R_3 + R_4)}\right) V \tag{6.6}$$

The bridge will be balanced when, $E = 0$,

$$R_1 R_3 = R_2 R_4 \text{ and } R_1 = \frac{R_2 R_4}{R_3}$$

If the bridge is initially balanced, $E = 0$, and the resistance values of R_1, R_2, R_3 and R_4 are changed by values of ΔR_1, ΔR_2, ΔR_3 and ΔR_4, then neglecting second order terms, ΔE can be given by

$$\Delta E = V \left(\frac{R_1 R_2}{(R_1 + R_2)^2}\right) \left(\frac{\Delta R_1}{R_1} - \frac{\Delta R_2}{R_2} + \frac{\Delta R_3}{R_3} - \frac{\Delta R_4}{R_4}\right) \tag{6.7}$$

When $R_2/R_1 = \gamma$

$$\Delta E = V \left(\frac{\gamma}{(1 + \gamma)^2}\right) \left(\frac{\Delta R_1}{R_1} - \frac{\Delta R_2}{R_2} + \frac{\Delta R_3}{R_3} - \frac{\Delta R_4}{R_4}\right) \tag{6.8}$$

If only the value of one of the resistances changes, say R_1 by a value of ΔR_1, while R_2, R_3 and R_4 remain constant, then the change in ΔE can be written as

$$\Delta E = V \left(\frac{\gamma}{(1 + \gamma)^2}\right) \left(\frac{\Delta R_1}{R_1}\right) \tag{6.9}$$

Since gauge factor

$$F = \frac{\Delta R_1}{R_1} / \varepsilon$$

$$\Delta E = V \left(\frac{\gamma}{(1 + \gamma)^2}\right) \varepsilon F \tag{6.10}$$

or sensitivity

$$S = \frac{\Delta E}{\varepsilon} = V \left(\frac{\gamma}{(1 + \gamma)^2}\right) F \tag{6.11}$$

Equation (6.11) gives the sensitivity (S) of the bridge assuming that the gauge current is independent of the applied voltage V and that the applied voltage remains constant.

When more than one active gauge is employed, i.e. a multiple gauge circuit, the circuit sensitivity is then dependent upon the number of gauges and is given by

$$S = \frac{\Delta E}{\varepsilon} = V\left(\frac{\gamma}{(1+\gamma)^2}\right) \quad nF \tag{6.12}$$

where n = number of active gauges.

Unbalanced bridge as shown above, is employed usually for dynamic measurements where the bridge voltage ΔE is measured directly and related to the strain level. In static applications, a null-balance bridge is usually adopted where the resistance of one of the arms is changed to match the effect of change in resistance of the active arm to bring the bridge into balance. A highly sensitive galvanometer is placed in the bridge.

At the initial stage when the bridge (Fig. 6-18a) is balanced, $R_1 R_3 = R_2 R_4$ and $R_5 = R_6$. If the resistance of R_1 changes by a certain amount, the values of R_5 and R_6 are changed so that balance is again restored in the bridge. The equivalent resistances are related to R_2, R_5 and R_3, R_6 such that

$$R_{2e} = \frac{R_2 R_5}{R_2 + R_5} \tag{6.13}$$

$$R_{3e} = \frac{R_3 R_6}{R_3 + R_6} \tag{6.14}$$

$$\Delta R_5 = -\Delta R_6$$

and $R_5 + R_6 = \text{constant}$

The change in resistance $\dfrac{\Delta R_1}{R_1}$ can be given by

$$\frac{\Delta R_1}{R_1} = F \cdot \varepsilon = \frac{2(1 + R_5/R_2)}{1 + 2(R_5/R_2) + (R_5/R_2)^2 [1 - (\Delta R_5/R_5)^2]}\left(\frac{\Delta R_5}{R_5}\right) \tag{6.15}$$

If $\Delta R_5/R_5$ is very small (< 0.1), the nonlinear term $(\Delta R_5/R_5)^2$ will be negligible and Eq. 6.15 can be rewritten as

$$\frac{\Delta R_1}{R_1} = F \cdot \varepsilon = \frac{2(1 + R_5/R_2)}{(1 + R_5/R_2)^2}\left(\frac{\Delta R_5}{R_5}\right) \tag{6.16}$$

or

$$\varepsilon = \frac{2}{F(1 + R_5/R_2)}\left(\frac{\Delta R_5}{R_5}\right)$$

The range of strain measurement possible with this circuit is dependent upon R_5/R_2. Putting $\Delta R_5/R_5 = 0.1$, and $F = 2$,

$$\varepsilon = \frac{0.1}{(1 + R_5/R_2)} \tag{6.17}$$

The higher the value of R_5/R_2, the lower the strain that can be measured with the circuit. However the sensitivity of the circuit is given by

$$S_\beta = \frac{\Delta R_5/R_5}{\varepsilon}$$

Substituting the value of ε from Eq. 6.16

$$S_\beta = \frac{F}{2}(1 + R_5/R_2) \tag{6.18}$$

S_β is higher the higher the R_5/R_2 ratio. In practice, therefore, compromise has to be made between the sensitivity of the circuit and the strain to be measured.

The circuits of commercial strain indicators, however, are more complicated. Usually these employ a reference bridge and a strain gauge bridge. More details shall be found in several standard books (PERRY and LISSNER, 1962, HETENYI, 1960).

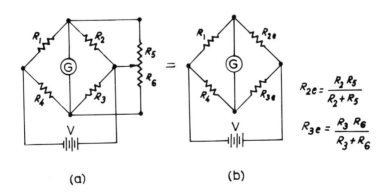

Fig. 6–18: Parallel-balance circuit and an equivalent circuit for the null-balance Wheatstone bridge.

(7) Limitations of electrical resistance strain gauges

Though the bonded electrical strain gauges are useful under a variety of conditions, these have the following limitations:

1. Because of the heterogeneous nature of the rocks, it is necessary to test a large number of specimens in order to obtain statistically reliable results. The cost of strain gauges may be prohibitive.
2. The strain measured by a gauge at a particular location on a specimen may differ greatly from the average strain over a greater length. Strain values obtained from these measurements may thus be inaccurate. It may be overcome by using long gauges or a number of gauges and statistically averaging the results.
3. Bonding of gauges to the specimen is time consuming.
4. In rocks which start flaking at load much lower than the fracture strength the strain gauges may be damaged.
5. The direct bonding is not suitable for measurement of strains right up to fracture.
6. The use of bonding cement may strengthen the rock locally giving higher modulus. This is particularly important for soft rocks.

6.4.4. Extensometers Using Electrical Gauges

To overcome some of the disadvantages of electrical resistance strain gauges, gadgets have been devised which can be used over and over again. Though these are not superior to the use of bonding the gauge onto the specimen surface and have their own disadvantages, they are useful when a large number of tests are to be done on specimens and where costs may be prohibitive or where tests are required to be conducted beyond the failure point.

1. Cantilever-type total contraction indicator

Electric resistance strain gauges have been used in the determination of the total contraction of the specimens in an apparatus developed at the University of Minnesota (FAIRHURST, 1961). The equipment consist of two spring steel cantilevers (Z) (Fig. 6-19) to each of which is attached a pair of strain gauges sensitive to cantilever deflection. The cantilevers indicate the upward movement of the lower platen during loading of specimen (Y). The cantilevers are pretensioned by adjusting the screw contacts resting on the upper platen. The platen is loosely fastened to the machine by light chain which allows the platen to fall from the cantilever on fracture of the specimen. The apparatus is calibrated by loading steel and aluminium cylinders of known modulus and of the same dimensions as the rock specimens to be tested.

Fig. 6–19: Cantilever gadget for measuring axial deformation
(after FAIRHURST, 1961).

The results obtained by this indicator show curvature at smaller loads due to compression of surface irregularities of the specimen. The indicator is not suitable for use with specimens having low strength and specimens have to be prepared extremely carefully and lapped accurately to avoid errors at initial loads.

2. LEEMAN *and* GROBBELAAR *Compressometer*

LEEMAN and GROBBELAAR (1957a) have designed a compressometer for measurement of strain continuously to fracture point using electrical resistance strain gauges for 28 mm (1–$^1/_8$ in) diameter specimens. The compressometer falls away, undamaged, at the fracture point and hence can be used again and again.

The compressometer consists of two clamping rings A and B (Fig. 6-20a) with 3 clamping screws placed at 120°.

On ring B is fitted a nut C which screws on a thread turned on ring B. When the rings are assembled, ring A rests concentrically on the nut C, and by screwing C, any desired distance between the clamping screw points can be achieved. To ensure the same gauge length for every specimen, C is screwed into the necessary position and a pin inserted in hole D which is drilled through the flange of ring A, through C and the flange of ring B. By inserting a well-fitting pin through this hole with the rings in contact, the gauge length between the clamping screws can be ensured.

The strain sensitive elements consist of four U-pieces, F, G, H and a fourth which is not shown in Fig. 6-20a. To these, wire resistance strain gauges are cemented and connected in circuit as shown in Fig. 6-20b.

The U-pieces are placed in position and located on the rings by steel balls soldered on the flanges of the rings. An equal pre-strain is then applied to each U-piece by tightening the Allen screws M_1, M_2, etc., in the one arm of each U-piece. On applying compression to the specimen the pre-strain in the U-pieces will tend to become relieved and the relief of the U-piece pre-strain will be proportional to specimen strain, provided, of course, the stress in the U-piece has not exceeded the proportional limit of the material of which they are made. Any bending of the specimen is compensated for by the spacing of the four U-pieces. When the specimen fractures, the U-pieces fall off the rings and suffer no damage.

An improvement in the sensitivity of the gauges can be obtained by reducing the cross-section at the centre of the U-pieces. This also reduces lateral loads on clamping screws. The compressometer can be used in conjunction with the lateral extensometer shown in Fig. 6-21.

3. Clip-on gauges

A number of simple clip-on gauges have been developed which make use of either the strain gauges or an LVDT as the sensing element. One such gauge using an LVDT is shown in Fig. 6-22. An LVDT is mounted between two symmetrical halves separated by a leaf spring. The two halves are provided with knife edges which act against the specimen wall using either a rubber band or a steel spring passing round the specimen and joined to the gauge at two ends. A rubber band or a steel spring offering a force of 1.5N (0.34 lbf) is sufficient to hold the gauge on to the specimen.

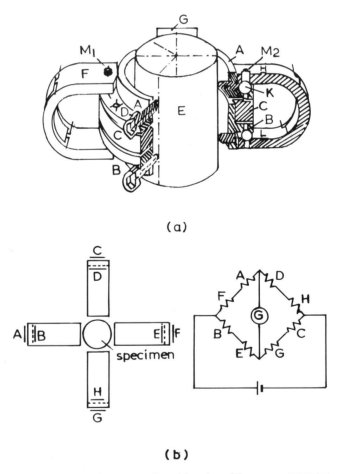

(a)

(b)

Fig. 6–20 (a): A partly cross-sectioned drawing of the compressometer
(b): The circuit diagram used in the compressometer
(after LEEMAN and GROBBELAAR, 1957a).

A number of clip-on gauges are used in crack opening detection and go by the name COD. These use strain gauges mounted onto the spring holding the two knife edges. The two knife edges are sometimes replaced by special shaped ends (Fig. 6-23a) which can be used for measuring the circumferential expansion of a ring containing a number of rollers placed round the specimen (Fig. 6-23b). The device is very useful and very quick to instal. The weight of these crack open detectors is of the order of a few decagrams. They are capable of measuring displacements up to half the gauge length. These are well adapted to use in servo-control systems.

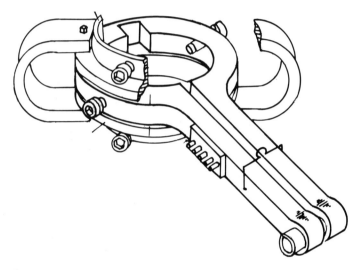

Fig. 6–21: The lateral extensometer used in conjunction with the compressometer (after LEEMAN and GROBBELAAR, 1957b).

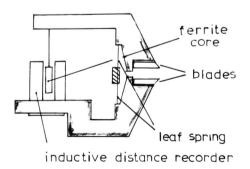

Fig. 6–22: Inductive axial-strain-measurement device (after WILHELM, 1975).

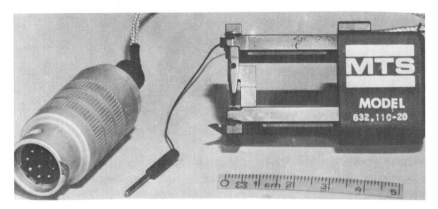

Fig. 6–23 (a): COD detector.

Fig. 6–23 (b): Circumferential strain incasing system, MTS, Minneapolis
(after LAMA, 1976).

6.5. Calculation of Elastic Constants from Tests

6.5.1. Simple Compression and Direct Tension

Referring to Fig. 6-24, if a uniform prismatic specimen is loaded with a compressive force F uniformly distributed over the cross-section A, there shall be contraction in its length parallel to the load axis and expansion at right angles to the load axis. The axial strain is then given by

$$\varepsilon = \frac{\Delta L}{L}$$

and lateral strain

$$\varepsilon' = \frac{\Delta B}{B}$$

If the area of cross-section is A, the stress is given by

$$\sigma = \frac{F}{A}$$

Assuming material to behave according to HOOKE's law (linearly elastic material) the modulus of elasticity in compression (E) can be given by

$$E = \frac{\sigma}{\varepsilon} = \frac{F/A}{\Delta L/L} = \frac{F \cdot L}{\Delta L \cdot A} \qquad (6.19)$$

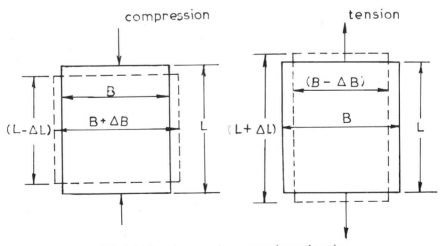

Fig. 6–24: Specimens under compression and tension

and Poisson's ratio
$$v = \frac{\varepsilon'}{\varepsilon} = \frac{\Delta B \cdot L}{B \cdot \Delta L} \tag{6.20}$$

When the specimen is loaded in tension, there is an increase in the length of the specimen and decrease in the width of the specimen but the above relationships are equally valid. The length L or B along which change in length is measured is called the base length.

When change in length is not measured along the full height (width) of the specimen but only along a limited portion, then only this limited length should be taken as the base length in the calculation of strain.

The bending of specimens under simple compression may result in serious errors in the measurement of strain at the specimen surface. The bending of the specimens may be caused due to non-parallelism of the specimen ends, non-verticality of the specimen, non-alignment of the specimen axis and the loading platen axis especially when spherical seats are used, non-uniform stiffness of the specimen or due to the excessive slenderness ratio of the specimen (particularly at points near its failure). Tests were con-

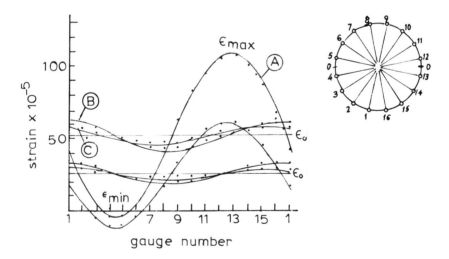

Fig. 6–25: Axial strain around the cylindrical specimen using different loading methods at load values of 146 bars and 291 bars
(A) Spherical cup of 20 cm diameter
(B) Rigid seat
(C) Loading through a 1 cm diameter ball
(after GUSTKIEWICZ, 1975a).

ducted by GUSTKIEWICZ (1975a) using limestone specimens of 50 mm (2 in) dia. ($h/d = 2$) with the following types of loadings:

A – Loading using platens with a spherical cup of 20 cm (7.9 in) diameter

B – Rigid seat

C – Loading through a 1 cm (0.4 in) diameter steel ball

Strains were measured using 16 strain gauges placed at mid-height of the specimen. The results are given in Fig. 6-25. Rigid seats and loading through a 1 cm (0.4 in) diameter ball give far more uniform strain and reduce bending moments. If only individual values of different strain gauges are taken and plotted, the modulus values calculated shall be very different (Fig. 6-26). The errors can be reduced by measuring strain at two or preferably three points placed at $0°$, and $180°$ or $0°$, $120°$, and $240°$ on the circumference of the specimen.

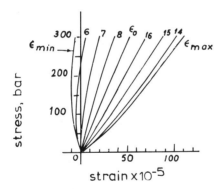

Fig. 6–26: Stress-strain curves obtained from a specimen in uniaxial compression (measured at different points at its mid circumference) as a result of bending of the specimen ($h/d = 2$, dia. $= 50$ mm). Numbers on figures refer to gauges placed around the mid-circumference of the specimen (see Fig. 6–25).
(after GUSTKIEWICZ, 1975a).

The bending and axial strains at a point can be separated in specimens using flexagauges. The flexagauge consists of two strain gauges mounted back to back on a plastic separator. When this assembly is bonded on to the specimen (Fig. 6-27), by virtue of triangles, it is clear that

$$\frac{\varepsilon_b}{D/2} = \frac{\varepsilon_1 - \varepsilon_2}{h} \tag{6.21}$$

or

$$\varepsilon_b = \frac{D}{2h}\,(\varepsilon_1 - \varepsilon_2) \qquad (6.22)$$

$$\varepsilon_a = \varepsilon_2 - \varepsilon_b$$

and

$$\varepsilon_a = \varepsilon_2 - \frac{D}{2h}\,(\varepsilon_1 - \varepsilon_2) \qquad (6.23)$$

where ε_a = true axial strain
  ε_b = bending strain at the specimen surface
  ε_1 = strain in uppermost gauge
  ε_2 = strain in gauge adjacent to the surface
  D = diameter of the specimen   and
  h = thickness of flexagauge plastic separator

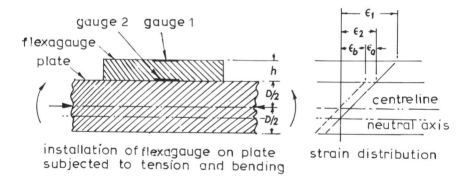

installation of flexagauge on plate subjected to tension and bending

strain distribution

Fig. 6–27: Strain distribution in flexagauge installed on plate subjected to combined compression and bending.

6.5.2. Bending

It is a very simple method of determining YOUNG's modulus values in tension and compression. The beam or a cylindrical core can be loaded at 3 or 4 points (Fig. 6-28) and YOUNG's modulus value can be calculated using the equations:

For 3 point loading (mid-point loading)

$$E = \frac{WL^3}{48y\,I_z} \qquad (6.24)$$

For 4-point loading (loading at 2 points as shown in Fig. 6-28)

$$E = \frac{WL^3}{56y\,I_z} \qquad (6.25)$$

where I_z = moment of inertia of the section with respect to centroid
y = deflection at the centre   and
L = length of the beam

The 4-point loading is superior to 3-point loading since there is no shear stress and maximum tensile stress is uniform in the central portion of the specimen.

The moment of inertia for the rectangular specimens and circular specimens is given by,

for *rectangular section*

$$I_z = \frac{bh^3}{12} \qquad (6.26)$$

where b = width
h = thickness;

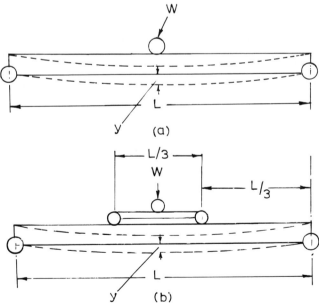

Fig. 6–28: Loading of beams in bending
(a) 3 point loading
(b) 4 point loading.

for *circular section*

$$I_z = \frac{\pi d^4}{64} \tag{6.27}$$

where d = diameter

Modulus values obtained by the bending test are different than those obtained in compression and tension tests because rocks do not have the same modulus values in compression and tension. The effect of the difference in modulus of elasticity in compression and tension modifies the values obtained from bending tests and can be related by the equation (ADLER, 1970) (assuming linear stress-strain behaviour)

$$E_r = \frac{4 E_c \cdot E_t}{(\sqrt{E_c} + \sqrt{E_t})^2} \tag{6.28}$$

where E_r = modulus of elasticity in bending, or reduced modulus of elasticity
   E_t = modulus of elasticity in tension   and
   E_c = modulus of elasticity in compression.

Eq. 6.28 is applicable only to rectangular specimens.

The modulus of elasticity in tension can be obtained from tests in bending using the equivalent section concept used in mechanics (MILLER and DOERINGSFELD, 1955). ADLER (1970) has discussed the cases of both rectangular and circular sections. The relationship between the modulus of elasticity in tension and compression is given by

$$E_t = n E_c \tag{6.29}$$

The constant n varies depending upon the cross-section of the specimen and is called the section multiple. ADLER (1970) has analysed the case of rectangular and circular sections using principle of equivalent section and has given nomograms for the two cases. The following examples make these clear.

Rectangular Section

Given

1. E_c = 5×10^6 lbf/in² (from compression testing)
2. W/y = 6×10^5 lbf/in (form flexural testing; 3 point midspan loading)
3. b = 1 in (width)
4. h = 1 in (depth)
5. L^3 = 10 in³ (span³)

Find

E_t

Solution

1. $f(n) = \dfrac{3\,WL^3}{48\,yE_c\,bh^3} = \dfrac{3 \times 6 \times 10^5 \times 10}{48 \times 5 \times 10^6 \times 1 \times 1^3} = 0.075$

2. Solve for n on Fig. 6-29 curve B:
   (a) Enter ordinate at $f(n) = 0.075$
   (b) Read abscissa at $n = 0.135$

3. Use Eq. 6.29:

   $E_t = E_c n = 5 \times 10^6 \times 0.135 = 0.675 \times 10^6 \text{ lbf/in}^2$

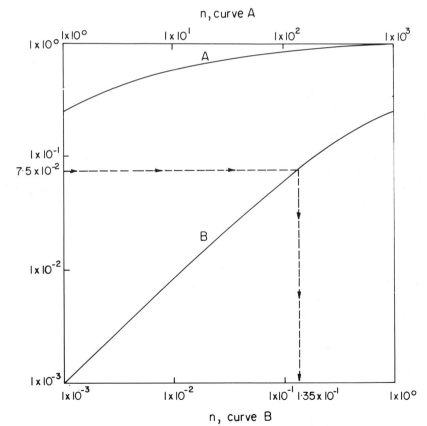

Fig. 6–29: Relationships between $f(n)$ and n; for rectangular sections;
sample problem 1
(after ADLER, 1970).

Circular Section

Given

1. E_c = 5×10^6 lbf/in^2 (from compression testing)
2. W/y = 6×10^5 lbf/in (from flexural testing; midspan 3 point loading)
3. d = 2 in (diameter)
4. L^3 = 10 in^3 (span^3)

Find

E_t

Solution

$$f(a) = \frac{64 L^3}{48 \pi d^4 E_c} \cdot \frac{W}{y} = \frac{64 \times 10 \times 6 \times 10^5}{\pi \times 48 \times 2^4 \times 5 \times 10^6} = 0.0318$$

2. Solve for n on Fig. 6-30, curve B:
   (a) Enter ordinate at $f(a) = 0.0318$
   (b) Read abscissa at $n = 5 \times 10^{-5}$
3. Use Eq. 6.29
   $E_t = n E_c = 5 \times 10^6 \times 5 \times 10^{-5}$
   $E_t = 250$ lbf/in^2

An alternate method is to measure the strain at the upper and lowermost fibres. For rectangular beams, the neutral axis of the section can be located at a distance (y_t) above the lower fibre using the relationship (Fig. 6-31).

$$y_t = \frac{\varepsilon_t}{\varepsilon_c} \cdot h \tag{6.30}$$

The value of n can be calculated from the Eq. 6.31.

$$n y_t A_t - y_c \cdot A_c = 0 \tag{6.31}$$

where A_t and A_c = areas of the respective tensile and compressive sections.

For circular section the distance $\bar{y}$ is related to the angle subtended by the secant of the circle given by the Eq. 6.32.

$$\bar{y} = \frac{D \cos \alpha}{2} \tag{6.32}$$

The value of n can be calculated using the Eq. 6.33.

$$n = \frac{\pi \cos \alpha}{2/3 \sin^3 \alpha - \alpha \cos \alpha + \sin \alpha \cos^2 \alpha} + 1 \tag{6.33}$$

When applying loads, simple supports in the form of round steel bars are provided across the beam which should be free to rotate relative to one another about the axis of the beam to avoid torsional loading of the beam. This is more easily achieved while using cores, but becomes more difficult when rectangular cross-sections are used.

6.5.3. Brazilian Test

The YOUNG's modulus and POISSON's ratio in Brazilian test can be calculated by measuring strains at the centre of the disc in both vertical and horizontal directions. The 90° strain gauge rosette placed at the centre of the disc is best suited for the purpose with the rosette axis placed parallel and at right angles to the load axis (Fig. 6-32). The relationships for discs (plane stress case) (HONDROS, 1959) are given by

$$E = -\frac{6P\,(1 - \nu^2)}{\pi\,Dt\,(\varepsilon_y + \nu\varepsilon_x)} \tag{6.34}$$

and

$$\nu = -\left(\frac{3\,\varepsilon_x + \varepsilon_y}{3\,\varepsilon_y + \varepsilon_x}\right) \tag{6.35}$$

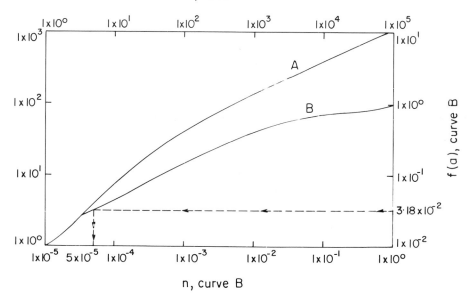

Fig. 6–30: Relationships between $f(a)$ and n, for circular section
(after ADLER, 1970).

where E = Young's modulus
  v = Poisson's ratio
  P = applied load
  D = diameter of the disc
  t = thickness of the disc
  ε_y = strain at the centre along the vertical axis   and
  ε_x = strain at the centre along the horizontal axis

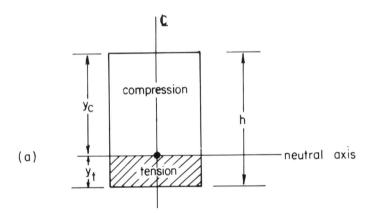

(a)

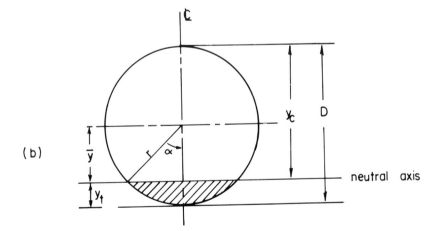

(b)

Fig. 6–31: Location of neutral axis
(a) Rectangular section
(b) Circular section.

Under plane strain conditions, the relationships are:

$$E = -\left(\frac{6P(1-\nu)(1-2\nu)}{\pi Dt [\nu\varepsilon_x + (1-\nu)\varepsilon_y]} \right) \tag{6.36}$$

and

$$\nu = -\left(\frac{\varepsilon_y + 3\varepsilon_x)}{2(\varepsilon_y - \varepsilon_x)} \right) \tag{6.37}$$

The relationships derived above assume that the properties are the same in compression and tension which is however not true for most of the rocks. Such values may be acceptable when stresses are low.

Since the stress and strain vary throughout the specimen, it is essential that gauges be as small as possible. The gauge length should not be more than $0.07D$ for an accuracy of 5%. When strain gauge rosettes are not used and it is not possible to place both the strain gauges at the centre, it is advisable to displace the gauge measuring the radial strain ε_y along the loaded axis of the disc keeping the gauge measuring tangential strain ε_x at the centre since the radial strain is fairly constant whereas the tangential strain varies more rapidly.

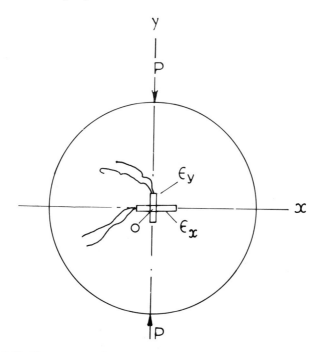

Fig. 6–32: Measurement of strains at the centre of a disc in Brazilian test.

6.5.4. Compression of Square Plates

The arrangement of tests is given in Section 3.3.4. of Vol. I (Fig. 3-39). The modulus value and POISSON's ratio can be calculated (assuming rock to be linearly elastic) by measuring the horizontal and vertical strains at the centre of the square plate and using the relationship (DAVIES and STAGG, 1970).

$$\nu = \frac{\beta - \alpha}{\beta\alpha - 1} \qquad (6.38)$$

and

$$E = \frac{\sigma_h}{\varepsilon_h} (1 + \nu\alpha) \qquad (6.39)$$

where σ_h = horizontal tensile stress

    ε_h = horizontal strain at the centre

    $\alpha = \left(-\dfrac{\sigma_v}{\sigma_h}\right)$ (ratio of vertical to horizontal stress)

    $\beta = \left(-\dfrac{\varepsilon_v}{\varepsilon_h}\right)$ (ratio of vertical to horizontal strain)

The value of σ_h is related to the load applied and is dependent upon the method of loading. A detailed discussion is given in Section 3.3.4. (Vol. I).

6.5.5. Triaxial Test (Solid and Hollow Cylinders)

Elastic constants can be determined from triaxial tests using strain gauges mounted to the exterior surface of the specimens and taking out the leads through specially designed bearing plates.

The relationship between the applied axial stress σ_1 and confining pressure p_3 and the axial strain ε_1 is given by

$$E = \frac{(\sigma_1 - 2\nu p_3)}{\varepsilon_1} \qquad (6.40)$$

If p_3 is maintained constant, the incremental change in axial stress and axial strain then can be related to give the modulus of elasticity as follows

$$E = \Delta\sigma_1/\Delta\varepsilon_1 \qquad (6.41)$$

When hollow cylinders are tested in a triaxial cell with internal pressure

$p = 0$, the axial strain (ε_1) is given by

$$\varepsilon_1 = \left[\frac{\sigma_1}{E} - \frac{2\, v p_3\, b^2}{E\,(b^2 - a^2)} \right]$$

or

$$E = \frac{1}{\varepsilon_1} \left[\sigma_1 - \frac{2\, v p_3\, b^2}{(b^2 - a^2)} \right] \qquad (6.42)$$

where b = outside radius of the cylinder   and
         a = radius of hole.

The modulus values can also be determined from the testing of hollow cylinders by measuring the deformation of the central hole, U,

$$U = -\frac{4\,(1 - v^2)\, ab^2\, p_3}{E\,(b^2 - a^2)} - 2a\, v\varepsilon_1$$

or

$$U = -\frac{4\, p_3\, b^2 a}{E\,(b^2 - a^2)} + \frac{2\, va\sigma_1}{E} \qquad (6.43)$$

For $p_3 = 0$,

$$U = -2\, va\varepsilon_1 \text{ or } v = \frac{-U}{2\, a\varepsilon_1} \qquad (6.44)$$

For a constant value of p_3

$$v = -\frac{1}{2a} \left(\frac{\varDelta U}{\varDelta \varepsilon_1} \right) \qquad (6.45)$$

Eliminating ε_1 from Eqs. 6.42 and 6.43

$$p_3 = -\frac{E\,(b^2 - a^2)\, U}{4\, ab^2} + \frac{v\,(b^2 - a^2)\, \sigma_1}{2\, b^2} \qquad (6.46)$$

For $\sigma_1 = 0$

$$E = -\frac{4\, ab^2\, p_3}{U\,(b^2 - a^2)} \qquad (6.47)$$

For $\sigma_1 = $ constant

$$E = -\frac{4\, ab^2\, \varDelta p_3}{(b^2 - a^2)\, \varDelta U} \qquad (6.48)$$

The measurement of changes in the internal diameter of a hollow cylinder is possible by measuring the change in the internal volume of the fluid and allowing for the compressibility of the fluid and the axial strain of the specimen. Gauges may be used which can be inserted into the specimen with the detector placed in the specially designed body of the upper or lower platens. With larger internal diameter specimens, LVDTs could be used. Specially designed LVDTs have been found to function well at pressures up to 4000 bars (58013 lbf/in²) (FUNG, 1975).

The measurement of lateral or radial strain of specimens in a constant pressure triaxial test can also be accomplished by measuring the change in volume of the confining pressure cell using an LVDT mounted on a small diameter pipe connected to the pressure vessel. Assuming that the diameter of the pipe to which the small piston is mounted on the end of an LVDT core is constant, the volumetric change of the vessel is given by

$$\Delta U = \Delta l \times a$$

where a = area of piston   and
         Δl = displacement

If the diameter of the loading piston and the diameter of the specimen are the same and it is assumed that the specimen deforms uniformly laterally along its length (WAWERSIK, 1975) then,

$$\varepsilon_2 = \varepsilon_3 = \frac{1}{(1 - \varepsilon_1)} \left[\frac{\Delta U}{C_1} - (C_2 \varepsilon_1 + C_3) F \right] \tag{6.49}$$

where ε_1 = axial strain

         $\varepsilon_2 = \varepsilon_3$ = lateral strain

            ΔU = cumulative volume change of the system

            C_1 = $2 A_t l_t$

$$C_2 = \frac{v_s}{E_s A_t}$$

$$C_3 = C_2 \left(\frac{l_t}{l_p - l_t} \right)$$

            A_t = area of the specimen

            l_t = length of the specimen

            l_p = effective internal length of pressure vessel

       E_s, v_s = elastic constants of the loading piston and

            F = axial force

If the diameter of piston and the specimen are not same, then

$$\varepsilon_2 = \varepsilon_3 = \frac{1}{(1 - \varepsilon_1)} \left[\frac{\Delta U}{C_1} - (C_2 \varepsilon_1 + C_3) F + C_4 (\varepsilon_1 + C_5 F) \right] \tag{6.50}$$

where $\quad C_4 = \dfrac{A_t - A_s}{2 A_t}$

$\quad\quad C_5 = \dfrac{L_c}{E_s A_t l_t}$

where $\quad A_s$ = area of loading piston
$\quad\quad L_c$ = length of the loading piston.

This method has been successfully used by CROUCH (1970).

6.6. Deformation of Rock

Deformation of rock in a uniaxial, biaxial or triaxial test is represented by plotting the difference between the maximum and minimum principal stresses $(\sigma_1-\sigma_3)$ against strain. Fig. 6-33 shows the typical stress-strain curves obtained depending on the deformational mode and fracture. Brittle fracture (Fig. 6-33a) usually occurs in uniaxial compression or uniaxial tension tests while ductile fracture (Fig. 6-33c) occurs in tests conducted under high lateral pressures and higher temperatures. Fig. 6-33b shows the transition stage. Important deformational properties are defined in Fig. 6-33d.

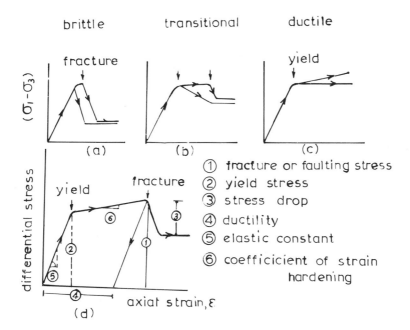

Fig. 6–33: Generalised stress-strain curves and definitions of some important deformational properties of rock.

The stress-strain curves for rocks differ markedly depending upon the method of testing, stress level, rate of loading, size of specimen etc. For a uniaxial compression test, stress-strain curves obtained for different rocks may be classified into the following three categories (Fig. 6-34).

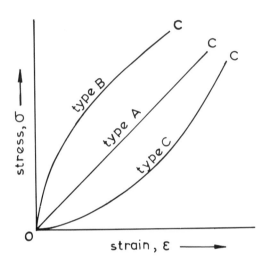

Fig. 6–34: Idealised stress-strain curves in uniaxial compression.

Type *A* (straight line) shows a linear elastic behaviour indicating a constant value of the modulus of elasticity till the point of failure. This type of curve is exhibited by most of the igneous rocks, such as basalt, diabase, gabbro and other very strong rocks such as quartzite, quartz, very strong sandstones, limestones, durain (coal), etc.

Type *B* (convex towards the stress axis) shows a relationship where there is pronounced strain with every increment in load. The value of the modulus of elasticity is highest at the early stages of loading, but continuously decreases. Such a behaviour is exhibited by softer rocks such as shales, siltstones, tuff, softer limestones, well cleated and bedded coal when tested parallel to the bedding planes, etc. Such a type of behaviour is very often called as strain-softening behaviour.

Type *C* (convex towards the strain axis) indicates a stress-strain curve where there is decreased strain with every increment in load. The value of the modulus of elasticity is the lowest at the start of loading but continuously increases. Such a behaviour is exhibited by sandstones, coals and other rocks when loaded perpendicular to the bedding planes, rock salt and certain metamorphic rocks and is termed as strain hardening behaviour.

The variations in the stress-strain curves of rocks are attributed to the following reasons:

1. At higher stress, compaction of rocks occurs and pore spaces and cracks are closed giving increased value of elasticity (Type *C*).
2. Strain hardening of material (salt) giving decreased strain (Type *C*).
3. Breakdown of specimen at high stresses giving higher strain (Type *B*).

In practice, when stating modulus of elasticity of any rock, either secant value (i.e. average value) or tangent value is reported. These are explained in Fig. 6-35. Except when otherwise mentioned, average value is taken as the tangent value at 50% of the compressive strength. The secant value calculated by taking the maximum deformation of the specimen at the time of failure or at any stress level is many times referred as the deformation modulus of the material.

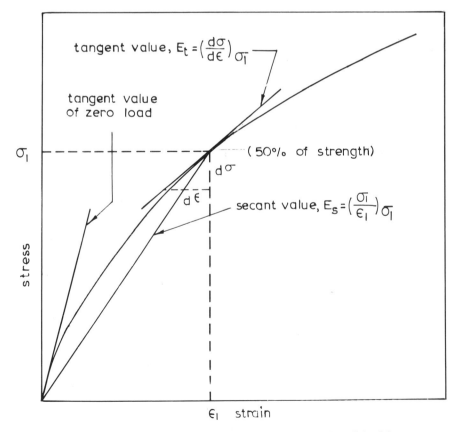

Fig. 6–35: Representation of three types of modulus of elasticity.

When tests are conducted in a stiff machine, where a complete stress-strain curve can be obtained in region beyond failure, post-failure modulus is measured as the tangent value at different stages of the post-failure curve. The tangent value (Fig. 6-36) at the point of failure is an indication of the brittleness of the rock. The higher this value the more brittle is the rock. The modulus value in the post-failure regions is many times termed as the post-failure stiffness of the material.

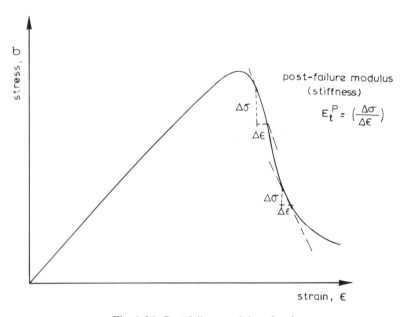

Fig. 6–36: Post-failure modulus of rock.

When rocks have a curve of the Type A, (Fig. 6-34) ending abruptly at the fracture point C, the rock is described as linear elastic and HOOKE's law is applicable to the rock. The rock may show the curve Type B or C and may be still perfectly elastic if on unloading the specimen the path followed is exactly the same as on loading. In that case the rock does not have any definite modulus value, but it changes from point to point.

If the path followed by the curve on unloading is different than that followed on loading but the net strain (residual strain or permanent strain) on unloading is zero (Fig. 6-37, $\varepsilon_0 = 0$) the rock is termed as elastic. The effect is called hysteresis which means that more work is done on loading than that recovered during unloading. The modulus value on unloading at different load values is different when compared to that during loading.

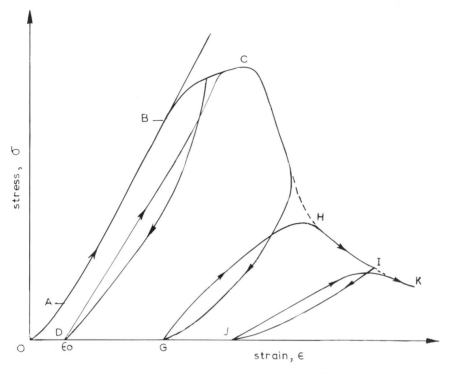

Fig. 6–37: Complete stress-strain curve for rock.

The actual curve for most of the rocks is a combination of the idealised curves. This can be divided into four distinct regions, OA, (Type C), AB (Type A), BC (Type B) and CK – (post-failure curve). In most of the rocks, the regions OA and AB are very much elastic with very small residual strain on unloading. Loading in this region does not produce any marked changes in the structure of the rock unless a number of cycles are repeated or load kept constant for a long time at elevated temperatures. The point B corresponds to the stress level at which changes in the structure start and the modulus value decreases rapidly as the point C which corresponds to the highest stress sustained by the specimen approaches. Unloading at any point lying between B and C always produces a permanent strain. At point C, failure of the specimen occurs and the curve followed is dependent upon the mode of fracture. The modulus value in the post-failure region is highest at the point near C and then drops as the specimen deteriorates with increased deformation. The point K represents a very much advanced stage of failure where the resistance offered by the specimen is due to friction between the various fractured portions of the specimen as these get displaced with increased deformation.

Considering the mechanics, the process of deformation consists of closing of cracks in the region OA, linear elastic deformation (AB), stable crack growth (BC) and unstable crack propagation starting at some point near C and extending to some point before H. These points also correspond to change in the gradient of stress-volumetric strain curve.

Strain-strain curve in tension does not show these four regions but is either basically linear with a sudden failure point C as in curve of Type A or shows curve of type B where the modulus value decreases continuously between O and C.

Certain materials maintain this maximum stress level C for a considerable further strain. Such materials are called ductile. Some softer rocks or rocks under high pressures or at high temperatures show similar behaviour.

6.7. Factors Influencing Stress-Strain Curves for Rocks

The stress-strain curves obtained for rocks in practice are influenced by a number of factors such as specimen geometry, platen conditions, rate of loading, temperature, confinement, cyclic loading etc. The influence of these factors is discussed below.

6.7.1. Specimen Geometry

The height of the specimen may or may not influence the measured modulus values substantially. Modulus values measured with very low (h/d) ratios for rocks which are more or less homogeneous (rocks which do not show any substantial structure in laboratory specimens) do not vary substantially at low loading rates. Tests conducted by HUDSON, BROWN and FAIRHURST (1971) on marble for (h/d) ratios of 3 to 0.33 did not show any marked influence on the stress-strain curve in the pre-failure range (Fig. 2-48, Vol. I).

Similarly, the diameter of the specimens does not influence the modulus values in the pre-failure range for homogeneous rocks.

PERKINS, GREEN and FRIEDMAN (1970) found very slight variations for specimens of (h/d) ratio equal to 2 and 1.5 at strain rates varying from 10^{-4} to 10^3.

In non-homogeneous rocks, where any change in the dimensions of the specimens is likely to change the number or the type of defect, both h/d ratio and the size may have a marked influence on the stress-strain curve and hence the modulus value of the rock. Results obtained from tests conducted by LAMA (1970) on coal which is a typical example of non-homogeneous discontinuous rock are given in Fig. 6-38. The deformation modulus value drops by a factor of 1.28 as (h/d) ratio increases from 0.25 to 1.0 for specimens of 10×10 cm (3.94×3.94 in) base. The modulus value drops by about 27 % as the size of the test specimen is increased from 5 cm (1.97 in) cube to 10 cm (3.94 in) cube. The drop in value for increased (h/d) ratio is slightly more marked for specimens of 10 cm $\times$ 10 cm (3.94 in $\times$ 3.94 in) base than that for specimens of 5 cm $\times$ 5 cm (1.97 in $\times$ 1.97 in) base. SYN-DZAO-MIN (1965) found the difference in modulus value for bedded coal in the ratio of 1.0:1.18:1.27 for (h/d) ratios of 1:0.5:0.25 in unconfined tests (base dimensions 12 cm $\times$ 12 cm (4.72 in $\times$ 4.72 in)).

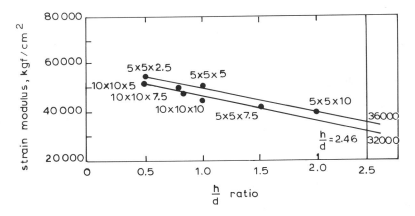

Fig. 6–38: Influence of h/d ratio and specimen size on deformation
modulus of coal (Seam 504, Silesian coal field)
(after LAMA, 1970).

The influence of shape of the specimen becomes marked when (h/d) ratio greatly increases (3 and beyond) and buckling becomes more dominant. Some results for cylinders and prisms on Coconino sandstone for (h/d) ratio of 3 at different confining pressures are given in Fig. 6-39. Very similar results have been obtained for Indiana and Lueders limestones (HANDIN et al., 1972).

In the post-failure range, the influence of h/d ratio using rigid platens is quite marked. The specimens maintain remarkably well the maximum stress when h/d ratio is small and as this value is increased post-failure modulus increases rapidly. The effect vanishes for the h/d ratio 2.0–3.0 (Fig. 6-40). Results obtained by HUDSON, CROUCH and FAIRHURST (1972) for marble are given in Fig. 6-41 which clearly indicate the influence of both h/d ratio and size. The influence of h/d ratio is more pronounced than the volume of the specimen.

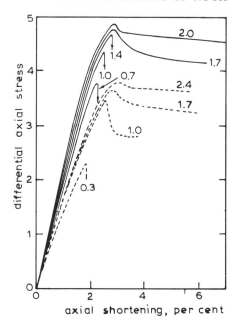

Fig. 6–39: Stress-strain curves for cylinders (dashed line) and prisms (solid line) of Coconino sandstone with h/d ratio of 3 at different confining pressures (kilobars) (after HANDIN et al, 1972).

When brush platens are used, the effect of h/d ratio on the post-failure modulus vanishes (GONANO, 1974). Results of tests conducted on marble and plaster with and without brush platens are given in Figs. 6-42 and 6-43. These clearly indicate that the post-failure modulus is effected by the end constraints; when the constraint is removed, the post-failure curve is independent of h/d ratio. Low h/d ratios of specimens result in high confining stresses and influence the mode of fracture of the specimens (LAMA and GONANO, 1976). The high confinement results in ductile failure while low confinement in brittle failure. That is why post-failure modulus is low for low (h/d) ratio and high for high (h/d) ratio.

6.7.2. Platen Conditions

The type of the loading platens (rough, smooth, relative diameter of platen to specimen, rigid or brush platens) influences markedly the stress distribution in the specimen. A detailed discussion on this aspect is given in Section 2.2. (Vol. I). Very rough platen surfaces induce high friction and high confinement regions extending deeper into the specimen. Deformation measured on the specimen at various heights of the specimen are not uniform and are usually high at the centre and low near the ends. ZNANSKI (1954) observed that deformation modulus values obtained for sapropolite specimens using rough platens were about 5% higher than those obtained using specially designed paraffin filled cups giving uniform loading.

The lateral strain distribution at different heights of the specimen is very uneven particularly when loading is not uniform (SELDENRATH and GRAMBERG, 1958) and special care should be taken to achieve uniform loading. A number of methods are discussed in Section 2.2. (Vol. I).

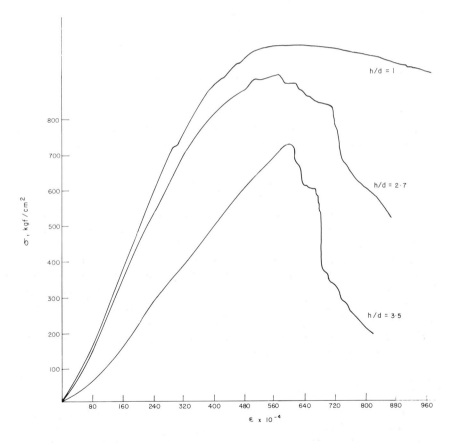

Fig. 6–40: Influence of *h/d* ratio on the complete stress-strain curve
of sapropolite shale
(after LAMA, 1975).

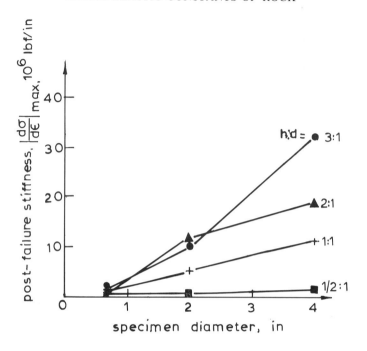

Fig. 6–41: Post-failure stiffness of Georgia Cherokee marble in uniaxial compression as a function of specimen diameter for various height: diameter ratios
(after HUDSON et al, 1972).

6.7.3. Rate of Loading

The rate of loading in any test refers to either the rate of change of strain (in strain controlled tests) or the rate of change of stress (in stress controlled tests). The strain rate and stress rate are related to each other and can be converted into each other using the relationship (assuming rock to be linearly elastic),

$$\text{stress rate} = \text{strain rate} \times \text{YOUNG's modulus}$$

If the rock does not behave linearly elastic, and a test is conducted at a particular stress rate, the corresponding strain rate shall be different at different stages of strain. It shall increase for a rock with a stress-strain curve type B (Fig. 6-34) and decrease for a rock with a stress-strain curve type C.

Most of the machines are capable of giving strain rates of the order of 10^{-3} to 10^{-4} and this is the range of strains at which most of rock testing is done.

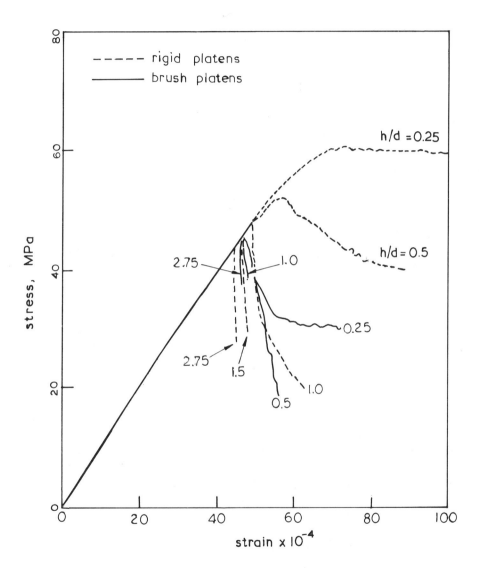

Fig. 6–42: Complete stress-strain curve for plaster (brush and solid
platens) in compression for different (h/d) ratios
(after LAMA and GONANO, 1976).

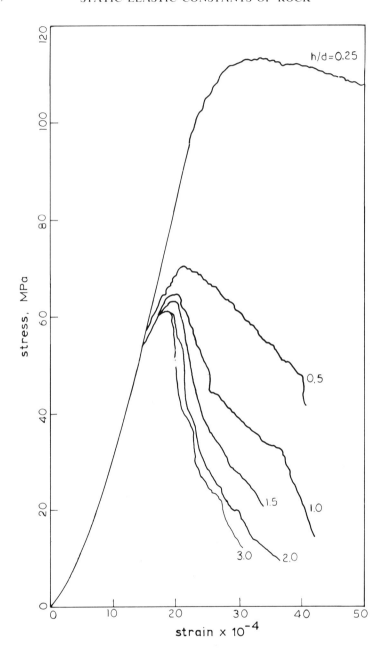

Fig. 6–43(a): Complete stress-strain curve for Wombeyan marble
(solid platens) in uniaxial compression
(after Lama and Gonano, 1976).

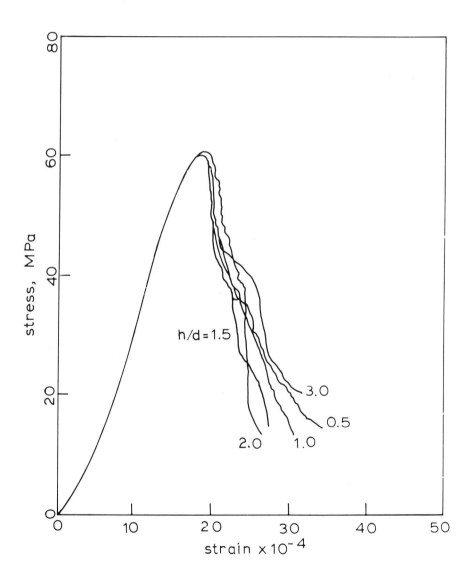

Fig. 6–43 (b): Complete stress-strain curve for Wombeyan marble
(brush platens) in uniaxial compression
(after Lama and Gonano, 1976).

The influence of rate of loading (stress or strain) on rocks is different depending upon the rock type. PHILLIPS (1948 a,b) conducted tests on sandstone beams in bending and reported an increase in modulus value of specimens by 50% when the loading rate was increased from 0.18 to 0.70 MPa/s (1568 to 6075 lbf/in²/min), a factor of only 4 (Fig. 6-44). WUERKER (1959) reported a rise in modulus by 25% for change in test duration from 10 minutes to half a minute. HORIBE and KOBAYASHI (1965) tested a number of rocks in compression and tension at stress rates of 10^{-2}–10^5 MPa/s (1.5–14.5 × 10⁶ lbf/in²/s) and found an increase in modulus value with increased strain rate.

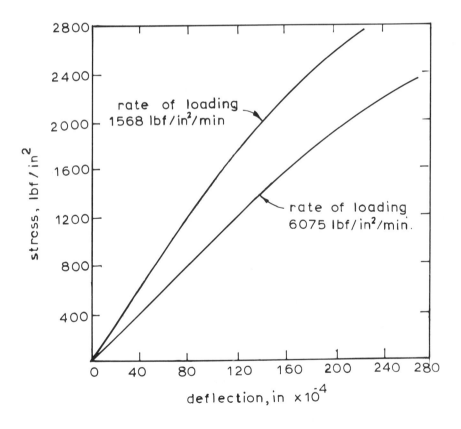

Fig. 6–44: Load-deflection curves for sandstone at two different rates of loading
(after PHILLIPS, 1948 a, b).

Modulus of elasticity increases more rapidly in tension than in compression. GREEN and PERKINS (1969) conducted tests on three different rock types (Solenhofen limestone, Westerly granite and volcanic tuff) at loading rates varying from 6 to 3 × 10⁷ MPa/s (870 to 435 × 10⁷ lbf/in²/s). They found that effect of loading rate is different for different materials. For Solenhofen limestone (an extremely fine grained very strong homogeneous material), the modulus of elasticity remained unchanged while both strength and strain increased with increased strain rate. For Westerly granite, modulus of elasticity increased but strain at failure remained constant. For volvanic tuff, there was no clear influence of strain rate except an increase in strength. Tests conducted by STOWE and AINSWORTH (1968) on basalt at stress rates from 7 × 10⁻² to 7 × 10⁸ MPa/s (10 to 1015 × 10⁸ lbf/in²/s)

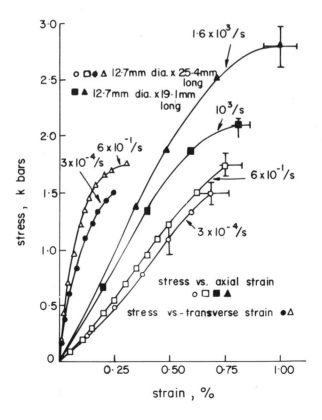

Fig. 6–45: Porphyritic tonalite at 25 °C loaded parallel to bulk core axis.
Each curve represents the average of 4–8 tests
(after PERKINS, GREEN and FRIEDMAN, 1970).

showed an increase in modulus of elasticity and strain at failure while a decrease in horizontal strain. Very similar results have been reported by STOWE (1969) on granite, limestone and tuff. Tests conducted by PERKINS, GREEN and FRIEDMAN (1970) on porphyritic tonalite at strain rates in the range of 10^{-4} to 10^3 have shown that at room temperature (25 °C) increase in strain rate increases both the tangent modulus and the strain at failure. The increase in modulus for this rock for strain rates from 3×10^{-4} to 6×10^{-1} was found to be about 15% and strain at failure increased by 10%. At low temperature, (—78 °C) the tangent modulus value increases by 34% (and strain at failure dropped by 2.5% (Figs. 6-45 and 6-46). The influence on POISSON's ratio and volumetric strain (Fig. 6-47) is very similar. BIENIAWSKI's (1970) tests on sandstone show that while modulus increases, strain at failure with higher loading rates decreases. This decrease

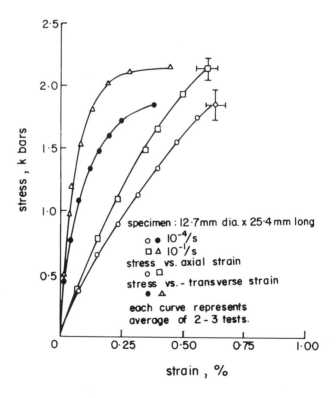

Fig. 6–46: Porphyritic tonalite at –78 °C loaded parallel to the bulk core axis (after PERKINS, GREEN and FRIEDMAN, 1970).

in strain (at room temperature) is in agreement with tests on concrete (RUSCH, 1960) but is not in agreement with those tests on rock conducted by other investigators (Chapter 9, Section 9.12.).

Tests conducted by PENG (1975) at the strain rates from 2.2×10^{-4} to 2.2×10^{-8} on Berea sandstone, Barre granite, Tennessee marble and Valder limestone show that except for Valder limestone, there is an indication of increase in modulus with strain rate, while the modulus value remains almost constant for Valder limestone. The strain at failure increases for increased strain rate as the strength increases. Tests on tuff at strain rates from 10^{-2} to 10^{-7} showed no effect on modulus except an increase in strain at failure (PENG and PODNIEKS, 1972). Results of tests by JOHN (1972) on norite, dolerite, Carrara marble and strong sandstone show

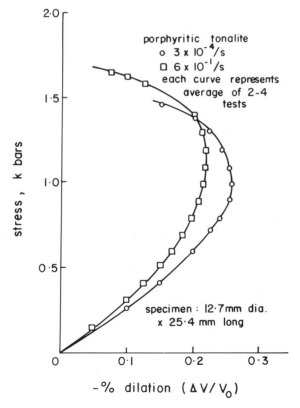

Fig. 6–47: Volume change as a function of stress for specimens loaded
parallel to bulk core axis
(after PERKINS, GREEN and FRIEDMAN, 1970).

no marked effect of strain rate (range 10×10^{-3} to 10×10^{-5}). For quartzite and weaker sandstone, modulus of deformation increased slightly with increased stress rate. Similar studies on Syrian marble by LAMA (1973) showed that for coarse grained marble, both modulus and strain at failure increased with increase in strain rate (Fig. 9-54).

Deformation under extreme loading conditions (shock loading – strain rate 10^5 to 10^6 s^{-1}) without any lateral strain has been conducted by JONES and FROULA (1969), GREEN and PERKINS (1969), GREEN et al. (1970). The modulus values under such conditions are much higher (Fig. 6-48) and the difference compared to quasi-static loading increases as the rate of loading increases. It shall be pertinent to mention tests conducted by WATSTEIN (1953) on weak (12 MPa) (1800 lbf/in²) and strong (45 MPa) (6600 lbf/in²) concrete at stress rates of 5×10^{-2} to 3×10^5 MPa/s (7 to 435×10^5 lbf/in²/s). He found an increase in modulus of elasticity for both cements of 10%. The strength increased by 1.8 for both and strain at failure by 1.3. Extensive tests have been done by RUSCH (1960) on concrete at strain rates

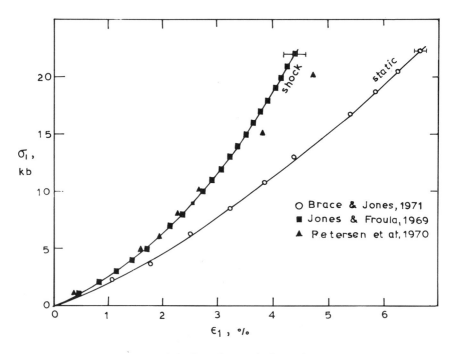

Fig. 6–48: Uniaxial deformation of tonalite. Typical error bars are shown; other data points have the same fractional error.

of 2×10^{-6} to 8×10^{-10} per second and for sustained load at different stress levels. He concluded that the limiting lines in Fig. 6-49 enclose all possible relationships between stress and axial strain. The limiting line on the left will become close to a straight line for extremely short duration of loading, until it coincides with the straight line E_c representing modulus under extremely fast loading. On the right side, it is limited by the creep deformation corresponding to an infinite duration of loading. The influence of strain rate on the post-failure modulus of the rocks varies from rock to rock. BIENIAWSKI's (1970) tests on sandstone showed that influence of higher strain rate increases post-failure stiffness and rock behaves more brittle (Fig. 9-53). PENG and PODNIEKS (1972) on tuff (Fig. 6-50), PENG (1975) on Berea sandstone, Tennessee marble, Barre granite, Valder limestone and LAMA (1973) (Fig. 9-54) on Syrian marble did not find any significant difference.

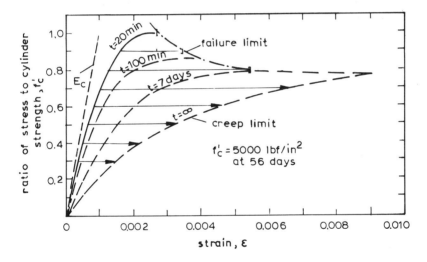

Fig. 6–49: Influence of sustained load and rate of loading on
stress-strain behaviour of concrete
(after RUSCH, 1960).

The effect of strain rate on post yield-curve of evaporites is quite marked and relative influence increases with increase in temperature. The coefficient of strain hardening increases rapidly which is clearly seen by comparing Figs. 6-51 and 6-52. Very similar results are obtained for softer rocks. The initial increase in modulus almost disappears at higher temperatures for such rocks and strain rate is the dominating factor. Summarising, the influence of rate of loading on the stress-strain curve in uniaxial compression for rocks is given in Fig. 6-53. It looks however, that homogeneous strong low porosity rocks showing linear elastic behaviour will not be affected by increase in rate of loading. In other rocks the modulus value will increase. The rocks having very brittle failure will show increased strain at failure, simply because of higher strength. Some other rocks may show increased modulus and increased strain (marbles, rock salt, sandstones) or only increased modulus but no increased strain. This is dependent upon the comparative increase in strength and modulus value as affected by loading rate.

WILLARD and McWILLIAMS (1969) have found a linear relationship between the (transgranular/intergranular[1]) fracture ratio R and the loading rate for Charcoal granite which is a homogeneous rock showing no tensile strength anisotropy. Tests were conducted using discs of 51 mm (2 in) dia. and 25 mm (1 in) thick at loading rates of 1.2, 5.0, 50.4 and 453.6 kgf/s (2.6, 11.0, 111.1 and 1000 lbf/s). The relationship is given as

$$R = k_1 + k_2 \ln (L_r) \tag{6.51}$$

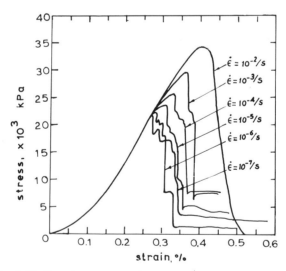

Fig. 6–50: Complete stress-strain curves for tuff at various strain rates
(after PENG and PODNIEKS, 1972).

---

[1] Transgranular-intergranular fracture is described by means of distance increments within mineral grains and along grain boundaries that are measured along a fracture trace in a thin-sectioned disc.

where

$$R = \text{transgranular-intergranular fracture factor}$$
$$L_r = \text{loading rate}$$
$$k_1 \ \& \ k_2 = \text{constants} \ (k_1 = 9.93, \ k_2 = -0.57)$$

At high strain rates, the propagating fractures have less tendency to include transgranular defect (or other defects) and there is no dissipation of stored energy due to rapid deformation; the rocks show both higher strength and higher modulus. The effect will become more marked as the loading rate increases and comes closer to the velocity of sound in the material.

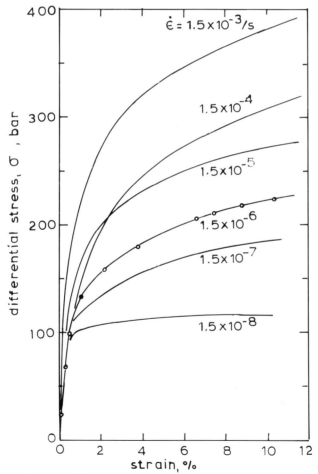

Fig. 6–51: Differential stress-strain curves for polycrystalline halite extended at 2 kb, $\dot{\varepsilon} = 1.5 \times 10^{-3}$ to 1.5×10^{-8}/s, and 100 °C (after HEARD, 1972).

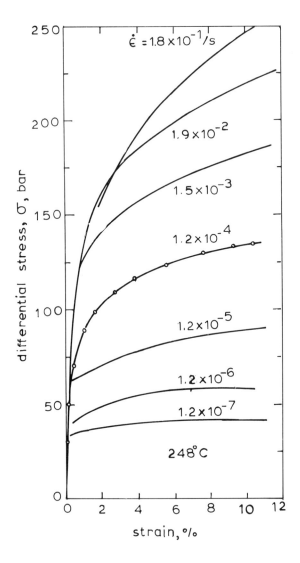

Fig. 6–52: Differential stress-strain curves for polycrystalline halite extended at 2 kb, $\dot{\varepsilon} = 1.8 \times 10^{-1}$ to 1.2×10^{-7}/s, and 248 °C (after HEARD, 1972).

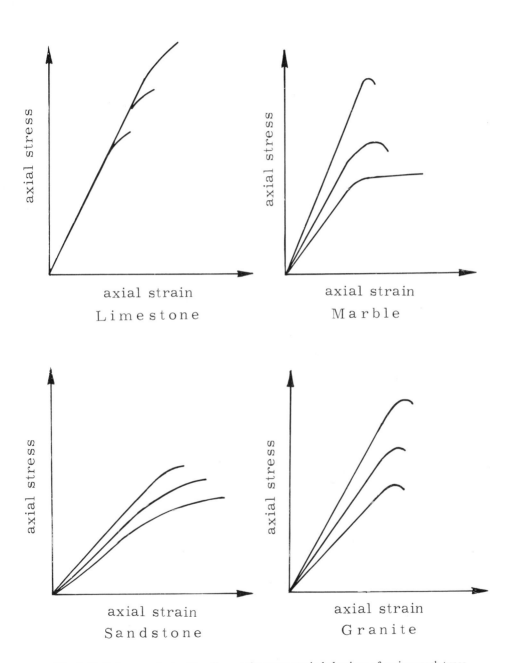

Fig. 6–53: Influence of rate of loading on the stress-strain behaviour of various rock types.

6.7.4. Temperature, Pressure and Brittle-Ductile Transition

The effect of temperature on modulus of elasticity is dependent upon the rock type. Lowering of temperature increases its value while rise in temperature lowers not only its value, but also changes the shape of the curve. The rocks which are linearly elastic at lower temperature show pronounced creep and tend to give the stress-strain curve of the type B at elevated temperatures (Fig. 6-34). The yield point C on the curve is lowered and the linearity span is decreased.

HANDIN and HAGER (1958) tested specimens of Barns sandstone at temperatures from room to 300 °C and found that decline in modulus value with rise in temperature is very small while results obtained by HUGHES and MAURETTE (1956) on Caplen Dorne sandstone at temperatures from 25° to 200 °C at constant confining pressure of 500 atm showed a decline in the value of E by 20%. Results of tests by GRIGGS, TURNER and HEARD (1960) on granite at various temperatures and at 5 kbars confining pressure are given in Fig. 6-54. At room temperature the rock shows brittle fracture. As the temperature is raised, decrease in strength and modulus value occurs. The amount of strain that is introduced without substantial loss in strength may be substantial. The transition from brittle to ductile fracture can be clearly noticed.

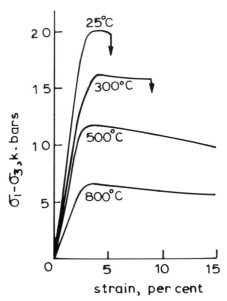

Fig. 6–54: Stress-strain curves for granite at a confining pressure
of 5 kilobars and various temperatures
(after GRIGGS, TURNER and HEARD, 1960).

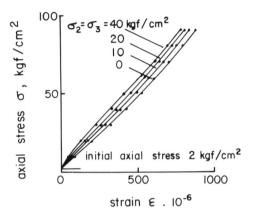

Fig. 6–55: Influence of lateral pressure on stress-strain curves of weathered granite (after SERAFIM and LOPES, 1961).

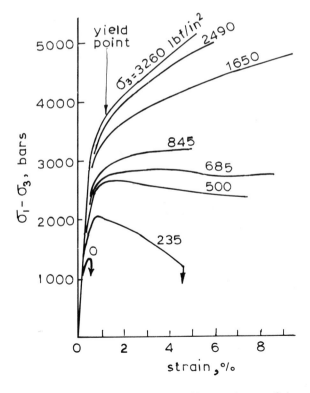

Fig. 6–56: Stress-strain curves for Carrara marble at various confining pressures (after VON KARMAN, 1911).

The influence of confining pressure on elastic modulus for strong low porosity rocks is small but for partially cracked, high porosity and weathered rocks is high. HOFFMANN's (1958) tests on sandstone showed that modulus of elasticity increased by 20% with rising pressure and dropped by 20 to 40% as fracture approached. Similar results were obtained on bituminous clay and coal (HOFFMANN, 1958) and sandstone (HANDIN and HAGER, 1957, 1958). SERAFIM and LOPES (1961) conducted tests on weathered granite from Alto Rabagao Dam at lateral pressures 0 to 3.92 MPa (0 to 40 kgf/cm²) (0 to 569 lbf/in²) and their results are shown in Fig. 6-55. Tests show an increase in modulus with lateral pressure. It is, however, doubtful if these curves have been corrected for the influence of lateral stress on the axial deformation observed. Another classic example are the tests conducted by VON KARMAN (1911) on Carrara marble (Fig. 6-56) which show clearly that the rock can undergo increased deformation while retaining the strength. At higher pressures, the resistance even increases with strain showing strain hardening. Very similar results on Solenhofen limestone have been obtained by HEARD (1960). Tests by SCHWARTZ (1964) on Stone Mountain granite, WAWERSIK and FAIRHURST (1970) on Tennessee marble, WAWERSIK and BRACE (1971) on Westerly granite, and CROUCH (1972) showed extremely small or no influence of confining pressure on modulus values. The influence of effective stress on modulus in Westerly granite both in tension and compression is quite marked and is shown in Fig. 6-57. The modulus values increase as the effective normal stress is increased. Modulus value in tension is lower than in compression. FAIRHURST (1961) found, for fine-grained granite, the ratio between the two moduli determined in bending tests to be 1.25. Tests in bending and compression on coal by EVANS and POMEROY (1966) showed that modulus in bending is about 5% higher than in compression. This result is influenced by the volume effect and hence is not reliable. The influence of confining pressure is to increase the initial tangent modulus value of rocks. The relationship can be expressed in the form (DUNCAN and CHANG, 1970):

$$E_t = E_i \left[1 - \frac{(\sigma_1 - \sigma_3)\, R_f}{(\sigma_1 - \sigma_3)_f} \right]^2 \qquad (6.52)$$

where

$$
\begin{aligned}
E_t &= \text{tangent modulus at stress level } \sigma_1 \\
E_i &= \text{initial tangent modulus} \\
\sigma_1 - \sigma_3 &= \text{deviator stress} \\
(\sigma_1 - \sigma_3)_f &= \text{failure deviator stress at confining pressure } \sigma_3 \\
R_f &= (\sigma_1 - \sigma_3)_f / (\sigma_1 - \sigma_3)_{ult} \quad \text{where} \\
(\sigma_1 - \sigma_3)_{ult} &= \text{asymptotic value of stress difference (Fig. 6-58).}
\end{aligned}
$$

The value of R_f is always less than one and is independent of confining pressure.

The initial tangent modulus is also dependent upon the confining pressure. Fig. 6-59 represents the results of some tests. Its value can be given by,

$$E_i = KP_a \left(\frac{\sigma_3}{P_a} \right)^n \tag{6.53}$$

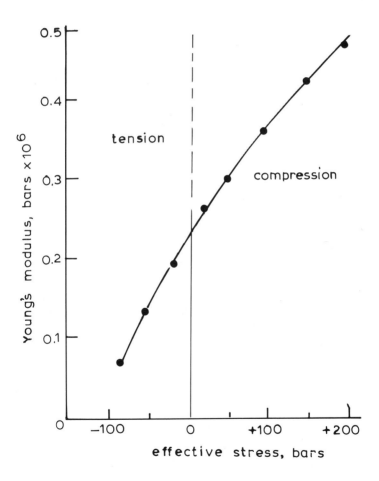

Fig. 6–57: Dependence of elastic constants of Westerly granite on effective normal stress near zero stress (after Brace, 1971).

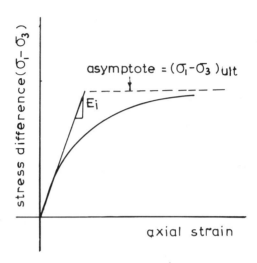

Fig. 6–58: Stress difference vs. axial strain.

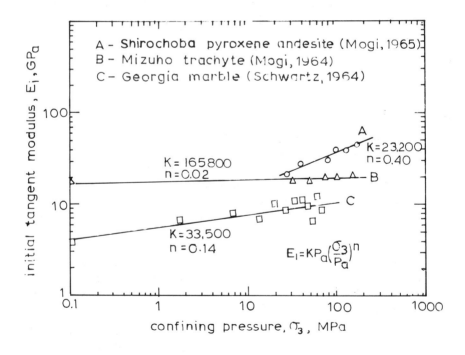

Fig. 6–59: Variations of initial tangent modulus with confining pressure
(after KULHAWY, 1975).

where

K = modulus number
n = modulus exponent
σ_3 = confining pressure   and
P_a = atmospheric pressure

Atmospheric pressure, σ_3 and E_i are expressed in the same units to ensure that K and n are prime numbers. These parameters for a variety of rocks are given in Tables 4, 5 and 6, and are summarised in Table 7.

The Tables 4, 5 and 6 and the summary shown in Table 7 indicate a number of general trends for the modulus parameters K, n and R_f:

1. Hard, crystalline or homogeneous rocks of low porosity tend to have high values of K and low values of n and R_f. The low n-value indicates that the modulus is not greatly affected by confining pressure while the low R_f-value indicates stress-strain curves which are close to being linear.
2. Porous, elastic or closely jointed rocks tend to have relatively low values of K and relatively high values of n and R_f, indicating substantial stress-dependency and non-linearity.
3. Anisotropic variations are most clearly shown by the value of K while, for the most part, the values of n and R_f tend to be fairly consistent with sample orientation (e.g. samples MFT-4, SCT-21, SCT-22).
4. Based upon the average values shown in Table 7 it can be seen that the average values of n and R_f are fairly consistent with each other, while there is substantial variation in the K values.

Pressure greatly influences strain at which brittle or ductile fracture or faulting occurs. The amount of deformation that a rock undergoes before fracture increases with confining pressure. The results of tests on Dunham dolomite are given in Fig. 6-60. This straight line relationship is valid for $\sigma_2 = \sigma_3$ conditions, but when $\sigma_2 \neq \sigma_3$ (multiaxial), the influence of inter-mediate principal stress becomes apparent (Fig. 6-61). The deformation before fracture decreases exponentially with increase in σ_2 (Fig. 6-61b). Fig. 6-62 shows the influence of σ_2 and σ_3 on deformation before fracture. The contour lines of yield of failure (ductility) are nearly parallel lines and as such the yield at failure can be approximately expressed as a function of the parameter $(\sigma_3 - a\sigma_2)$. The coefficient of strain hardening measured as slope of the stress strain curve in the post yield region is dependent on the intermediate principal stress (Fig. 6-61a) and monotonically increases with increase in intermediate principal stress.

RICE (1975) has related strain hardening to suction induced pore fluid dilatancy of the deforming rock and the effective stress principle. But this concept cannot account fully for the strain hardening observed in dry and drained rocks at different confining pressures.

TABLE 4
Triaxial stress-strain parameters for sedimentary (clastic and chemical) rock types
(after KULHAWY, 1975)

Rock number	Description	Density (gm/cm³)	Specific gravity	Porosity (%)	Modulus number K (in 10³)	Exponent N	Failure ratio, Rf	Cohesion, C (MN/m²)	Angle of friction, Ø	Range of confining pressure (MN/m²)	Reference
SCT-1	Stockton Shale Breccia (waxy to earthy)	2.48	–	–	19.4	–	–	1.45	36–50	0.1–12.4	DEKLOTZ et al. (1966b)
SCT-2	Berea Sandstone (medium-grained, well-cemented)	–	2.66	18.2	43.6	0.26	0.73	27.2	27.8	0–200.	HANDIN et al. (1963)
SCT-3	Week's Island Sandstone (massive, hard, friable, fine-grained, well-cemented)	–	–	–	5.96	0.25	0.66	55.2	27.5	0–203.	HANDIN & HAGER (1957)
SCT-4	Oil Creek Sandstone (massive, very hard, very fine-grained, well-cemented)	–	–	–	161.8	0.39	0.42	22.1	44.5	0–203.	HANDIN & HAGER (1957)
SCT-5	Bartlesville Sandstone (massive, fine-grained, well-cemented)	–	–	–	58.6	0.07	0.64	8.0	37.2	0–203.	HANDIN & HAGER (1957)
SCT-6	Pottsville Sandstone (unweathered, almost pure silica)	2.28	2.64	14.0	106.0	0.27	0.25	14.9	45.2	0–68.9	SCHWARTZ (1964)
SCT-7	Boise Sandstone (well-cemented)	1.90	–	27.0	73.6	0.07	–	–	–	1.4–34.4	KING (1968)
SCT-8	Mase Sandstone (uniform, medium-grained)	2.69	–	0.9	135.4	0.12	0.62	41.9	41.5	0.1–150.	HOSHINO & KOIDE (1970)
SCT-9	Mutenberg Sandstone	–	–	–	127.0	0.08	–	–	–	0–243.	HEARD (1967)
SCT-10	Barnes Sandstone (massive, fine-grained, well-cemented)										HANDIN & HAGER (1957)
-10a	(// to bedding)	–	–	–	48.0	0.07	0.60	8.6	34.0	0–203.	
-10b	(⊥ to bedding)	–	–	–	62.4	0.10	0.52	8.2	34.0	0–203.	

No.	Description										Reference
SCT-11	Repetto Siltstone (hard, fissile, dry)	–	2.58	5.6	76.8	0	0.55	34.7	32.1	51.-203.	HANDIN & HAGER (1957)
SCT-12	Repetto Siltstone (hard, fissile, saturated)	–	2.58	5.6	25.8	0.16	0.72	34.7	32.1	0-200.	HANDIN et al. (1963)
SCT-13	Stockton Northview Shale (dense, silty, fine-grained)	2.47	–	–	2.07	0.23	–	0.69	21.0	0.8-4.1	DEKLOTZ et al. (1966b)
SCT-14	Stockton Shale (soft, waxy)	2.38	–	–	4.06	0.08	–	0.34	22.0	0.8-4.1	DEKLOTZ et al. (1966b)
SCT-15	Muddy Shale (hard, fine-grained, dry)	–	2.67	4.7	3.57	0.43	0.36	35.4	55.5	0-203.	HANDIN & HAGER (1957)
SCT-16	Muddy Shale (hard, fine-grained, saturated)	–	2.67	4.7	85.6	0.01	0.36	38.4	14.4	0.-200.	HANDIN et al. (1963)
SCT-17	"5900 Foot Sands Formation" Shale (hard, fissile)	–	–	–	94.6	0.12	0.78	24.7	23.8	0-203.	HANDIN & HAGER(1957)
SCT-18	Edmonton Clay Shale (w = 20 %)	–	–	–	0.18	0.80	0.55	0.1	28.0	0.1-0.8	SINCLAIR & BROOKER (1967)
SCT-19	Edmonton Bentonitic Shale (w = 30 %)	–	–	–	0.22	0.68	0.78	0.3	7.5	0.1-3.1	SINCLAIR & BROOKER (1967)
SCT-20	Green River Shale (hard, calcareous)										
-20a	(// to bedding)	–	–		0.06	1.22	0.84	0	46.6	0-203.	HANDIN & HAGER (1957)
-20b	(⊥ to bedding)	–	–		54.9	0.11	0.81	24.9	22.6	0-203.	
SCT-21	Green River Shale-1 (fine-grained, brittle, calcitic & dolomitic, interbedded with kerogen)										
-21a	(// to bedding)	–	–		131.2	0.10	0.66	73.1	29.0	6.9-172.	McLAMORE (1966)
-21b	(15° to bedding)	–	–		92.5	0.15	0.64	62.0	30.0	6.9-172.	
-21c	(20° to bedding)	–	–		115.6	0.08	0.46	53.8	30.0	6.9-172.	
-21d	(30° to bedding)	–	–		85.0	0.14	0.50	44.1	30.5	6.9-172.	
-21e	(45° to bedding)	–	–		97.9	0.12	0.57	55.8	30.5	6.9-172.	
-21f	(60° to bedding)	–	–		117.0	0.11	0.49	59.3	30.0	6.9-172.	
-21g	(75° to bedding)	–	–		61.8	0.21	0.45	59.3	30.5	6.9-172.	
-21h	(⊥ to bedding)	–	–		85.6	0.13	0.38	62.7	30.5	6.9-172.	

TABLE 4 (continued)

Rock number	Description	Density (gm/cm³)	Specific gravity	Porosity (%)	Modulus number K (in 10³)	Exponent N	Failure ratio, R_f	Cohesion, C (MN/m²)	Angle of friction, ∅	Range of confining pressure (MN/m²)	Reference
SCT-22	Green River Shale-2 (fine-grained, plastic, calcitic & dolomitic, interbedded with kerogen)										McLamore (1966)
-22a	(// to bedding)	—	—	—	50.3	0.13	0.64	44.5	21.0	6.9-172.	
-22b	(10° to bedding)	—	—	—	67.3	0.04	0.74	41.4	20.5	6.9-172.	
-22c	(20° to bedding)	—	—	—	56.5	0.07	0.75	34.5	19.9	6.9-172.	
-22d	(30° to bedding)	—	—	—	29.9	0.11	0.63	29.0	18.0	6.9-172.	
-22e	(40° to bedding)	—	—	—	27.9	0.15	0.65	31.7	19.0	6.9-172.	
-22f	(60° to bedding)	—	—	—	13.9	0.25	0.69	36.5	20.6	6.9-172.	
-22g	(⊥ to bedding)	—	—	—	78.9	0.03	0.80	38.6	20.7	6.9-172.	
SCHT-1	Devonian Limestone (heterogeneous, coarse-grained)	—	—	—	123.1	0.12	0.96	20.6	33.6	0-203.	Handin & Hager (1957)
SCHT-2	Fusselman Limestone (heterogeneous, coarse-grained)	—	—	—	63.3	0.20	0.60	11.1	32.7	0-203.	Handin & Hager (1957)
SCHT-3	Wolf Camp Limestone (heterogeneous, fine-grained)	—	—	—	77.8	0.28	0.63	23.6	34.8	0-203.	Handin & Hager (1957)
SCHT-4	Marianna Limestone (massive, friable, dry)	—	2.70	13.0	107.5	0	0.34	26.8	26.4	0-203.	Handin & Hager (1957)
SCHT-5	Marianna Limestone (massive, friable, saturated)	—	2.70	13.0	59.9	0.09	0.80	12.6	37.6	0.3-4.4	Handin et al. (1963)
SCHT-6	Wells Station Limestone (heterogeneous, fine-grained)	—	—	—	455.	0	—	—	—	20.6-98.	Paterson (1958)
SCHT-7	Solenhofen Limestone (homogeneous)	—	—	—	594.	0	—	—	—	0.1-1013.	Griggs (1936)
SCHT-8	Solenhofen Limestone (homogeneous, at 25 °C.)	—	—	—	315.	0.06	—	—	—	0-500.	Heard (1960)

		–	–	–	–	–	–	–	–	–	
SCHT-9	Limestone	–	–	–	544.	0.02	–	–	–	0.1-100.	Birch (1966)
SCHT-10	Indiana Limestone (oolitic)	2.20	2.64	19.4	44.5	0.18	0.54	6.72 / 29.6	42.0 / 7.0	0-9.6 / 9.6-68.9	Schwartz (1964)
SCHT-11	Crown Point Limestone	–	2.70	–	52.1	0.25	0.58	86.0	21.3	20.-180.	Donath (1970)
SCHT-12	AEC Nevada Site Limestone (dense, fine-grained)	2.70	2.72	0.5	742.	0.0	0.33	14.5	44.0	0-27.6	Stowe (1969)
SCHT-13	Blair Dolomite (homogeneous, fine-grained)	–	–	–	168.6	0.16	0.50	35.9	39.0	0-203.	Handin & Hager (1957)
SCHT-14	Clear Fork Dolomite (coarse to fine-grained)	–	–	–	196.5	0.22	0.39	73.1	35.0	0-203.	Handin & Hager (1957)
SCHT-15	Fusselman Dolomite (heterogeneous, fine-grained, calcitic)	–	–	–	86.9	0.26	0.60	48.4	39.5	0-203.	Handin & Hager (1957)
SCHT-16	Glorieta Dolomite (heterogeneous, med.-grained, calcitic)	–	–	–	60.5	0.29	0.74	25.8	35.0	0-203.	Handin & Hager (1957)
SCHT-17	Luning Dolomite (fine-grained, calcitic)	–	–	–	101.3	0.21	0.88	23.7	34.0	0-203.	Handin & Hager (1957)
SCHT-18	Hasmark Dolomite (homogeneous, dry, coarse-grained)										Handin & Hager (1957)
-18a	(// to foliation)	–	2.91	3.5	176.2	0.11	0.86	23.1	32.1	101.-203.	
-18b	(⊥ to foliation)	–	2.91	3.5	153.6	0.13	0.81	45.6	30.6	0-203.	
SCHT-19	Hasmark Dolomite (homogeneous, coarse-grained, saturat.)	–	2.91	3.5	88.0	0.17	0.61	22.8	35.5	0.8-5.9	Handin et al. (1963)
SCHT-20	Stockton Dolomite & Dolomite Breccia (calcareous, medium to fine-grained)	2.56	–	–	231.0	0.02	–	3.45	61.0	0.05-12.4	Deklotz et al. (1966b)
SCHT-21	Stockton Dolomite with Shale Seams (laminated)	2.56	–	–	56.6	0.27	–	0.69	51.0	0.4-12.4	Deklotz et al. (1966b)
SCHT-22	Stockton Dolomite with Stylolites (clay-filled)	2.56	–	–	68.9	0.32	–	0.76	56.0	0.8-4.1	Deklotz et al. (1966b)
SCHT-23	Chalk (95 % CaCO$_3$)	1.62	2.72	40.0	0.13	0.67	0.54	0	31.5	10.-90.	Dayre et al. (1970)
SCHT-24	Blaine Anhydrite (fine-grained)	–	–	–	93.2	0.10	0.78	43.4	29.4	0-203.	Handin & Hager (1957)

TABLE 5

Triaxial stress-strain parameters for igneous (plutonic and volcanic) rock types

(after KULHAWY, 1975)

Rock number	Description	Density (gm/cm³)	Specific gravity	Porosity (%)	Modulus number K (in 10³)	Exponent N	Failure ratio, R_f	Cohesion, C (MN/m²)	Angle of friction, ϕ	Range of confining pressure (MN/m²)	Reference
IPT-1	Inada Biotite Granite (medium-grained)	2.61	–	–	988.	0	0.72	55.2	47.7	0.1-98.	MOGI (1964)
IPT-2	Westerly Granite	–	2.61	0.4	544.	0.03	–	–	–	0.1-100.	BIRCH (1966)
IPT-3	Stone Mt. Granite (medium-grained)	2.61	2.66	0.2	63.2	0.19	0.18	55.1	51.0	0-68.9	SCHWARTZ (1964)
IPT-4	AEC Nevada Site Granite (dense, coarse-grained, unweathered)	2.69	2.69	0.3	694.	0.01	0.19	22.1	52.0	0-27.6	STOWE (1969)
IPT-5 / -5a	Quartz Monzonite (medium-grained, porphyritic) (mutually perpendicular axes)	2.68	2.70	0.2	789.	0.01	0.26	36.9	53.0	0-20.7	DEKLOTZ & HECK (1965)
-5b		2.69	2.69	0.2	884.	0.01	0.29	40.3	53.0	0-20.7	
-5c		2.68	2.68	0.2	878.	0.01	0.27	36.2	56.0	0-27.7	
IPT-6	AEC Nevada Site Quartz Monzonite	2.66	–	0.2	822.	-0.01	–	27.6	56.0	1.0-9.3	CORDING (1967)
IPT-7	Ukigane Diorite (very dense, medium-grained)	3.09	–	0.2	1101.	0	0.70	56.2	41.5	0.1-98.	MOGI (1964)
IPT-8	Orikabe Quartz Diorite (fine-grained)	2.78	–	0.4	168.7	0.14	0.62	176.0	23.8	18.-250.	MOGI (1965)
IPT-9	Nabe-Ishi Hornblende Peridotite (medium-grained)	3.16	–	0.02	690.	0	0.48	63.0	37.6	18.-250.	MOGI (1965)
IPT-10	Cedar City Tonalite	–	–	–	–	–	–	16.5	45.0	0-69.	MOGI (1965)
	(friable, medium to fine-grained)	–	2.60	4.9	242.	0.03	0.32	88.2	31.0	69.-248.	SAUCIER (1969)
IPT-11	Omega Diabase	–	3.04	–	1027.	0.01	–	–	–	0.1-100.	BIRCH (1966)

No.	Description											Reference
IVT-1	Howard Prairie Dam Basalt	–	2.82	–	756.	0.01	–	–	–	0.1-100.		BIRCH (1966)
IVT-2	AEC Nevada Site Basalt	2.70	2.83	4.6	286.	0.05	0.39	27.6	64.0	0-3.4		STOWE (1969)
	(dense, fine-grained, unweathered)							66.2	31.0	3.4-34.5		
IVT-3	Panguna Andesite	–	–	–	1.7	1.15	0.68	0	36.6	0-6.5		JAEGER (1969)
	(closely jointed)							3.72	33.4	6.5-41.4		
IVT-4	Honkomatsu Augite Andesite	2.23	–	9.9	154.	0	0.80	20.6	28.2	0.1-145.		MOGI (1964)
	(dense, coarse to fine-grained)											
IVT-5	Shirochoba Pyroxene Andesite	2.08	–	5.1	269.	0.03	0.65	36.1	35.3	0.1-147.		MOGI (1964)
IVT-6	Shirochoba Pyroxene Andesite	2.45	–	5.1	23.2	0.40	0.27	35.0	39.5	28.-130.		MOGI (1965)
IVT-7	Shinkomatsu Pyroxene Andesite	2.17	–	12.6	388.	0	0.59	72.4	0	13.-100.		MOGI (1965)
	(very porous)											
IVT-8	Mizuho Trachyte	2.24	–	8.5	43.	0.26	0.55	56.0	18.4	15.-200.		MOGI (1965)
	(massive, holocrystalline)											
IVT-9	Mizuho Trachyte	2.24	–	8.7	165.8	0.02	0.79	27.4	27.7	0.1-147.		MOGI (1964)
	(massive, holocrystalline)											
IVT-10	Saku-Ishi Andesitic Welded Tuff	1.95	–	21.6	100.4	0	0.23	25.0	0	10.-50.		MOGI (1965)
	(porous)											
IVT-11	Ao-Ishi Dacite Pumice Tuff	2.00	–	24.0	35.5	0	0.54	17.6	19.4	0.1-147.		MOGI (1964)
	(dense, finely laminated)											
IVT-12	Tatsuyama Dacite Pumice Tuff	2.23	–	11.2	241.	0	0.62	38.7	28.5	0.1-147.		MOGI (1964)
	(dense)											
IVT-13	Ooya-Ishi Dacite Pumice Tuff	1.45	–	30.0	20.2	-0.08	0.57	2.6	15.5	0.1-147.		MOGI (1964)
	(massive, slightly altered)											
IVT-14	Tatsuyama Dacite Pumice Tuff	2.26	–	10.2	274.	0	0.42	77.4	19.5	28.-200.		MOGI (1965)
	(dense, laminated)											
IVT-15	Ao-Ishi Tuff	2.01	–	17.3	110.6	0.07	0.51	38.0	0	6.5-200.		MOGI (1965)
	(massive, finely laminated)											
IVT-16	AEC Nevada Site Tuff	1.92	2.39	19.8	36.0	0.08	0.26	3.75	22.5	0-10.3		STOWE (1969)
	(fairly welded ash, w = 21.1 %)											

TABLE 6

Triaxial stress-strain parameters for metamorphic (foliated and non-foliated) rock types

(after Kulhawy, 1975)

Rock number	Description	Density (gm/cm³)	Specific gravity	Porosity (%)	Modulus number K (in 10³)	Exponent N	Failure ratio, R_f	Cohesion, C (MN/m²)	Angle of friction, ∅	Range of confining pressure (MN/m²)	Reference
MFT-1	Schistose Gneiss (fine-grained)	2.79	2.80	0.5	—	0.04	0.47	20.7	43.0	0-20.7	Deklotz et al. (1965)
	(⊥ to foliation)							46.9	28.0	20.7-70.	
MFT-2	Schistose Gneiss (30° to foliation)	—	2.75	1.9	434.	0.02	0.40	14.8	27.6	0-69.	Deklotz et al. (1966a)
MFT-3	Mettawee Slate (fine-grained)	—	—	—	281.	0.17	0.55	46.6	47.6	0-203.	Handin & Hager (1957)
	(⊥ to foliation)				82.7						
MFT-4	Texas Slate (fine-grained)										McLamore (1966)
-4a	(// to foliation)	—	—	—	166.2	0.17	0.18	60.7	33.5	34.5-276.	
-4b	(10° to foliation)	—	—	—	258.0	0.09	0.33	51.7	28.5	34.5-276.	
-4c	(20° to foliation)	—	—	—	25.7	0.34	0.35	29.0	25.0	34.5-276.	
-4d	(30° to foliation)	—	—	—	98.6	0.18	0.68	26.2	21.0	34.5-276.	
-4e	(40° to foliation)	—	—	—	116.6	0.14	0.52	35.2	19.0	34.5-276.	
-4f	(50° to foliation)	—	—	—	95.8	0.18	0.68	52.4	15.0	34.5-276.	
-4g	(60° to foliation)	—	—	—	50.5	0.26	0.59	53.8	20.0	34.5-276.	
-4h	(70° to foliation)	—	—	—	23.8	0.36	0.50	64.1	22.5	34.5-276.	
-4i	(80° to foliation)	—	—	—	50.6	0.27	0.46	67.0	24.7	34.5-276.	
-4j	(⊥ to foliation)	—	—	—	70.2	0.28	0.61	70.3	26.9	34.5-276.	

		Reference										
MNFT-1	Sioux Quartzite (fine-grained)	Handin & Hager (1957)	0-203.	48.0	70.6	0.57	0.06	360.	–	–	–
MNFT-2	Yamaguchi Marble (pure, fine-grained)	Mogi (1964)	0.1-147.	29.2	30.4	0.91	0	538.	0.3	–	2.62
MNFT-3	Mito marble (little quartz, medium-grained)	Mogi (1964)	0.1-147.	30.6	22.8	0.84	0	730.	0.2	–	2.69
MNFT-4	Yamaguchi Marble (pure, coarse-grained)	Mogi (1964)	0.1-147.	27.8	22.2	0.83	0	272.	0.1	–	2.48
MNFT-5	Wombeyan Marble (coarse-grained)	Paterson (1958)	0-100.	–	–	–	0	348.	–	–	–
MNFT-6	Carrara Marble	Heard (1967)	0-320.	–	–	–	0	473.	–	–	–
MNFT-7	Georgia Marble (calcitic, very dense)	Schwartz (1964)	5.6-68.9	60.0	0 / 21.2	0.43	0.14	39.5	0.3	2.76	2.69
MNFT-8	Yule Marble (medium-grained)	Griggs (1936)	0.1-1013.	–	–	–	0	240.	–	–	–
MNFT-9	Yule Marble (medium-grained)										
-9a	(⊥ to foliation)	Handin & Hager (1957)	0-101.	35.3	1.12	0.67	0	512.	–	–	–
-9b	(// to foliation)	Handin & Hager (1957)	0-203.	36.4	14.6	0.92	0	287.	–	–	–
MNFT-10	Cabramurra Serpentine	Raleigh & Paterson (1965)	100.-500.	–	–	–	0	585.	–	–	–

TABLE 7
Summary of triaxial stress-strain parameters
(after KULHAWY, 1975)

| Parameter | Rock type | igneous | | metamorphic | | sedimentary | | all |
		plutonic	volcanic	non-foliated	foliated	clastic	chemical	
Density (gm/cm³)	No. values	10 (8)	14	4	1	6	6	41 (39)
	maximum	3.16	2.70	2.69	–	2.69	2.70	3.16
	minimum	2.61	1.45	2.48	–	1.90	1.62	1.45
	average	2.76	2.14	2.62	2.79	2.37	2.37	2.45
Specific gravity	No. values	8 (6)	3	1	2	6	9 (8)	29 (26)
	maximum	3.04	2.83	–	2.80	2.67	2.91	3.04
	minimum	2.60	2.39	–	2.75	2.58	2.64	2.39
	average	2.71	2.68	2.76	2.78	2.63	2.77	2.72
Porosity	No. values	11 (9)	14	4	2	8	8 (7)	47 (44)
	maximum	4.9	30.0	0.3	1.9	27.0	40.0	40.0
	minimum	0.02	4.6	0.1	0.5	0.9	0.5	0.02
	average	0.7	13.5	0.2	1.2	10.1	12.0	8.0
Cohesion, C (MN/m²)	No. values	12 (9)	17 (15)	8 (6)	14 (4)	35 (20)	22 (20)	108 (74)
	maximum	176.0	77.4	70.6	70.3	73.1	96.0	176.0
	minimum	16.5	0.0	0.0	14.8	0.0	0.0	0.0
	average	56.1	32.2	22.9	45.7	31.7	26.3	34.5
Angle of friction, ∅	No. values	12 (9)	17 (15)	8 (6)	14 (4)	35 (20)	22 (20)	108 (74)
	maximum	56.0	64.0	60.0	47.6	55.5	61.0	64.0
	minimum	23.8	0.0	25.3	15.0	7.5	7.0	0.0
	average	45.6	24.7	36.6	27.3	29.2	35.9	32.0
Modulus number K (in 10³)	No. values	13 (11)	16	11 (10)	13 (4)	37 (22)	25 (24)	115 (87)
	maximum	1101.0	756.0	730.0	434.0	161.8	742.0	1101.0
	minimum	63.2	1.4	39.5	23.8	0.1	1.0	0.1
	average	683.9	181.4	398.6	134.9	62.2	186.4	216.5
Exponent, n	No. values	13 (11)	16	11 (10)	13 (4)	37 (22)	25 (24)	115 (87)
	maximum	0.19	1.15	0.14	0.36	1.22	0.67	1.22
	minimum	−0.01	−0.08	0.00	0.02	0.00	0.00	−0.08
	average	0.03	0.12	0.02	0.19	0.20	0.17	0.14
Failure ratio, R_f	No. values	10 (8)	15	7 (6)	13 (4)	32 (17)	18 (17)	95 (67)
	maximum	0.72	0.80	0.92	0.68	0.84	0.96	0.96
	minimum	0.18	0.23	0.43	0.18	0.25	0.33	0.18
	average	0.40	0.52	0.74	0.49	0.57	0.64	0.56

Numbers in parentheses indicate number of different rock types.

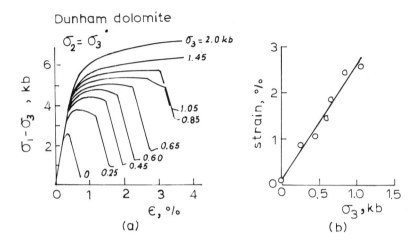

Fig. 6–60: Stress-strain curves of Dunham dolomite in the conventional
triaxial test (a)
and strain at failure as a function of confining pressure (b)
(after Mogi, 1971).

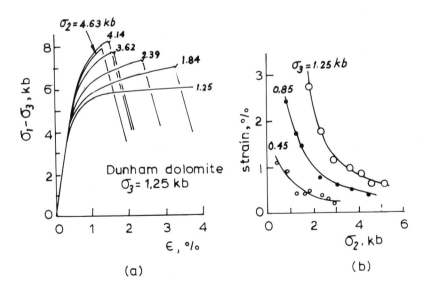

Fig. 6–61: Stress-strain curves at various values of the intermediate principal stress (σ_2)
and the minimum principal stress of 1.25 kb for Dunham dolomite (a) and strain at failure
at various values of σ_3 as a function of σ_2 (b)
(after Mogi, 1972).

The strain at failure and the stress drop at fracture are related with each other. The effect of σ_2 and σ_3 on the stress drop in Yamaguchi marble is shown in Fig. 6-63. The curves are very much similar to those of Fig. 6-62. The transition from brittle to ductile behaviour occurs not only with increase in σ_3 but also with decrease in σ_2. With high values of σ_2 fracture can be brittle with larger drop in stress and lower deformation before failure. HANDIN et al. (1963) have given the amount of permanent strain that the rocks will undergo before fracture as a function of depth for some dry and

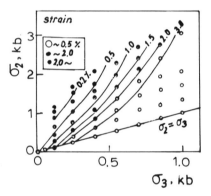

Fig. 6–62: Strain just before fracture as a function of the minimum principal stress (σ_3) and the intermediate principal stress (σ_2) in Yamaguchi marble
(after MOGI, 1972).

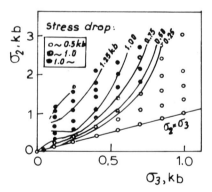

Fig. 6–63: Stress drop at fracture as functions of the minimum principal stress (σ_3) and the intermediate principal stress (σ_2) in Yamaguchi marble
(after MOGI, 1972).

salinated rocks (Fig. 6-64). The curves are derived from earlier tests (HANDIN and HAGER, 1958) and give the average values. The shaded area gives the spread of the results. The comparison of these figures clearly reveals that both strength and ductility depend upon effective confining pressure. For certain materials (e.g. halite) temperature is more important while for others pore pressure effects will be smaller (e.g. quartzite and slate or other crystalline rocks of very low porosity).

6.7.5. Stress Level

When the stress-strain curve for a rock is non-linear, the modulus values (tangent or secant) are dependent upon the stress level to which the specimen has been subjected. The modulus values may decrease or increase depending upon the convexity or concavity of the stress-strain curve (Fig. 6-34). Rocks tending to creep give increased deformation as the load is increased and the time (elapsed) between the attaining of a particular load value and read out of strain. The modulus values so calculated will show lower values. Certain investigators cyclically load the specimens until complete linearity is obtained or the hysteresis loop is extremely small. The values calculated using this method are higher for strain hardening materials and lower for strain softening materials. Most of the harder rocks show a higher value of modulus calculated this way. Results of some tests by DIETZE (1958) and DREYER (1972) are given in Fig. 6-65 where the values were calculated using the incremental stress and the corresponding increment in strain. BAULE and MÜLLER (1956), EVERLING (1960) and a number of other investigators attribute this modulus increase to collapse of pores and/or closure of cracks. The collapse of the pores is not possible at lower stress levels. DREYER (1972) determined the porosity of samples of sandstone before and after the tests (sample SN_5, Fig. 6-65) and found that its value changed only from 7.1 % (error $\pm$ 0.3 %) to 7.3 % (error $\pm$ 0.5 %) showing a possible increase in porosity. Basalt which has a lower porosity (0.5 %) shows much smaller increase in modulus with stress while the sandy shale has porosity of 1.3 % and the increase in modulus is somewhere over hundred percent. Any collapse in pore or crack closure should give rise to decrease in porosity on unloading. It looks therefore that the increase in modulus with stress is due to elastic changes in the structure of these rocks, e.g. compressibility, elastic sliding of crack surfaces, and much less due to any permanent changes. (Influence of pores and cracks on stress-strain curve is discussed in detail in Section 6.7.6.).

In other rocks (e.g. rock salt) this may be due to extensive crystal deformation by cleavage and twinning (BRAGG and FRYE, 1947). With increased strain, the disturbance extends leading to considerable size reduction of the

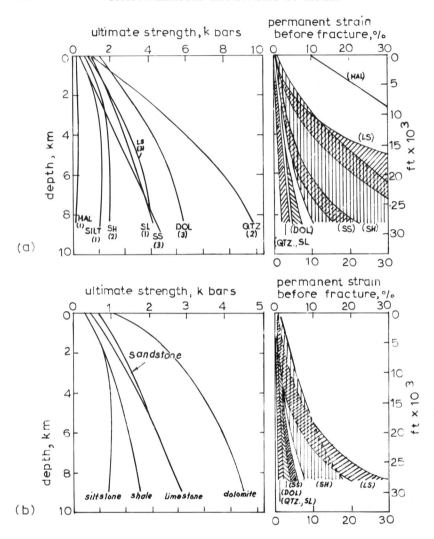

Fig. 6–64 (a): Ultimate compressive strengths and strains before fracture of dry rocks as functions of depth. Effects of confining (overburden) pressure and temperature (30 °C/km) included.

(b): Ultimate compressive strengths and strains before fracture of water-saturated rocks as functions of depth. Effects of confining (overburden) pressure, temperature (30 °C/km) and "normal" formation (pore) pressure included

(after HANDIN et al, 1963).

Nos. in brackets refers to number of rocks tested.

HAL	–	Halite	SL	–	Slate
SILT	–	Siltstone	DOL	–	Dolomite
SH	–	Shale	QTS	–	Quartzite
LS	–	Limestone	SS	–	Sandstone

mosaic blocks in the affected grains resulting in increase in strain hardening. As fracture approaches, the structure becomes increasingly disturbed due to large scale cracking, partial destruction of contacts between the grains and pulverisation of grains. Materials having slow failure show this effect quite pronounced (e.g. sandstone, shales, etc.) while those failing suddenly do not show any drop in value (e.g. Solenhofen limestone).

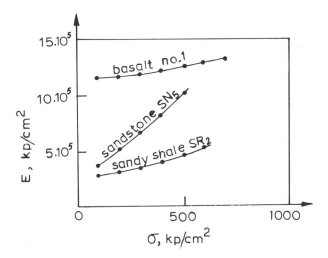

Fig. 6–65: Elastic moduli of sandstone, shale and basalt as a function of stress level (after DREYER, 1972).

6.7.6. Influence of Pores and Cracks

The non-linear stress-strain curve and increase in modulus and POISSON's ratio with stress as well as the phenomenon of hysteresis in cyclic loading have been explained due to presence of pores and cracks. Studies on elasticity and plasticity of rocks (starting from ADAMS and NICOLSON, 1901; ADAMS and WILLIAMSON, 1923; and a host of others; for more details see Section 9.2.) have shown that rocks and minerals do not show plasticity at room temperature. ADAMS and WILLIAMSON (1923) were perhaps the first to suggest that the non-linearity was due to cracks or narrow fissures in rock. This concept has been now widely accepted and has been expanded to explain the various deformational properties such as modulus (WALSH, 1965a, COOK, 1965, WALSH and BRACE, 1966a & b; BRADY, 1970, 1973) POISSON's ratio (WALSH, 1965b), compressibility etc. (WALSH, 1965c; WARREN, 1973).

WALSH (1965a) has analysed the influence of narrow elliptical open and closed cracks in both plane strain and plane stress conditions using reciprocal theorem (LOVE, 1927).

For low crack concentration and open cracks, the effective modulus at the start of the loading cycle is given by

$$E_{\text{eff}} = \frac{E}{\left(1 + \dfrac{4\pi\,\bar{c}^3}{3\,\bar{v}}\right)} \qquad \text{(plane stress)} \tag{6.54}$$

$$E_{\text{eff}} = \frac{E}{\left[1 + \dfrac{4\pi}{3}\left(1 - v^2\right)\dfrac{\bar{c}^3}{\bar{v}}\right]} \qquad \text{(plane strain)} \tag{6.55}$$

For closed cracks, the effective modulus in the linear portion of the stress strain curve on loading is given by

$$E_{\text{eff}} = \frac{E}{\left[1 + \dfrac{4\pi\bar{c}^3}{15\bar{v}}\left\{\dfrac{2 + 3\mu^2 + 2\mu^4}{(1 + \mu^2)^{3/2}} - 2\mu\right\}\right]}$$
$$\text{(plane stress)} \tag{6.56}$$

$$E_{\text{eff}} = \frac{E}{\left(1 + \dfrac{4\pi\bar{c}^3(1 - v^2)}{15\bar{v}}\left\{\dfrac{2 + 3\mu^2 + 2\mu^4}{(1 + \mu^2)^{3/2}} - 2\mu\right\}\right)}$$
$$\text{(plane strain)} \tag{6.57}$$

where

E_{eff} = effective modulus of rock on loading

E = real modulus of the rock substance when no cracks are present (intrinsic modulus of material)

v = POISSON's ratio of the material

$\bar{c}$ = average crack length $[\bar{c}^3 = \sum_{N} c^3/N]$

$\bar{v}$ = average volume of the region containing a single crack $(N\bar{v} = V)$

V = total volume of the specimen

N = total number of cracks and

μ = coefficient of friction between two faces of the crack

Thus, as shown in Eqs. 6.54 to 6.57, the effective modulus of the rock in the presence of cracks is less than that of its intrinsic modulus. The use of these equations requires knowing the average crack length and average crack concentration expressed by $\bar{c}$ and $\bar{v}$ which are not available. However, a qualitative idea can be obtained during the unloading of the curve.

The sliding of the closed cracks that occurs on application of load is reversed only when the stress drops by a certain value ($\Delta\sigma$) which is given by

$$\Delta\sigma \geqslant \frac{2\,\sigma\mu\,\tan\beta}{[1 + \mu\tan\beta]} \tag{6.58}$$

where

σ = applied stress at which unloading occurs
μ = coefficient of friction between the two surfaces of the crack and
β = inclination of the crack to the applied stress σ.

The value of $\Delta\sigma$ is positive as long as $\beta \neq 0$. When $\beta = 0$, such cracks do not decrease the effective modulus. Therefore, the modulus value obtained immediately on loading should give the intrinsic modulus of the material. However, when rock contains cavities other than cracks which deform under pressure, the unloading modulus is not necessarily the intrinsic modulus value. Tests conducted by MOGI (1959) on marble specimens which were subjected successively to higher stresses before unloading showed that loading modulus decreased markedly as the fracture stress was approached, the initial slope during unloading was the same for all samples. Similarly, tests by NISHIHARA (1957) on saturated and dry sandstone showed that while the average modulus value for saturated samples was $^2/_3$ that of dry samples, the modulus value at initial loading was the same for both dry and saturated specimens. This is in accordance with the theory that while the average modulus value on loading depends upon the frictional coefficient of cracks, the modulus in initial stages of loading is independent of these parameters and is solely dependent upon crack density. Another test of the validity of these expressions can be made by comparing the friction coefficient calculated from the stress-strain curve with the actually measured value.

By combining Eqs. 6.54 and 6.56,

$$\frac{E_f^{-1} - E_g^{-1}}{E_0 - E_g} = \frac{1}{5}\left[\frac{2 + 3\mu^2 + 2\mu^4}{(1 + \mu^2)^{3/2}} - 2\mu\right] \tag{6.59}$$

where

E_0 = modulus when all cracks are open
E_g = unloading modulus
E_f = modulus in the linear part of the curve.

The E_o can best be measured by applying a certain tensional load to the specimens. By substituting the values of E_f, E_g and E_o from the measured values, the frictional coefficient of the rock can be calculated. The so calculated frictional values are not in agreement with the measured values of the friction coefficient in direct sliding. These equations are derived with the assumption of elliptical cracks and that crack density is low with no interaction between the cracks. When interaction is assumed, the modulus value may increase by about 10%. The penny-shaped cracks give about 15% less value and the difference between plane strain and plane stress may be about 10%. Such differences, therefore, are negligible in comparison with the overall level of approximation. BUDIANSKY and O'CONNELL (1976) have given a method to obtain elastic moduli of bodies containing randomly distributed cracks with or without fluids.

Schematically, the loading and unloading curve is given in Fig. 6-66. For a specimen containing a single inclined closed crack, the E_{eff} is constant on loading (OA) and on unloading the curve follows the line AB (Fig. 6-66a).

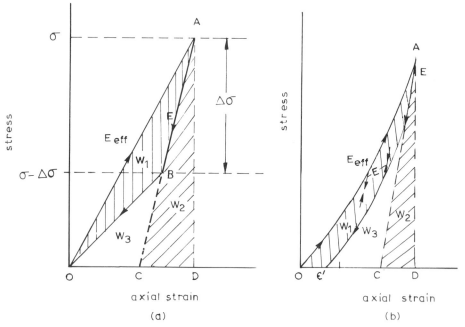

Fig. 6–66 (a): Stress-strain curve for a body in uniaxial compression containing a single inclined closed crack in loading and unloading
(b): Actual curve during loading and unloading containing a number of differently oriented cracks.

At point B, reverse sliding of crack takes place and if this line is extended to intersect the strain axis, the work $(W_1 + W_3)$ would be stored in the body in the presence of the crack. If complete reversal of crack sliding occurs, strain is fully recovered, and the area between the loading and unloading line gives the work done during sliding (W_1). The modulus value during reverse sliding is lower than that during loading and W_3 is the additional strain energy released due to crack.

In a multiple crack body, certain cracks on unloading may not slide back to their original position because of the absence of necessary stress drop required, thus giving permanent strain ε' (Fig. 6-66b).

The difference in modulus values on loading and unloading is the cause of hysteresis observed in rock specimens. When all cracks have been closed under high stresses, the effective modulus becomes the intrinsic modulus and no hysteresis effect shall be observed.

The effect of cracks on Poisson's ratio and compressibility of the rock has been similarly examined by WALSH (1965b,c) and is given by the following relationships:

Poisson's ratio, when all cracks are closed (high stress)

$$\nu_{\text{eff}} = \nu + \frac{1-2\nu}{2}\left(1 - \frac{E_{\text{eff}}}{E}\right) \qquad \text{(for plane strain)} \qquad (6.60)$$

$$\nu_{\text{eff}} = \nu + \left[\frac{2\pi(1-2\nu)(1-\nu^2)N\bar{c}^3}{15\,V}\right]\left[\frac{2+3\mu^2+2\mu^4}{(1+\mu^2)^{3/2}} - 2\mu\right]$$

$$\text{(for plane stress)} \qquad (6.61)$$

When all cracks are open, (initial stage of loading) initial Poisson's ratio (ν_0);

$$\nu_{\text{eff}} = \nu_0 = \nu\,(E_0/E) \qquad (6.62)$$

or

$$\nu_{\text{eff}} = \nu_0 = \nu\left(1 - \frac{4\pi N\bar{c}^3}{3\,V}\right) \qquad (6.63)$$

Compressibility, when all cracks are open (WALSH, 1965c)

$$K_{\text{eff}} = K\left[1 + \frac{4\pi}{3}\left(\frac{1-\nu^2}{1-2\nu}\right)\frac{\bar{c}^3}{\nu}\right] \qquad \text{(plane strain)} \qquad (6.64)$$

$$K_{\text{eff}} = K\left[1 + \frac{4\pi}{3(1-2\nu)}\frac{\bar{c}^3}{\bar{\nu}}\right] \qquad \text{(plane stress)} \qquad (6.65)$$

When all cracks are closed

$$K_{\text{eff}} = K \tag{6.66}$$

As indicated by Eq. 6.60, at low stress POISSON's ratio shall be less than the intrinsic value and at high stresses greater than the intrinsic value. This behaviour agrees with the increase in POISSON's ratio with increasing stress as reported by SCHMIDT (1926), ZISMAN (1933), NISHIHARA (1957) and SALUSTOWICZ (1965).

Eq. 6.62 indicates that where (E_0/E) approaches zero, i.e. start of loading of a highly cracked rock, the POISSON's ratio shall be zero.

This also makes clear that rocks having lower E_{eff} (highly cracked) will show lower POISSON's ratio and those highly compact (high E_{eff}) will show higher POISSON's ratio. ZISMAN (1933) found that at low stress the rocks having lower E have lower v and those having higher E have higher v. This also explains the influence of stress on the measured values of POISSON's

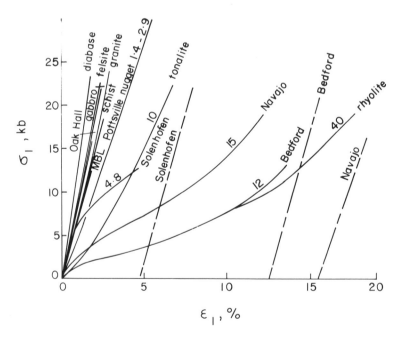

Fig. 6–67: Stress in the axial direction as a function of axial strain. The small number on some of the curves is porosity in per cent. The dotted lines are stress-strain curves which would be followed if porosity were zero
(after BRACE and RILEY, 1972).

ratio which show an increase with increased stress level. At higher stress levels, opening of cracks parallel to axis of maximum stress will result in higher measured values of lateral strain. This gives rise to lateral strain curve convex towards the stress axis as is usually observed in practice and the POISSON's ratio calculated in extreme cases may exceed 0.5 (LAMA, 1966).

It shall be seen that while the expressions for POISSON's ratio and axial modulus contain frictional term (μ), the volumetric strain (compressibility) does not. As such no hysteresis shall be expected in the volumetric strain curve. Experimental observations show that this is very nearly true (PAULDING, 1965). Tests (with lateral strain – 0) by BRACE and RILEY (1972) on rocks of different porosities (porosity range from almost zero to about 40%) at extremely high stress values (up to 30 kb) are given in Fig. 6-67. The porosity values at zero stress are given in Table 8. The stress-strain curves are of all the three types as indicated earlier and the difference is due to difference in crack porosity. Rhyolite, Bedford limestone, Solen-hofen limestone have low porosity and no crack porosity whereas Pottsville sandstone has high crack porosity. The stress-strain curve in the absence of any porosity would be quite different and theoretical values are given by dotted lines. The position where these intersect the abcissa gives the porosity of the rock (e.g. Solenhofen limestone porosity = 4.8%). The stress at which the measured curves intersect these dotted lines is the stress at which porosity reaches zero. For the two limestones, this stress appears to be 15–20 kb. For the Navajo sandstone this stress probably exceeds 30 kb. For the tonalite it may be a great deal higher. It may be mentioned that permanent strain produced in the rock on unloading is not necessarily directly related to the initial porosity (Fig. 6-68) though stress concentration around the pores determines the stress required for collapse of the pores and this value must be exceeded at which the specimen is to be unloaded to achieve the acceptable value.

YOUNG's modulus for soft rocks is very sensitive to weathering and variation of YOUNG's modulus with apparent specific gravity for marlstone obtained by BLAIR (1955) is shown in Fig. 6-69.

6.7.7. Rock Fabric and Modulus-Anisotropy

Rock being a composite material consisting of a large number of different minerals, the effective modulus values therefore are dependent upon the contribution of these individual minerals, their relative proportion, shape and orientation. Theoretical analysis of treating rock as a composite material has been limited only to a two-phase system where it is assumed that it consists of elastic granular inclusions surrounded by a matrix having

TABLE 8
Porosity of rocks at zero stress
(Ref. Fig. 6-67)
(after BRACE and RILEY, 1972)

Diabase, II . . . . Frederick, Md.	0.1
Gabbro . . . . San Marcos, Cal.	0.2
Schist . . . . source unknown	0.3
White marble . . . . source unknown	0.3
Lynn felsite . . . . Saugus, Mass.	0.3
Limestone II . . . . Oak Hall, Pa.	0.6
Granite . . . . Barre, Vt.	0.6
Granite . . . . Westerly, R.I.	0.9
Nugget sandstone . . . . Utah	1.9
Pottsville sandstone . . . . Tennessee	2.9
Limestone . . . . Solenhofen	4.8
Tonalite . . . . Cedar City, Utah	7
Limestone . . . . Bedford, Ind.	12
Navajo sandstone . . . . source unknown	15.5
Rhyolite tuff . . . . Colorado	40

different elastic properties. A number of equations have been developed by various investigators (PAUL, 1960, HASHIN, 1962, GRESZCZUK, 1966, LAMA, 1969, Ko and HAAS, 1972). Their application to rocks has not been proved because of obvious reasons, but these do help to demonstrate the modulus anisotropy in rocks which have preferred orientation of grains of elongated shapes.

PAUL (1960) has shown that for a two components material, the effective lower and upper bounds of the YOUNG's modulus are given by

$$E_l = \frac{1}{(f/E_1) + (1-f)/E_2} \tag{6.67}$$

$$E_u = \left[\frac{1 - \nu_1 + 2m(m - 2\nu_1)}{1 - \nu_1 - 2\nu_1^2}\right] E_1 f +$$
$$\left[\frac{1 - \nu_2 + 2m(m - 2\nu_2)}{1 - \nu_2 - 2\nu_2^2}\right] E_2 (1-f) \tag{6.68}$$

where E_l and E_u = lower and upper bounds of the composite

ν = POISSON's ratio with subscripts 1 and 2 referring to matrix and inclusion respectively

E = YOUNG's modulus with subscripts 1 and 2 referring to matrix and inclusion respectively

f = ratio of volume of matrix to volume of elemental cube

$m = \dfrac{\nu_1(1 + \nu_2)(1 - 2\nu_2)fE_1 + \nu_2(1 + \nu_1)(1 - 2\nu_1)(1-f)E_2}{(1 + \nu_2)(1 - 2\nu_2)fE_1 + (1 + \nu_1)(1 - 2\nu_1)(1-f)E_2}$

The upper bound assumes uniform stress in the material and the lower bound uniform strain in the material. The upper and lower bounds have been developed using the minimum potential energy theorem and the theorem of least work, respectively, of the theory of elasticity. Although the bounds are theoretically exact the variation between them is quite large to allow a good estimate to be made. PAUL (1960) has further shown that an approximate effective YOUNG's modulus can be given by

$$\frac{1}{E_{\text{eff}}} = \int_0^1 \frac{dz}{E_1 + (E_2 - E_1) A_{2(z)}} \tag{6.69}$$

where E_{eff} = effective YOUNG's modulus of rock

E_1 and E_2 = modulus of the matrix and solid inclusions

$A_{2(z)}$ = cross-sectional area of the inclusion expressed as a function of coordinate Z representing the direction of applied stress.

This equation clearly shows the influence of the shape of the inclusions on the effective modulus of the composite. In a rock, if the grain sizes are spherical and uniformly distributed, the YOUNG's modulus value will be the same in the various directions. But if the shape of the particles deviates from that of a sphere, the value of $\int_0^1 A_{2(Z)}$ will be different and so the value of effective YOUNG's modulus. For a multi-phase system such as rocks are, it is more convenient to use the averaging method put forward by VOIGHT or REUSS. According to VOIGHT (1910), the YOUNG's modulus of the composite can be given by

$$\frac{1}{E_{eff}} = \frac{V_a}{E_a} + \frac{V_b}{E_b} + \frac{V_c}{E_c} + - - - - \tag{6.70}$$

where V_a, V_b, V_c are the percentages of the minerals in the rock as determined from modal analysis and E_a, E_b and E_c are the respective YOUNG's modulus values. According to REUSS (1929) the effective modulus is given by

$$E_{eff} = V_a E_a + V_b E_b + V_c E_c + - - - - - \tag{6.71}$$

When the shape of the grains is spherical, the volumetric percentage of the minerals is taken in calculations. When the shape is not spherical and the

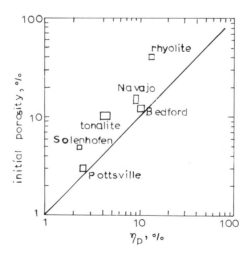

Fig. 6–68: Comparison of permanent volumetric compaction with initial porosity for high-porosity rocks. Size of boxes indicates uncertainty (after BRACE and RILEY, 1972).

particles have a preferred orientation, the percentage surface area of the minerals along a section parallel to the direction of application of the stress be used in calculations. This permits to determine the modulus anisotropy in different directions.

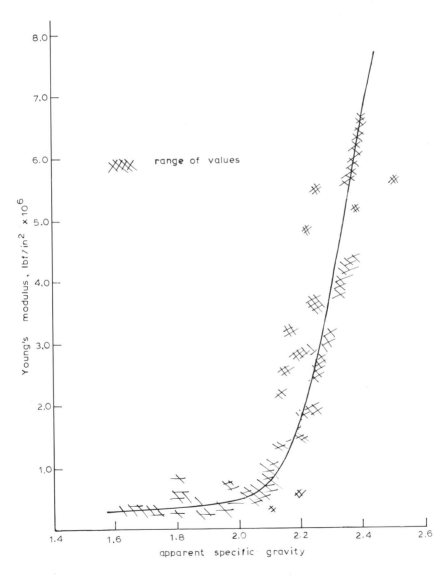

Fig. 6–69: Variation of YOUNG's modulus with apparent specific gravity of marlstone (after BLAIR, 1955).

It should be noted that VOIGHT's average is based on the assumption of uniform strain or deformation of the mineral aggregates in the rock and the REUSS average is based upon the assumption of uniform stress or pressure in the aggregate. VOIGHT's average therefore gives the upper limit and the REUSS's average gives the lower limit. The actual value shall be somewhere in between the two limits. HILL (1952) has suggested the use of arithmetic or the geometric average of the VOIGHT and REUSS values. REUSS value is more suitable for isotropic rocks and for high amplitude stress or high rate of compression (KUMAZAWA, 1969). At low strain rates or low static load values and when highly anisotropic mineral constituents are present in a rock, HILL's geometric value is better suited. Calculated and measured values of a number of elastic constants such as compressibility, wave velocity, LAMÉ's constant are in fair agreement with the above.

Bedding planes and joints influence the modulus anisotropy greatly. Rocks show higher modulus values when stressed in the direction parallel to the bedding planes or joints than when loaded at right angles to them.

If we consider a layered rock body consisting of n number of beds of thickness S_1, S_2, S_3 .... S_n and subjected to a stress σ at right angle to their bedding plane, the deformation of each bed can be given by

$$\Delta S_1 = \frac{\sigma \cdot S_1}{E_1}$$

$$\Delta S_2 = \frac{\sigma \cdot S_2}{E_2}$$

$$\Delta S_3 = \frac{\sigma \cdot S_3}{E_3}$$

$$- - - - -$$

$$- - - - -$$

$$\Delta S_n = \frac{\sigma \cdot S_n}{E_n}$$

The total deformation of the rock body ΔS is the sum of the deformation of each bed such that

$$\sum_1^n \Delta S = \left(\frac{\sigma S_1}{E_1} + \frac{\sigma S_2}{E_2} + \frac{\sigma S_3}{E_3} + - - - - \frac{\sigma S_n}{E_n} \right)$$

or $$\frac{\sum_1^n \Delta S}{\sigma} = \left(\frac{S_1}{E_1} + \frac{S_2}{E_2} + \frac{S_3}{E_3} + - - - - \frac{S_n}{E_n} \right)$$

$$E_{eff} = \frac{\sum_{1}^{n} S}{\left(\dfrac{S_1}{E_1} + \dfrac{S_2}{E_2} + \dfrac{S_3}{E_3} + \cdots \dfrac{S_n}{E_n}\right)} \tag{6.72}$$

If $S_1 = S_2 = S_3 = S$, then $\sum_{1}^{n} S = ns$

$$E_{eff} = \frac{n}{\left(\dfrac{1}{E_1} + \dfrac{1}{E_2} + \dfrac{1}{E_3} + \cdots + \dfrac{1}{E_n}\right)} \tag{6.73}$$

Similarly, if a body is intersected by horizontal joints with a density of J (i.e. J joints per unit length), and if the modulus value of the rock substance is E_r and normal stiffness of the joint is K_j, the total strain of the rock mass (ε_{rm}) under the action of stress σ acting at right angles to the joint plane is equal to the sum of the strain of the rock substance (ε_r) and joint closure (ε_j) (joint strain) and can be given by

$$\varepsilon_{rm} = \varepsilon_r + J\,\varepsilon_j$$

or

$$\frac{\sigma}{E_{eff}} = \frac{\sigma}{E_r} + J\frac{\sigma}{K_j}$$

or

$$E_{eff} = \frac{1}{\left(\dfrac{1}{E_r} + \dfrac{J}{K_j}\right)} \tag{6.74}$$

Similarly the shear modulus of a bedded deposit is given by

$$G_{eff} = \frac{\sum_{1}^{n} S}{\left(\dfrac{S_1}{G_1} + \dfrac{S_2}{G_2} + \dfrac{S_3}{G_3} \cdots \dfrac{S_n}{G_n}\right)} \tag{6.75}$$

and for a rock intersected with horizontal joints

$$G_{eff} = \frac{1}{\left(\dfrac{1}{G_r} + \dfrac{J}{K_s}\right)} \tag{6.76}$$

where G_{eff} = effective shear modulus of the rock mass
         $G_1, G_2, G_3, \ldots, G_n$ = shear modulus of beds 1 to n
         G_r = shear modulus of rock substance
         K_s = shear stiffness of joints.

When a specimen is loaded parallel to the bedding plane, stress distribution in the specimen is non-uniform but each bed undergoes uniform strain. Under these assumptions, the total force F per unit width acting on the specimen is equal to the sum total carried by each bed such that

$$F = F_1 + F_2 + F_3 + \ldots F_n$$

If the average stress is σ, and the displacement caused in each bed is Δl, and the thickness of each bed is $S_1, S_2, S_3, \ldots S_n$, then

$$F = \frac{\Delta l \, E_{eff} \sum_1^n S}{l} = \frac{\Delta l \, S_1 E_1}{l} + \frac{\Delta l \, S_2 E_2}{l} + \frac{\Delta l \, S_3 E_3}{l} + \ldots \frac{\Delta l \, S_n E_n}{l}$$

or $$E_{eff} \sum_1^n S = S_1 E_1 + S_2 E_2 + S_3 E_3 + \ldots S_n E_n$$

or $$E_{eff} = \frac{S_1 E_1 + S_2 E_2 + S_3 E_3 + \ldots S_n E_n}{\sum_1^n S} \tag{6.77}$$

If the thickness of these beds is uniform $S_1 = S_2 = S_3 \ldots S_n$, then

$$E_{eff} = \frac{E_1 + E_2 + E_3 \ldots E_n}{n} \tag{6.78}$$

Similarly, the shear modulus can be proved to be

$$G_{eff} = \frac{S_1 G_1 + S_2 G_2 + S_3 G_3 + \ldots S_n G_n}{\sum_1^n S} \tag{6.79}$$

or when $$S_1 = S_1 = S_3 \ldots S_n$$

$$G_{eff} = \frac{G_1 + G_2 + G_3 + \ldots G_n}{n} \tag{6.80}$$

When the direction of loading is not parallel to any of the two orthogonal directions, the modulus values will be different. If the elastic constants in the two directions in the same plane are different and their values are known, then the constants in any direction in the plane (Fig. 6-70) can be calculated using the relationship (LEKHNITSKII, 1968).

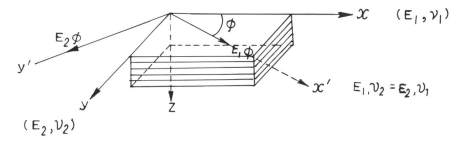

Fig. 6–70: Anisotropy of elastic constants.

$$\frac{1}{E_1 \, \varphi} = \frac{\cos^4 \varphi}{E_1} + \left(\frac{1}{G_1} - \frac{2\nu_1}{E_1}\right) \sin^2 \varphi, \cos^2 \varphi + \frac{\sin^4 \varphi}{E_2} \qquad (6.81)$$

$$\frac{1}{E_2 \, \varphi} = \frac{\sin^4 \varphi}{E_1} + \left(\frac{1}{G_1} - \frac{2\nu_1}{E_1}\right) \sin^2 \varphi, \cos^2 \varphi + \frac{\cos^4 \varphi}{E_2} \qquad (6.82)$$

$$\frac{1}{G'} = \frac{1}{G_1} \left(\frac{1+\nu_1}{E_2} + \frac{1+\nu_2}{E_2} - \frac{1}{G}\right) \sin^2 2\varphi \qquad (6.83)$$

$$\nu_1\varphi = E_1\varphi \left[\frac{\nu_1}{E_1} - \frac{1}{4}\left(\frac{1+\nu_1}{E_2} + \frac{1+\nu_2}{E_2} - \frac{1}{G}\right)\right] \sin^2 2\varphi \qquad (6.84)$$

$$\nu_2\varphi = \nu_1\varphi \frac{E_2\varphi}{E_1\varphi} \qquad (6.85)$$

where E_1, E_2 = YOUNG's modulus (tension or compression) in the direction x and y respectively.

$E_1\varphi, E_2\varphi$ = YOUNG's modulus (tension or compression) in the new directions x' and y' respectively.

$\nu_1, \nu_2, \nu_1\varphi, \nu_2\varphi$ = respective values of POISSON's ratios

G_1 = shear modulus in the plan $x - y$

G' = shear modulus in the new direction

φ = angle of rotation between the two coordinates.

In practice the elastic constants a_{ij}[1] referred to an arbitrary system of

---

[1] a_{ij} are called the compliances and are the reciprocals of the moduli. The notations refer to the standards used in mechanics.

coordinates x and y are measured in a simple compression or tension test and we need to determine the principal elastic constants.

Let x' and y' be the principal axes, and a'_{ij} the principal elastic constants. The unknown angle φ, which is formed by axis x with one of the principal directions will be determined as the smallest angle which will satisfy the following conditions:

The angle φ through which the axis must be rotated is given by

$$tg2\varphi = \frac{a_{16} + a_{26}}{a_{11} - a_{22}}$$

$$tg4\varphi = 2\ \frac{a_{16} - a_{26}}{a_{11} + a_{22} - a_{12} - a_{66}} \tag{6.86}$$

The condition that these two equations be identically satisfied is

$$(a_{16} - a_{26})\,(a_{11} - a_{22} + a_{16} + a_{26})\,(a_{11} - a_{22} - a_{16} - a_{26})$$
$$= (a_{16} + a_{26})\,(a_{11} - a_{22})\,(a_{11} + a_{22} - 2a_{12} - a_{66}) \tag{6.87}$$

When this condition is not satisfied, then there are no principal directions in plane xy, i.e., the plate is not orthotropic. For example, let the elastic constants of an anisotropic plate which are referred to a certain system of coordinates xy, namely,

$$a_{16} = a_{26} \neq 0,\ a_{11} = a_{22}, a_{11} + a_{22} - 2a_{12} - a_{66} > 0.$$

The condition of the existence of principal directions (Eq. 6.87) is satisfied obviously. Equation (6-86) takes the form

$$tg\,2\varphi = \infty,\quad tg\,4\varphi = 0 \tag{6.88}$$

From the Eq. 6.88 we find $\varphi = \dfrac{\pi}{4}, \dfrac{3\pi}{4}, \dfrac{5\pi}{4}, \dfrac{7\pi}{4} \ldots$ and from the second, $\varphi = 0, \dfrac{\pi}{4}, \dfrac{\pi}{2}, \dfrac{3\pi}{4} \ldots$ The solutions of the first equation are applied to the second. Consequently, we can assume that $\varphi = \dfrac{\pi}{4}$. The principal directions of elasticity exist, they are the directions of the bisectors of the angles between axes x and y. The principal elastic constants can be determined from Eqs. 6.81 to 6.84 by substituting the value of φ.

In the case of bedded or schistose rocks, the deformability normal and parallel to these planes is measured and usually represents the limiting value. Results of some laboratory and in situ tests by PINTO (1970) on schists are given in Fig. 6-71. The type of curve necessarily remains the same, and depends upon the relative modulus values.

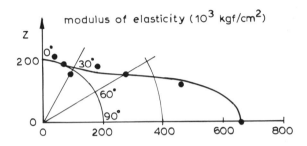

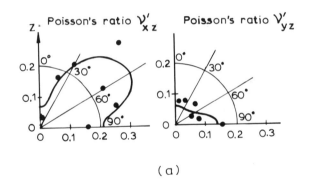

(a)

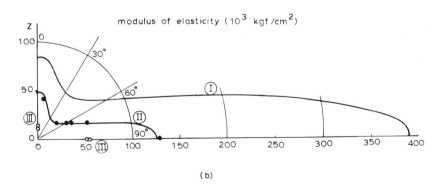

(b)

Fig. 6–71 (a): Modulus of elasticity and POISSON's ratio for a schistose
rock (laboratory tests).
(b): Elastic modulus variations for 3 different Graywacke schist in situ tests
(after PINTO, 1970).

The ratio between the extreme values is many times referred to as the degree of anisotropy (PANOV, SAPEGIN and KHRAPKOV, 1970). The results of various investigations on different rocks are given in Tables 9 and 10. In most cases the laboratory results show the ratio to be of the order of 1–2 but in situ values are greatly higher. PINTO (1970) found the modulus ratios in in situ tests to be 3.5. It is however, possible that the minimum value of the modulus may not be at right angle to the schistosity planes (LNEC, 1964).

The influence of orientation of bedding on the modulus of elasticity at different confining pressures obtained by GRAY (1967) is given in Fig. 6-72. Sensitivity of the modulus to confining pressure is different for different orientations. The cause of existence of the peak value is not clear.

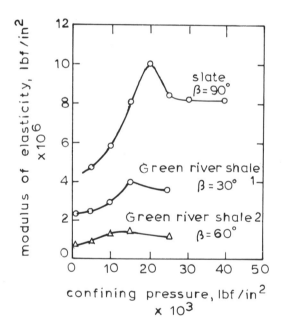

Fig. 6–72: Variation of the modulus of elasticity with respect to confining pressures. β – angle of orientation of bedding plane to the direction of maximum principal stress (after GRAY, 1967).

6.8. Poisson's Ratio of Rocks

POISSON's ratio is calculated from the lateral and longitudinal strains measured on specimens subjected to uniaxial compression or uniaxial tension and is given by

$$v = \frac{\varepsilon_{lat}}{\varepsilon_{long}} \qquad (6.89)$$

where v = POISSON's ratio

ε_{lat} = strain at right angle to the direction of application of load and

ε_{long} = strain parallel to the direction of application of load.

The reciprocal of it is called the POISSON's number and is denoted by m.

The values of POISSON's ratio so measured are affected by a number of factors such as stress level, presence and absence of cracks, temperature, rate of loading etc., in similar manner as the modulus values. Presence of cracks and pores decreases the POISSON's ratio, but cracks if oriented parallel to the direction of application of stress may tend to open up resulting in higher lateral strain and higher POISSON's ratio. That is why the POISSON's ratio calculated increases as the stress is raised. POISSON's ratio at small loads is very small and may be even negative when specimens are not carefully lapped and longitudinal strain is measured by gauges placed between the press platens. With increase in load its value rises and may even exceed the theoretical value of 0.5. HORIBE and KOBAYASHI (1960) found that, for Iwaki sandstones, POISSON's number varies from 18 at a stress of 1 MPa (142 lbf/in²) to 2 at a stress of 12.24 MPa (1775 lbf/in²). Similar results have been obtained by SERAFIM and LOPES (1961) in triaxial tests on granite. Results obtained by the Association des Ingenieurs Sartes de Liège (SA-LUSTOWICZ, 1965) on the value of m are given in Table 11.

The relationship between POISSON's number and stress obtained by BAR-SCHINGER for sandstones is represented in Fig. 6-73 (SCHMIDT, 1926). Results obtained by WALSH and BRACE (1966b) on Westerly granite are given in Fig. 6-74.

Tests on coal specimens conducted by LAMA (1966) show that in well cleaved specimens lateral strain far exceeds the vertical strain. This is due to opening out of the cleavage planes giving rise to high lateral strain, a major portion of which is permanent and irreversible (Fig. 6-75). This results in calculated value of POISSON's ratio exceeding the theoretical maximum value of 0.5 and sometimes gives values as high as 1.0 or 1.2.

TABLE 9
Measured cross-anisotropic parameters
(after Gerrard, Davis and Wardle, 1972)

Material	E_h/E_v	F_h/F_v	v_{vh}	v_h	References
Marl-Turonian	2.17	—	—	—	Lozińska-Stepień, 1966
Clay shale	1.36	—	—	—	Stepanov et al., 1967
Clay shale	2.00	—	—	—	Stepanov et al., 1967
Clay shale	2.86	—	—	—	Stepanov et al., 1967
Clay limestone – Maestrichtian	0.62	—	—	—	Lozińska-Stepień, 1966
Clay limestone – Turonian	0.84 – 1.69	—	—	—	Lozińska-Stepień, 1966
Opaka marl – Maestrichtian	0.85 – 1.00	—	—	—	Lozińska-Stepień, 1966
Opaka marl – Turonian	1.30 – 2.06	—	—	—	Lozińska-Stepień, 1966
Marly opaka – Maestrichtian	1.00	—	—	—	Lozińska-Stepień, 1966
Sandy shale	1.56	—	—	—	Stepanov et al., 1967
Sandy shale	1.28	—	0.19	0.10	Stepanov et al., 1967
Limestone	1.8 – 3.5	—	—	—	Khachikian, 1966
Opaka limestone – Turonian	0.92	—	—	—	Lozińska-Stepień, 1966
Sandstone	1.53	—	—	—	Stepanov et al., 1967
Sandstone	1.23	—	—	—	Stepanov et al., 1967
Sandy shale	1.68	—	—	—	Stepanov et al., 1967
Sandstone	1.33	—	—	—	Stepanov et al., 1967
Hard blue slate	1.50	1.50	0.43	0.43	Attewell, 1970

Basalt	1.50	—	—	—	Bamford, 1969
Slate	1.70	—	—	—	Bamford, 1969
"Phyllades de Revin"	3.90	—	0.16	0.22	Masure, 1970
Bandera sandstone	1.50	1.10	0.16	0.14	King, 1968
Berea sandstone	1.40	1.20	0.22	0.14	King, 1968
"Schist I"	1.30	1.40	0.21	0.27	Pinto, 1970
"Schist II"	1.90	1.50	0.14	0.22	Pinto, 1970
"Schist"	3.20	3.50	0.067	0.13	Pinto, 1970
Granite	1.40	1.15	0.14	0.17	Duvall, 1965
Aleurolite	1.09	1.12	0.26	0.215	Lekhnitskii, 1966
Coarse Phyllite	1.28	1.11	0.33	0.27	Lekhnitskii, 1966
Fine Phyllite	1.33	1.02	0.28	0.26	Lekhnitskii, 1966
Sylvinite (rich)	1.13	—	0.28	0.26	Stepanov et al., 1967
Chlorite slate	1.57	—	0.28	1.17	Stepanov et al., 1967
Tuffuceous sandstone	1.26	—	0.25	0.02	Stepanov et al., 1967
Siltstone	1.08	—	0.28	0.17	Stepanov et al., 1967
"Basalt I"	1.34	—	0.20	0.11	Stepanov et al., 1967
"Basalt II"	1.05	—	0.14	0.10	Stepanov et al., 1967
"Basalt III"	1.08	—	0.22	0.18	Stepanov et al., 1967
Peridotite	1.78	—	0.32	0.29	Stepanov et al., 1967
Gneiss	5.40	—	—	—	Stepanov et al., 1967
Barre granite	0.73	—	—	—	Douglass & Voight, 1969
Laurentian granite	1.24	—	—	—	Douglass & Voight, 1969
Granite	2.00	—	—	—	Serafim, 1963

TABLE 10
Stress-strain properties of earthern materials
(after GERRARD, 1975)

(a) Quartzofeldspathic Rocks

Material	Fabric description	Symmetry	Test method
1.1 Barre Granite			Longitudinal and shear wave velocities
		Isotropic	Uniaxial compression = 0
1.2		"n" fold axis parallel to direction of uniaxial stress (axis 3)	Uniaxial compression = 40 MPa
1.3 Barre Granite	Orthorhombic with the three planes of symmetry being those of least cleavage resistance (rift), intermediate (grain), and greatest resistance (hardway). The normals to these planes are respectively the directions of 3, 2 and 1. Concentration of microfractures in the rift plane.	Orthorhombic – planes of symmetry in stiffness properties similar to those for microfractures	Laboratory tests – Oriented samples uniaxially loaded to 67 MPa at a rate of 6.7 MPa/min. Strains recorded by bonded strain gauges. Results quoted are secant moduli.
1.4 Stanstead Granite			
1.5 Laurentian granite	One principal axis of stiffness coincides with rift axis but the other two are offset by 60° from the grain and hardway axes.	Orthorhombic?	
1.6 Alvarenga	Joint systems often in two or three inter-secting steeply dipping sets. In some cases the two intersecting sets are at right angles, suggesting an orthorhombic structure.	Orthorhombic in some cases. Axis 3 i.e. approximately vertical and approximately parallel to the joints is often direction of least stiffness. Axes 1 and 2 in some cases are approximately normal to the joint sets.	Laboratory tests – Oriented samples uniaxially loaded to 19.6 MPa. Strains recorded by bonded strain gauges. Results quoted are secant moduli.
1.7 Granite			
1.8 Altolindoso			
1.9 Granite			
1.10			
1.11 Vilarinho Gneiss			
1.12 Granite			
1.13 Granite Alto Rabagao Dam	Microfissures on horizontal plane. Axis 3 is vertical.	Orthorhombic?	Laboratory tests on prisms, together with field plate loading tests in galleries.
1.14 Granite Vilar Dam	Microfissures on horizontal plane. Axis 3 is vertical, axis 1 is horizontal in the direction of the river, axis 2 is normal to river.	Orthorhombic?	
1.15 Granite		Isotropic	Longitudinal and shear wave velocities on cube samples.

(a) Quartzofeldspathic Rocks

Test sample size (mm)	GPa						v_{12} (v_{21})	v_{13} (v_{31})	v_{23} (v_{32})	Reference
	E_1	E_2	E_3	G_{23}	G_{13}	G_{12}				
100 dia. x 100 long										
	36.7		36.7	18.3		18.3	0.036			Nur and Simmons (1969)
	43.6		56.2	24.2		20.1	0.086	0.070 (0.085)		
29 dia. x 43 long	46.8	36.6	33.7							Douglass and Voight (1969)
	47.5	35.5	25.9							
50 x 50 x 150 prisms	78.0	42.8	30.2							Rodrigues (1966)
	37.5	17.5	24.3							
	37.3	48.3	30.4							
	49.8	36.2	28.5							
	26.4	34.2	41.2							
	48.0	42.2	38.3							
	48.5	34.6	36.3							
	3.62	3.07	1.58							Serafim (1963)
	14.4	5.99	3.24							
	81.9			33.5						Alexandrov et al. (1969)

TABLE 10 (continued)

(a) Quartzofeldspathic Rocks

Material	Fabric description	Symmetry	Test method
1.16 Granite		Orthorhombic (axes 1, 2, 3)	Resonant frequency and ultra-sonic pulse tests on oriented cores
1.17 Granite		Orthorhombic?	
1.18 Granite Gneiss 1.19 (average of three)	Layering parallel to plane 12	Orthorhombic?	Field tests with radial jack. Pressure ≈ 4 MPa
1.20 Granite Gneiss 1.21 (compact)			Pressure ≈ 5 MPa
1.22 Gneiss with · Shale 1.23 (average of two)			Pressure ≈ 2.5 MPa
1.24 Gneiss (slaty) 1.25			Pressure ≈ 4 MPa
1.26 Mylonite Gneiss 1.27			Pressure ≈ 1.5 MPa
1.28 Gneiss 1.29	Layering parallel to plane 12. Slightly fissured	Orthorhombic	Laboratory compression tests on oriented samples | Pressure = 0−2 MPa Pressure = 49−59 MPa
1.30 Gneiss		Orthorhombic?	
1.31 Gneiss		"n" fold axis?	
1.32 Gneiss	Layering parallel to plane 12	"n" fold axis (axis 3)	Uniaxial unconfined tensile tests of oriented samples. Tangent moduli taken at half strength

(b) Basic Lithic

2.1 Basalt		Isotropic	Longitudinal and shear wave velocities on cube sample
2.2 Basalt		Isotropic	Resonant frequency and ultrasonic pulse methods

(a) Quartzofeldspathic Rocks

Test sample size (mm)	GPa						v_{12} (v_{21})	v_{13} (v_{31})	v_{23} (v_{32})	Reference
	E_1	E_2	E_3	G_{23}	G_{13}	G_{12}				
54 dia. x 305 long	42.4	44.5	30.3	17.2	16.9	19.6	0.168 (0.176)	0.203 (0.146)	0.209 (0.141)	DUVALL (1965)
	27.4	44.2	30.4							OBERT and DUVALL (1967)
2200 dia. x 2000 long	53. (41.)		15. (11.)							LAUFFER and SEEBER (1961), SEEBER (1970)
	32. (28.)		23. (21.)							Unbracketed figures correspond to recovered deformation, bracketed figures correspond to total deformations
	37. (28.)		6.1 (3.1)							
	14.6 (9.8)		6.0 (2.0)							
	6.9 (5.9)		2.1 (0.93)							
10 x 10 x 30 Prisms	17.3	15.1	7.3	6.2	6.4	7.6	0.005	0.058	0.087	TREMMEL and WIDMANN (1970)
	40.7	42.7	36.2	12.1	15.2	18.5	0.318	0.367	0.351	
	23.1	12.4	18.6							OBERT and DUVALL (1967)
	48.9		9.0							STEPANOV and BATUGIN (1967)
29.7 dia. x 68 long	$5E_3$		E_3							BARLA and GOFFI (1974)

(b) Basic Lithic

	62.9			31.4						ALEXANDROV et al. (1969)
75 x 76 x 54	98.0						0.28			KAARSBERG (1968)

TABLE 10 (continued)

(b) Basic Lithic

	Material	Fabric description	Symmetry	Test method
2.3 2.4 2.5	Basalt	Preferred orientation of mineral grains with optical axes oriented perpendicular to direction of tectonic compression, parallel to direction of tension	"n" fold axis parallel to the direction of tectonic compression (axis 3)	Laboratory tests — longitudinal and shear wave velocities
2.6	Peridotite			
2.7	Dunite	Strong concentration of Olivine a axes and girdles of b and c axes	"n" fold axis of symmetry parallel to a axes (i.e. axis 3)	Longitudinal and transverse wave velocities. Confining pressures from 0.1 GPa to 1.0 GPa
2.8	Dunite	Strong concentration of all three Olivine axes (a, b, c)	Orthorhombic a axis = axis 3 b axis = axis 2 c axis = axis 1	Results quoted are for a confining pressure of 0.4 GPa
2.9	Dunite		Orthorhombic	Longitudinal and shear wave velocities on cube samples
2.10	Olivinite		Orthorhombic	Longitudinal and shear wave velocities on cube samples
2.11	Enstatite		Orthorhombic	Longitudinal and shear wave velocities on cube samples
2.12	Andesite		Isotropic	Laboratory tests — uniaxial compression and tension tests, together with torsion tests. Strain gauges attached to samples
2.13	Hombendite		Orthorhombic	Longitudinal and shear wave velocities on cube samples
2.14	Eclogite	Preferred orientation of pyroxene grains relative to axis 3	"n" fold axis parallel to axis 3	Longitudinal and shear wave velocities on cube samples
2.15	Zoisitic prasinite		Orthorhombic	Longitudinal and shear wave velocities on cube samples

(b) Basic Lithic

Test sample size (mm)	GPa						v_{12} (v_{21})	v_{13} (v_{31})	v_{23} (v_{32})	Reference
	E_1	E_2	E_3	G_{23}	G_{13}	G_{12}				
50 dia. x 120 long	42. 53. 46.		40. 49. 34.		18. 21. 16.	18. 22. 20.	0.16 0.20 0.15	0.14 0.22 0.20		STORIKOVA (1965)
	94.		53.		21.	34.	0.38	0.31 (0.28)		STORIKOVA (1965)
25 dia. x 60 long	155.	155.	176.8	70.	70.	62.	0.240	0.400	0.308	CHRISTENSEN and RAMANANAN- TOANDRO (1971)
25 dia. x 60 long	190.	156.	216.	75.	80.	74.				CHRISTENSEN and RAMANANAN- TOANDRO (1971)
	101.	87.4	62.0	24.1	23.8	36.2	0.185	0.258	0.329	ALEXANDROV et al. (1969)
	171.	160.	151.	73.2	70.4	68.4	0.314	0.333	0.274	ALEXANDROV et al. (1969)
	157.	148.	134.	51.6	55.6	60.9	0.261	0.244	0.270	ALEXANDROV et al. (1969)
	32.			14.			0.153			NISHIMATSU (1970)
	115	106	98.	36.4	38.0	42.5	0.278	0.276	0.317	ALEXANDROV et al. (1969)
	142.	142.	178.	58.5	58.5	55.4	0.283	0.204	0.204	ALEXANDROV et al. (1969)
	135.5	125.	113.	63.5	51.1	45.5	0.228	0.354	0.364	ALEXANDROV et al. (1969)

TABLE 10 (continued)

(b) Basic Lithic

Material	Fabric description	Symmetry	Test method
			Ultrasonic longitudinal and shear wave velocities. Sample subject to range of hydrostatic pressures −
2.16 Bandera Sandstone			$\approx$ 0 MPa pressure
2.17	Preferred orientation of cracks parallel to bedding plane		68.9 MPa pressure
2.18 Berea Sandstone			$\approx$ 0 MPa pressure
2.19			68.9 MPa pressure
2.20 Sandstones (Average of 2)	Bedding plane is plane 12	Orthorhombic?	
2.21 Siltstone	Direction of bedding coincides with maximum wave velocity, minimum	"n" fold axis perpendicular to	Laboratory tests − longitudinal and
2.22 Tuffaceous Sandstone	velocity perpendicular to bedding (i.e. axis 3)	bedding?	shear wave velocities
2.23 Arkansas Sandstone	Breaks along bedding at iron oxide deposits spaced 50−150 mm apart. Bedding is plane 12.	Orthorhombic	Triaxial compression tests

(c) Pelitic Clay

Material	Fabric description	Symmetry	Test method
3.1 Sandy Clay Shale	Two fracture systems: (a) dip 30°−90°, strike SSE	Orthorhombic fabric but not aligned	Field tests − cube samples in the
3.2	(b) dip 30°−90°, strike SW Joint distance of the order of 30 mm.	with orthorhombic symmetry of test.	invert of test adit. Horizontal loads
3.3 Clay Shale		Plane 32 only common symmetry plane to both systems	by flat jacks in slots, vertical loads by hydraulic jacks
3.4 Colorado Oil Shale	Strong bedding in plane 12	"n" fold axis of symmetry perpendicular to bedding planes	Resonant frequency and ultrasonic pulse methods
3.5 Clay Shale (Average of 3)	Bedding plane is plane 12	"n" fold axis? (axis 3)	
3.6 Sandy Shale (Average of 3)	Bedding plane is plane 12	"n" fold axis? (axis 3)	

(b) Basic/Lithic

Test sample size (mm)	GPa						v_{12} (v_{21})	v_{13} (v_{31})	v_{23} (v_{32})	Reference
	E_1	E_2	E_3	G_{23}	G_{13}	G_{12}				
44 dia. x 89 long	18.6		12.4		7.3	8.1	0.14	0.25 (0.16)		KING (1968)
	28.9		26.2		12.0	12.7	0.14	0.19 (0.18)		
	19.3		14.5		7.0	8.5	0.14	0.30 (0.22)		
	34.4		34.4		14.6	14.6	0.18	0.18 (0.18)		
	8.7	10.1	6.5							OBERT and DUVALL (1967)
50 dia. x 120 long	42.		39.		17.	17.	0.23	0.28		STORIKOVA (1965)
	71.		57.		28.	28.	0.27	0.25 (0.02)		
19 dia. x 38 long	31.2	32.8	19.4				0.14 (0.15)	0.21 (0.14)	0.18 (0.11)	CHENEVERT and GATLIN (1965)

(c) Pelitic Clay

Test sample size (mm)	E_1	E_2	E_3	G_{23}	G_{13}	G_{12}	v_{12} (v_{21})	v_{13} (v_{31})	v_{23} (v_{32})	Reference
1000 x 1000 x 1000	0.93	0.15	1.37				0.04 (0.03)	1.07 (0.01)	2.77 (0.08)	LÖGTERS and VOORT (1974)
	0.73	0.69	9.80				0.05 (0.21)	0.50 (0.11)	1.00 (0.32)	
	0.48	0.83	1.22				0.06 (0.08)	0.49 (0.06)	1.00 (0.09)	
	49.9	49.9	36.2	15.3	15.3	19.7	0.266 (0.266)	0.144 (0.199)	0.273 (0.198)	KAARSBERG (1968)
	19.3		9.7							STEPANOV and BATUGIN (1967)
	45.3		29.6				0.10	0.29 (0.19)		STEPANOV and BATUGIN (1967)

TABLE 10 (continued)

(c) Pelitic Clay

	Material	Fabric description	Symmetry	Test method
3.7	Pennsylvanian Shale	Plane of laminations is plane 12	"n" fold axis normal to laminations?	Drained triaxial compression tests. Initial moduli quoted curve is concave upward
3.8	Shale	Sound, nonfissured with sub-horizontal stratification (i.e. plane 12)	"n" fold vertical axis?	Field test — Rocha flat jack method on cube sample
3.9	Shale	Layering (plane 12)	"n" fold axis?	Field tests — Rocha flat jack method
3.10	Marl	Layering (plane 12)	"n" fold axis?	
3.11	Opoka Marl	Pronounced stratification plane 12	"n" fold axis?	Laboratory tests — uniaxial compres-
3.12				sion tests oriented so that symmetry
3.13				elements were common to the
3.14	Marl			samples and the applied stress tensor. Stress rate 0.98 MPa/sec
3.15	Mudstone	Horizontal bedding — 3 main joint sets — (a) parallel to bedding (b) steep dip, strike SE (c) steep dip, strike NE	"n" fold axis perpendicular to bedding planes (i.e. planes 12)	Field test — seismic refraction — 91.5 m spread with geo- phones at 15.2 m spacing. Two shots beyond either end
3.16	Sandstone and Siltstone			of spread and one at centre of spread. Longitudinal and transverse wave velocities
3.17	Siltstone	Partially metamorphosed thin bedded	"n" fold axis of	Wave tests
3.18	Coarse Phyllite	sediments; parallel planes of	symmetry	
3.19	Fine Phyllite	stratification (planes 12)	(axis 3)	
3.20	Phyllite	Layering parallel to plane 12	Orthorhombic?	Field test with
3.21				radial jack pressure ≈ 3 MPa
3.22	Calcareous			pressure ≈ 5.5 MPa
3.23	Slate			

(c) Pelitic Clay

Test sample size (mm)	GPa						ν_{12} (ν_{21})	ν_{13} (ν_{31})	ν_{23} (ν_{32})	Reference
	E_1	E_2	E_3	G_{23}	G_{13}	G_{12}				
38 dia. x 76 long	0.062		0.034							MESRI and GIBALA (1971)
1000 x 1000 x 1000	10.8		4.9							DE BEER et al. (1974)
1000 x 1000 x 1000	12.2		7.3							ROCHA and SILVA (1970)
	57.8		30.8							
50 x 50 x 50	3.03		2.25							LOZINSKA-STEPIEN (1966)
	1.71		1.86							
	3.62		1.76							
	2.54		1.17							
	6.6	6.6	4.5	1.4	2.1		0.64	0.12 (0.08)		G_{12} assumed = $1.5G_{23}$
										WIEBENGA et al. (1964)
	12.9	12.9	7.2	3.0	4.5		0.44	0.53 (0.29)		G_{13} assumed = $.75G_{12}$
	60.7		55.5		22.4	25.0	0.215	0.28		LEKHNITSKII (1966)
	72.4		46.4		25.7	28.4	0.27	0.53		
	73.5		55.7		28.6	29.2	0.26	0.37		
2200 dia. x 2000 long	6.6 (3.4)		1.7 (0.7)							LAUFFER & SEEBER (1961) SEEBER (1970) Unbracketed figures correspond to recovered deformation, bracketed figures correspond to total deformation
	15.9 (13.4)		8.1 (3.8)							

TABLE 10 (continued)

(c) Pelitic Clay

Material	Fabric description	Symmetry		Test method
3.24 Slate	Laminations parallel to plane 12	"n" fold axis parallel to axis 3		Longitudinal and shear wave velocities on cube samples
3.25 Slate	Axis 3 is normal to the plane of foliation	Orthorhombic		Uniaxial compression and tension tests together with torsion tests. Strain gauges attached to samples
3.26 Hard Blue Pehrhyn Slate	Slaty cleavage associated with preferred orientation of micaceous and chloritic components in the rock. "Planes of elastic and velocity symmetry are coincident with the planes of fabric and tectonic symmetry."	Orthorhombic Stiffest along axis 1	direction of cleavage dip and max. fabric extension	Ultrasonic longitudinal and shear waves
3.27 Green Pehrhyn Slate		least stiff along axis 3	normal to planar features and direction of max. fabric shortening	
3.28 Chloritic Slate	Direction of bedding coincides with maximum wave velocity, minimum velocity perpendicular to bedding (i.e. axis 3)	"n" fold axis perpendicular to bedding?		Laboratory tests − longitudinal and shear wave velocities
3.29 Very Altered Slate	Axis 3 is normal to the plane of foliation	Orthorhombic?		Laboratory tests − oriented samples loaded to 9.8 MPa. Strains recorded by bonded strain gauges. Results quoted are secant moduli
3.30 Little Altered Slate				
3.31 Sound Slate				
3.32 Green River Shale	Thinly banded, finely laminated. Bedding is plane 12	Orthorhombic		Triaxial compression tests
3.33 Permian Shale	Thinly banded, finely laminated. Bedding is plane 12	Orthorhombic		

(c) Pelitic Clay

Test sample size (mm)	GPa						v_{12} (v_{21})	v_{13} (v_{31})	v_{23} (v_{32})	Reference
	E_1	E_2	E_3	G_{23}	G_{13}	G_{12}				
	114.0	114.0	53.0	19.8	19.8	53.5	0.067	0.328	0.328	ALEXANDROV et al. (1969)
	71.5	63.0	82.1	26.9	27.9	24.9	0.218	0.380	0.284	NISHIMATSU (1970)
5, 10, 15 x 64 x 90	46.3	44.1	30.1	10.6	15.4	16.0	0.426	0.426	0.439	ATTEWELL (1970)
	40.5		23.9	8.1	8.1	13.9	0.469		0.462	
50 dia. x 120 long	130.0		81.0		35.0	49.0	0.33	0.28 (0.17)		STORIKOVA (1965)
50 x 50 x 150 Prisms	34.8	55.4	7.8							RODRIGUES (1970)
	63.8	54.1	11.3							
	85.2	90.7	30.4							
19 dia. x 38 long	35.2	35.0	29.4				0.15 (0.16)	0.18 (0.28)	0.19 (0.18)	CHENEVERT and GATLIN (1965)
	35.4	31.6	24.4				0.17 (0.18)	0.21 (0.13)	0.21 (0.13)	CHENEVERT and GATLIN (1965)

TABLE 10 (continued)

(d) Pelitic (Mica)

	Material	Fabric description	Symmetry	Test method
4.1	Muscovite Mica Schist	Laminations parallel to plane 12	"n" fold axis parallel to axis 3	Longitudinal and shear wave velocities on cube samples
4.2	Schist	Foliation surfaces: strike NS (axis 1) dip 30° E Stratification surfaces: strike NS dip 30° W	Monoclinic, axis 1 is normal to the plane of symmetry	Laboratory tests — uniaxial compression of cylinders oriented with their axes parallel to NS, EW, and vertical stress up to 15 MPa
4.3	Schist (Phyllades de Revin)	Horizontal foliation surface lineation in one horizontal direction (axis 1)	Orthorhombic axes 1, 2, 3 are normals to planes of symmetry	Laboratory tests — uniaxial compression of cylinders oriented in the three orthogonal directions up to 30 MPa
4.4	Biotitic Mica Schist	Axis 3 is normal to the plane of schistosity	Orthorhombic?	Laboratory tests — oriented samples loaded to 9.8 MPa. Secant moduli quoted
4.5	Schist I	Pronounced stratification parallel to plane 12	"n" fold axis (axis 3)	Laboratory tests — uniaxial compression tests arranged so that one of the test planes was parallel to the axis of the samples while the other two test planes were inclined. Strain gauges attached to samples in various orientations
4.6	Schist II			
4.7	Schist III			
4.8	Serpentine Schist	Laminations parallel to plane 12	"n" fold axis (axis 3)	Uniaxial unconfined tensile tests of oriented samples. Tangent moduli taken at half strength
4.9 4.10	Mica Schist	Layering parallel to plane 12	Orthorhombic?	Field tests with radial jack pressure ≈ 2 MPa

(d) Pelitic (Mica)

Test sample size (mm)	GPa						v_{12} (v_{21})	v_{13} (v_{31})	v_{23} (v_{32})	Reference
	E_1	E_2	E_3	G_{23}	G_{13}	G_{12}				
	77.7	77.7	60.9	25.2	25.2	33.5	0.159	0.246	0.246	Alexandrov et al. (1969)
96 dia. x 192 long	49.0	26.5	14.5				0.29 (0.15)	0.33 (0.14)	0.35 (0.33)	Masure (1970)
96 dia. x 192 long	120.0	100.0	28.5				0.24 (0.14)	0.56	0.60	Masure (1970)
50 x 50 x 150 Prisms	69.2	75.6	43.1							Rodrigues (1966)
	95.4		74.5		27.2	37.6	0.268	0.270 (0.211)		Pinto (1970)
	76.9		41.0		20.5	31.4	0.219	0.271 (0.145)		
	63.4		20.0		7.9	27.9	0.134	0.212 (0.067)		
29.7 dia. x 68 long	2.5E_3		E_3							Barla and Goffi (1974)
2200 dia. x 2000 long	57.0[1] (53.0)[2]		3.8[1] (1.9)[2]							Lauffer and Seeber (1961) Seeber (1970)

[1] recovered deformation   [2] total deformation

TABLE 10 (continued)

(d) Pelitic (Mica)

Material	Fabric description	Symmetry	Test method
4.11 Greywacke Schist	Medium weathering; strongly folded	"n" fold axis perpendicular to schistosity planes (i.e. planes 12)	Field tests in adits — modulus measured at a range of angles. G_{23} calculated in the assumption that $v_{13} = 0$
4.12	Slight weathering; strongly folded		
4.13	Highly weathered; strongly folded		
4.14 Schist	Stratification parallel to planes 12	"n" fold axis? (axis 3)	Field tests in 2 m x 2 m gallery 30 m deep. Plate load tests parallel and perpendicular to stratification. Load to 3.4 MPa

(e) Saline Carbonate

Material	Fabric description	Symmetry	Test method
5.1 Clay Limestone	Pronounced stratification	"n" fold axis? (axis 3)	Laboratory tests — uniaxial compression tests oriented so that symmetry elements were common to the samples and the applied stress tensor. Stress rate 0.98 MPa/sec
5.2			
5.3 Clayey Limestone	Sub-horizontal strata (i.e. plane 12)	"n" fold axis normal to stratification?	Field tests (a) rigid plate (b) flexible plate
5.4			
5.5			(c) Goodman jack (d) microseismic velocities in the rock mass
5.6			
5.7 Limestones (Chirkei Damsite)	I. Almost horizontal dense bedding — cryptocrystalline limestone	"n" fold axis perpendicular to bedding planes? (axis 3)	Field tests — radial jacking device, 1.8 m dia. x 1.47 m long — internal pressures up to 9.78 MPa with up to three cycles
5.8	II. interbedded with marl and marl-clay — near vertical tectonic cracks opening to		
5.9			
5.10	III. 20 mm and filled with calcite and marly clay — near vertical lithogenic fissures		
5.11	IV. opening to 1 mm		
5.12 Limestones	Axis 3 is normal to bedding	Orthorhombic?	
5.13			
5.14 Marble		Orthorhombic	Longitudinal and shear wave velocities on cube samples

(d) Pelitic (Mica)

Test sample size (mm)	GPa						v_{12} (v_{21})	v_{13} (v_{31})	v_{23} (v_{32})	Reference
	E_1	E_2	E_3	G_{23}	G_{13}	G_{12}				
	38.0		8.1	1.3						Pinto (1970)
	12.2		4.7	1.0						
	5.1		1.2							
	≈ 20.		≈ 8.5							De Beer et al. (1968)

(e) Saline/Carbonate

Test sample size (mm)	GPa						v_{12} (v_{21})	v_{13} (v_{31})	v_{23} (v_{32})	Reference
	E_1	E_2	E_3	G_{23}	G_{13}	G_{12}				
50 x 50 x 50	1.27		2.06							Lozinska-Stepien (1966)
	3.13		1.86							
Plate diam. = 280	12.7		4.3							Aramburu (1974)
Plate diam. = 1000	12.3		5.5							
	5.6		4.8							
	18.6		14.7							
	13.9		2.96							Evdokimov and Sapegin (1967)
	13.6		5.96							
	13.5		6.43							
	5.62		2.30							
	6.24		1.83							
	40.9	37.2	33.4							Obert and Duvall (1967)
	56.4	61.8	69.4							
	84.7	80.1	74.8	29.7	30.7	32.6	0.314	0.353	0.306	Alexandrov et al. (1969)

TABLE 10 (continued)

(e) Saline Carbonate

Material	Fabric description	Symmetry	Test method
5.15 Marble	Bedding plane is plane 12	Orthorhombic?	
5.16 Yule Marble 5.17 5.18	Preferred orientation of the (c) axes of calcite indicates in "n" fold axis of symmetry. Thill et al. (1969) show that wave velocity patterns have the same symmetry.	"n" fold axis of symmetry parallel to c axes of calcite grains (axis 3)	Longitudinal wave velocities (C_{12} assumed 1.1 C_{13}). Static compression $\cdot 0.254$ mm/min strain Uniaxial compression tests
5.19 Rock Salt	Bedding plane parallel to plane 12	"n" fold axis parallel to axis 3	

(f) Sands

6.1 Natural Beach Sand	Bedding dips seaward at 5°. Sand contains some platy grains, one long axis parallel to coast while other dips landward at 10°. (axis 2 parallel to coast, axis 3 vertical)	Monoclinic. Symmetry plane is vertical and normal to coast	Drained triaxial compression tests (secant moduli taken at half failure)
6.2 Earlston Sand 6.3	Laboratory samples prepared by dropping in air. Short axes of sand grains vertical.	"n" fold vertical axis of symmetry	Triaxial test (low strain) Torsion triaxial test (low strain)
6.4 Leighton Buzzard Sand	Laboratory prepared samples by dropping in air	"n" fold vertical axis of symmetry (i.e. axis 3)	Compression tests on oriented cube samples. Confining stress = 0.055 MPa. Secant moduli at 1 % strain

(g) Clays

7.1 Krasnozem Silty Clay	Planar pore pattern relates to terrain slope which strikes N 29° E. (axis 1 is EW, axis 3 is vertical)	Monoclinic symmetry plane is vertical and normal to the strike of the slope	Undrained triaxial compression on partly saturated samples. (Secant moduli taken at half failure)
7.2 Foliated Clay 7.3	Layering parallel to plane 12	Orthorhombic?	Field test with radial jacks Pressure ≈ 2 MPa
7.4 Neo Comian Clays 7.5 Meso-Cenozoic Clays	Severe jointing coinciding with horizontal bedding direction. Symmetry of anisotropic modulus coincides with symmetry of anisotropic swelling	"n" fold vertical axis (axis 3)	Laboratory triaxial compression tests

(e) Saline Carbonate

Test sample size (mm)	GPa						v_{12} (v_{21})	v_{13} (v_{31})	v_{23} (v_{32})	Reference
	E_1	E_2	E_3	G_{23}	G_{13}	G_{12}				
	63.1	71.6	49.3							OBERT and DUVALL (1967)
12.7 x 12.7 x 25.4	74.2	74.2	53.9	16.5	16.5	29.0	0.270	0.36 (0.26)	0.36 (0.26)	RICKETTS and GOLDSMITH (1972)
	56.0	56.0	49.2	13.1	13.1	22.5	0.246	0.258	0.258	
25 x 25 x 50	59.7	59.7	38.8							LEPPER (1949)
	42.9	42.9	46.4	12.7	12.7	15.0	0.296 (0.296)	0.206 (0.218)	0.206 (0.218)	DREYER (1972)

(f) Sands

Test sample size (mm)	E_1	E_2	E_3	G_{23}	G_{13}	G_{12}	v_{12} (v_{21})	v_{13} (v_{31})	v_{23} (v_{32})	Reference
101 dia. x 203 long	0.035	0.040	0.054							LAFEBER and WILLOUGHBY (1971)
102 dia. x 203 long	0.017		0.034			0.0064	0.33	−0.41		MORGAN and GERRARD (1973) (E_1/E_3 and v_{12} obtained from cube sample tests)
	0.0041		0.0083	0.0037		0.0016	0.33	0.39		
100 x 100 x 100	0.015		0.018							ARTHUR and MENZIES (1972)

(g) Clays

Test sample size (mm)	E_1	E_2	E_3	G_{23}	G_{13}	G_{12}	v_{12} (v_{21})	v_{13} (v_{31})	v_{23} (v_{32})	Reference
101 dia. x 203 long	0.0043	0.0071	0.0097							LAFEBER and WILLOUGHBY (1971)
	1.28[1] (0.82)[2]		0.59[1] (0.29)[2]							SEEBER (1970)
	0.032		0.016						0.08 (0.04)	ROGATKINA (1967)
	0.121		0.062						0.23 (0.18)	

[1] recovered deformation   [2] total deformation

TABLE 10 (continued)

(g) Clays

	Material	Fabric description	Symmetry	Test method
7.6	London Clay (Ashford) − 20 m deep	Heavily overconsolidated clay with orientation of fissures predominantly horizontal together with some	"n" fold vertical axis (i.e. axis 3)	Undrained triaxial compression tests with oriented sam-
7.7	(Ashford) − 35 m deep	concentrations at high angle dips. Patterns vary from pseudo-orthorhombic to "n" fold axis (Skempton et al., 1969)		ples cut in various directions from blocks. Analysed assuming saturated and no volume change. Secant moduli taken at half ultimate.
7.8	London Clay (High Ongar Oxford Circus)			Drained triaxial compression tests with samples cut from blocks. In the analysis E_1/E_3 assumed 1.8
7.9	London Clay (Barbican Arts Centre)			Laboratory compres- sion. Drained and undrained triaxial and plane strain tests. Parameters quoted are effec- tive stress values
7.10 7.11 7.12 7.13	Kaolinitic Clay	Laboratory prepared samples from slurry. One dimensional consolidation to 0.39 MPa	"n" fold axis parallel to consolidation direction (i.e. axis 3)	UU tria- OCR = 1 axial tests. Constant OCR = 10 volume assum- OCR = 15 ed in analysis. Secant OCR = 20 moduli at 3 % strain
7.14	Florida Clay $\sigma = 0.55$ $K_0 = 0.48$	Test materials prepared by one directional consolidation from slurry with the pressure increasing from 0 to 0.23 MPa	"n" fold axis parallel to consolidation direction	Laboratory tests − 16 undrained hollow cylinder torsion
7.15	Florida Clay $\sigma = 0.41$ $K_0 = 0.47$	over 90 hours. Subsequent one directional consolidation under different cell pressures thereby producing different	(i.e. axis 3)	tests performed on each material in compression and
7.16	Florida Clay $\sigma = 0.275$ $K_0 = 0.49$	degrees of anisotropy. The first two clays are almost kaolinite, while the third contains some illite.		extension. For any test the rate of true stress and the prin-
7.17	Hydrite Clay $\sigma = 0.41$ $K_0 = 0.52$	σ in the first column indicates the cell pressure in MPa during the second stage of consolidation		cipal stress inclina- tion were fixed. Tests at constant volume with sam- ples fully saturated.

(g) Clays

Test sample size (mm)	GPa						v_{12} (v_{21})	v_{13} (v_{31})	v_{23} (v_{32})	Reference
	E_1	E_2	E_3	G_{23}	G_{13}	G_{12}				
38 dia. x 76 long	0.088		0.042		0.018	0.046	−0.04	1.04		WARD et al. (1959) WARD et al (1965) GIBSON (1974)
	0.145		0.084		0.037	0.064	0.13	0.87		
38 dia. x 76 long							−0.04	0.28		SIMONS (1971) GIBSON (1974)
	0.022		0.011			0.011	0.0	0.38 (0.19)		ATKINSON (1975)
	0.0022		0.0027		0.0006	0.0007	0.58	0.42		BHASKARAN (1975)
	0.0018		0.0022		0.0005	0.0006	0.60	0.40		
	0.0017		0.0017		0.0004	0,0006	0.50	0.50		
	0.0015		0.0012		0.0002	0.0003	0.49	0.61		
Hollow cylinder 71 (outside dia.) 51 (inside dia.) 142 (long)	0.903 1.80		1.		0.465 (1.14)	0.583 (1.59)	0.55 (0.115)	0.45 (0.885)		SAADA and OU (1973)
	0.567 1.36		1.		0.402 (1.235)	0.331 (1.03)	0.717 (0.322)	0.283 (0.678)		N.B.: Results for E_1, G_{13}, and G_{12} are given as ratios of E_3
	0.578 1.33		1.		0.354 (1.102)	0.337 (1.00)	0.712 (0.335)	0.288 (0.665)		Compression tests unbracketed.
	0.670 2.12		1.		0.279 (1.41)	0.402 (2.24)	0.665 (−0.057)	0.335 (1.06)		Extension tests bracketed.

TABLE 10 (continued)

(g) Clays

Material	Fabric description	Symmetry	Test method
7.18 Hydrite Clay $\sigma = 0.275$ $K_0 = 0.52$ 7.19 Grundite Clay $\sigma = 0.41$ $K_0 = 0.69$	Test materials prepared by one directional consolidation from slurry with the pressure increasing from 0 to 0.23 MPa over 90 hours. Subsequent one directional consolidation under different cell pressures thereby producing different degrees of anisotropy. The first two clays are almost pure kaolinite, while the third contains some illite. σ in the first column indicates the cell pressure in MPa during the second stage of consolidation	"n" fold axis parallel to consolidation direction (i.e. axis 3)	Laboratory tests − 16 undrained hollow cylinder torsion tests performed on each material in compression and extension. For any test the rate of true stress and the principal stress inclination were fixed. Tests at constant volume with samples fully saturated
7.20 Fulford Clayey Silt	Preconsolidation pressure 47 KPa exist. Overburden effective stress 20 KPa $\ast\ast\ast$ lightly overconsolidated K_0 about unity	"n" fold vertical axis (i.e. axis 3)	Multi stage compress. undrained triaxial compression and extension tests. Secant extens. moduli at 1 % axial strain

(h) Miscellaneous

8.1 Coal	Bedding parallel to plane 12, major cleats parallel to plane 23, minor cleats parallel to plane 13	Orthorhombic	Oriented cylinders loaded in triaxial compression to failure. (Confining pressures 0.34 to 4.1 MPa). Tests also on oriented cubes
8.2 Bitumious Concrete	Vertical kneading compaction of laboratory samples	"n" fold axis parallel with compaction direction	Triaxial compression, simple shear, and torsional tests. Confining pressure = 0.28 MPa Load time = 5 secs Temperature = 26.7 °C

(g) Clays

Test sample size (mm)	GPa						v_{12} (v_{21})	v_{13} (v_{31})	v_{23} (v_{32})	Reference
	E_1	E_2	E_3	G_{23}	G_{13}	G_{12}				
Hollow cylinder 71 (outside dia.) 51 (inside dia.) 142 (long)	0.534 1.75		1.		0.342 (1.30)	0.308 (1.55)	0.734 (0.127)	0.266 (0.873)		SAADA and OU (1973)
	0.845 1.084		1.		0.481 (0.838)	0.538 (0.745)	0.576 (0.458)	0.424 (0.542)		N.B.: Results for E_1, G_{13}, and G_{12} are given as ratios of E_3 Compression tests unbracketed. Extension tests bracketed.
38 dia. x 76 long	$0.32E_3$		E_3		$0.65E_3$	0.84	0.16			
	$2.3E_3$		E_3		$0.48E_3$	−0.15	1.15			

(h) Miscellaneous

Test sample size (mm)	E_1	E_2	E_3	G_{23}	G_{13}	G_{12}	v_{12} (v_{21})	v_{13} (v_{31})	v_{23} (v_{32})	Reference
55 dia. x 110 long (cylinders) 101 (cubes)	3.16	2.24	2.24	0.61	0.44	0.38	0.43 (0.21)	0.37 (0.16)	0.26 (0.21)	KO and GERSTLE (1972)
25 x 25 x 50 12.7 x 12.7 x 54 28.5 dia. x 12.7 long	0.65		0.98	0.13		0.22	0.49	0.43 (0.39)		BUSCHING et al. (1967)

TABLE 11
Poisson's number for some Belgian coal measure rocks
(after Salustowicz, 1965)

Depth in feet (metres)			
Rock type	400 (120)	2034 (610)	3936 (1180)
Sandstone	20	9	4
Shale	12	7	3.5
Coal	3	2.5	2

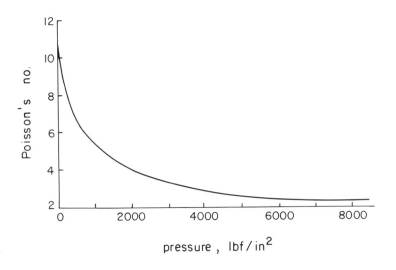

Fig. 6–73: Relationship between Poisson's number and stress level
(after Schmidt, 1926).

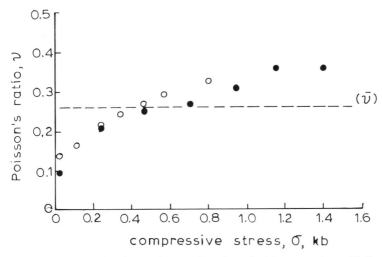

Fig. 6–74: Poisson's ratio v for granite sample under uniaxial compression σ. Notice that v is less than $\bar{v}$ (the intrinsic value) at low stress and greater than $\bar{v}$ at high stress. Open and closed circles refer to different samples of the same rock
(after Walsh and Brace, 1966).

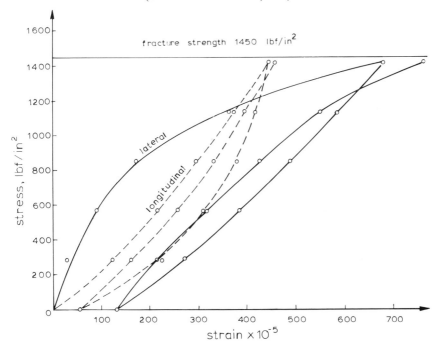

Fig. 6–75: Stress-strain curve for coal
(after Lama, 1966).

The influence of confining pressure on POISSON's ratio results in lowering of its value for weaker rocks, but for stronger rocks it may not have any influence.

For non-linear axial lateral strain curve, KULHAWY and DUNCAN (1972) and KULHAWY (1975) have represented the relationship between the tangential POISSON's ratio at any strain level and initial POISSON's ratio in the form

$$\nu_t = \frac{\nu_i}{(1 - d\varepsilon_a)^2} \qquad (6.90)$$

$$\nu_i = G - F \log\left(\frac{\sigma_3}{p_a}\right)$$

$$\varepsilon_a = \frac{\sigma_1 - \sigma_3}{E_i\left[1 - \frac{(\sigma_1 - \sigma_3)}{(\sigma_1 - \sigma_3)_f} R_f\right]}$$

where ν_t = tangential POISSON's ratio at any stage
         ν_i = initial tangential POISSON's ratio (Fig. 6-76)
         ε_a = axial strain at the stage
         $\sigma_1 - \sigma_3$ = deviator stress
         E_i = initial tangential YOUNG's modulus
         G, F = constants

$$R_f = \frac{(\sigma_1 - \sigma_3)_f}{(\sigma_1 - \sigma_3)_{ult}}$$

where $(\sigma_1 - \sigma_3)_f$ is the failure deviator stress at confining pressure and $(\sigma_1 - \sigma_3)_{ult}$ is asymptotic value of stress difference (Fig. 6-58).

Fig. 6-77 represents the relationship between experimental values and those obtained from Eq. 6.90 for Cedar City tonalite [compressive strength = = 101.5 MPa (14,600 lbf/in²)]. The POISSON's ratio parameters of Eq. 6.90 for various rock types are given in Table 12 and their variations are given in Table 13. Analysis of these limited data shows that the variations in the values of d is relatively small, indicating that the rate of increase of POISSON's ratio with strain or stress level is similar for these rock materials. The values of F range from —0.05 to 0.05 indicating that the initial tangent POISSON's ratio values may either decrease or increase 0.05 per log cycle of stress, a factor which may or may not be significant depending upon the magnitude of stress changes. The largest and most significant variations occur in the values of G the initial tangent POISSON's ratio at one atmosphere. These values range from 0.11 to 0.30.

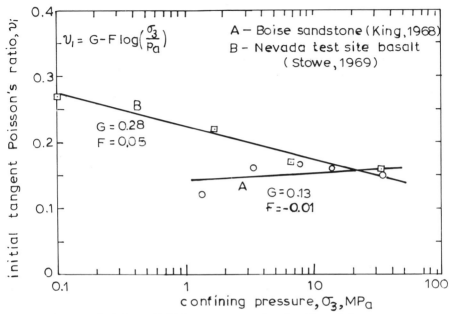

Fig. 6–76: Variations of initial tangent POISSON's ratio with confining pressure
(after KULHAWY, 1975).

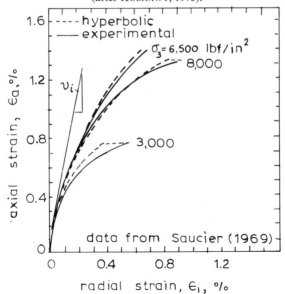

Fig. 6–77: Experimental and hyperbolic axial strain-radial strain curves for Cedar City
tonalite
(after KULHAWY, 1975).

TABLE 12
Triaxial POISSON's ratio parameters for various rock types
(after KULHAWY, 1975)

Rock Number	Description	Density (gm/cm³)	Specific gravity	Porosity (%)	Poisson's ratio parameters			Range of confining pressure (MN/m²)	Reference
					G	F	d		
IPT-4	AEC Nevada Site Granite (dense, coarse-grained, unweathered)	2.69	2.69	0.3	0.23	−0.01	115.0	0–27.6	STOWE (1969)
IPT-5	Quartz Monzonite (medium-grained, porphyritic)								DEKLOTZ & HECK (1965)
5a	(mutually perpendicular axes)	2.68	2.70	0.2	0.19	0	114.0	0–20.7	
5b		2.69	2.69	0.2	0.20	0.01	120.0	0–20.7	
5c		2.68	2.68	0.2	0.19	−0.01	128.0	0–20.7	
IPT-10	Cedar City Tonalite (friable, medium to fine-grained)	–	2.60	4.9	0.11	−0.05	62.0	0–248.	SAUCIER (1969)
IVT-2	AEC Nevada Site Basalt (dense, fine-grained, unweathered)	2.70	2.83	4.6	0.28	0.05	107.0	0–34.5	STOWE (1969)
IVT-16	AEC Nevada Site Tuff (fairly welded ash, w = 21.1 %)	1.92	2.39	19.8	0.24	0.03	114.0	0–10.3	STOWE (1969)
MFT-1	Schistose Gneiss (fine-grained, granular) (⊥ foliation)	2.79	2.80	0.5	0.11	−0.04	194.0	0–70.0	DEKLOTZ et al. (1965)
SCT-7	Boise Sandstone (well-cemented)	1.90	–	27.0	0.13	−0.01	–	1.4–34.4	KING (1968)
SCHT-12	AEC Nevada Site Limestone (dense, fine-grained, stylolite seams)	2.70	2.72	0.5	0.30	−0.01	77.0	0–27.6	STOWE (1969)

The increase in value of POISSON's ratio with stress in some rocks may be associated with plastic deformation in the specimen and in that case the rate of rise of lateral stress with depth will be higher and at greater depths hydrostatic stress conditions will approach.

For accurate calculation of the POISSON's ratio of a rock substance, it is suggested that the specimen be subjected to a lateral pressure to about 20–40% of its strength and then axially loaded and the axial and lateral strains measured at about 50% of the specimen strength under these test conditions. The values calculated at lower and higher stresses are likely to be influenced by cracks in the specimen.

While measuring the deformation for the purpose of calculating the POISSON's ratio, it should be kept in view that if a rock specimen is bedded, the values calculated will be affected if the measuring base for deformation extends to more than one bed. If the specimens are cut at right angle to the direction of bedding, and the longitudinal base extends to two beds with YOUNG's modulus values of E_1 and E_2 and POISSON's ratios of ν_1 and ν_2 and the base length in each bed is S_1 and S_2 while the lateral strain base is limited to one bed with POISSON's ratio value of ν_1, the longitudinal and lateral strains measured for a base length of $(S_1 + S_2)$ shall be

$$\Delta S_{\text{long}} = \sigma \left(\frac{S_1}{E_1} + \frac{S_2}{E_2} \right)$$

$$\Delta S_{\text{lat}} = \nu_1 \frac{\sigma}{E_1} (S_1 + S_2)$$

$$\nu_{\text{eff}} = \frac{\Delta S_{\text{lat}}}{\Delta S_{\text{long}}} = \frac{\nu_1 \dfrac{\sigma}{E_1} (S_1 + S_2)}{\sigma \left(\dfrac{S_1}{E_1} + \dfrac{S_2}{E_2} \right)}$$

$$\nu_{\text{eff}} = \frac{\nu_1 (S_1 + S_2) E_2}{(S_1 E_2 + S_2 E_1)} \tag{6.91}$$

Similarly, if the specimen is tested with bedding parallel to the direction of application of stress, then for thinly spaced beds, it is possible that the lateral strain gauge will extend to more than one bed and the longitudinal strain gauge will be limited to one bed. The POISSON's ratio calculated will obviously differ from that obtained if both the gauges were located in the same bed and can be given by

$$\nu_{\text{eff}} = \frac{(S_1 \nu_1 E_2 + S_2 \nu_2 E_1)}{E_2 (S_1 + S_2)} \tag{6.92}$$

where

ν_2, E_2 = elastic constants of the bed in which the axial strain gauge is located

while the lateral strain gauge is located in two beds of moduli ν_1, E_1 and ν_2, E_2 with respective gauge lengths of S_1 and S_2.

TABLE 13
Summary of non-linear triaxial POISSON's ratio parameters
(after KULHAWY, 1975)

Parameter	Number of values studied	Maximum	Minimum	Average
G	10	0.30	0.11	0.20
F	10	0.05	−0.05	0.00
d	9	194	62	115

6.9. Compressibility of Rock

Compressibility of rock is defined as either linear or volumetric. Volumetric compressibility is also defined as the inverse of bulk modulus. Compressibility K is expressed as

$$K_T = \frac{1}{V_0}\left(\frac{\partial V}{\partial P}\right)_T = \frac{1}{\varrho}\left(\frac{\partial \varrho}{\partial P}\right)_T \qquad (6.93)$$

where

K_T = compressibility at pressure P and temperature T
V_0 = specific volume and
ϱ = density at given pressure P and temperature T.
K is positive and has dimensions reciprocal of pressure.

Compressibility can be determined directly as a volume change under pressure or may be computed from changes of linear dimensions under pressure. For an anisotropic material measurements in linear change in 3 mutually perpendicular directions are required. The compressibility may then be computed from

$$K_T = k_1 + k_2 + k_3 + 2(k_1k_2 + k_1k_3 + k_2k_3) \qquad (6.94)$$

where k_1, k_2 and k_3 = linear compressibilities in three mutually perpendicular directions.

When compressibility is small,

$$K_T = k_1 + k_2 + k_3 \tag{6.95}$$

In cases where rock is shock loaded the value so determined is the adiabatic compressibility K_s rather than the isothermal compressibility K_T defined above. The relationship between the two quantities is

$$K_T = K_s + \frac{T\alpha^2}{\varrho c_p} \tag{6.96}$$

where

T = absolute temperature
a = volume thermal expansion
c_p = specific heat per gram at constant pressure.

The difference between K_T and K_s does not exceed a few percent for most solids at ordinary temperature.

The compressibility of rocks can be measured by the use of electrical resistance strain gauges attached to the jacketed rock specimens. Cores from samples are obtained and cut to suitable size, 2.5–5 cm (1–2 in) diameter and 5 to 10 cm (2–4 in) in length. The cores are jacketed using copper jacket (0.05 mm) (0.002 in) thick and an initial pressure of 1 kb is applied in a suitable pressure cell (a standard triaxial cell) to seat the jacket uniformly against the rock surface. Strain gauges are then applied on the jacket and the specimen subjected to hydrostatic compression using fluid as a medium. Rosette gauges may be used to measure compressibility in two mutually perpendicular directions and if measurements are conducted on the surface of the cylinder at places mutually perpendicular to each other, the compressibility in the different directions can be calculated.

In many measurements, compressibility results are reported in the form

$$\frac{\Delta v}{v_0} = ap - bp^2 \tag{6.97}$$

where a and b are constants depending upon temperature and $\Delta v = v_0 - v$.

When the $(\Delta v/v_0)/p$ is plotted against pressure, the values of a can be calculated by taking the intercept at $p = 0$ and the slope of the line gives the value of b. When the plot is non-linear, the value of a so determined is rather arbitrary. The values of a and b for some minerals are given in Table 14 and compressibility of some important minerals and rocks under different pressures is given in Tables 15, 16 and 17.

TABLE 14
Compressibility of minerals
(Extracted from CLARK, 1966)

$$K = \frac{\triangle v}{v_0} = ap\text{-}bp^2, p = \text{pressure in megabars}$$

Mineral	a mb^{-1}	b mb^{-2}	Remarks
Diamond	0.18-0.174		Pressure 12 kb
Graphite (hex) linear ‖	3.0 2.8	 45	 Pressure 16 kb
Anhydrite	1.84		
Barite	1.77-1.81 1.71 1.76	 7.6 11.9	 to 30 kb

Beryl
    linear, ‖ c axis
    linear, ⊥ c axis
	
	0.2075
	0.1664
		
		0.24
		0.24	

Calcite
    linear, ‖ c axis
    linear, ⊥ c axis
	
	0.822
	0.273
		
		2.9
		0.24	

Dolomite	1.22		
Fluorite	1.22-1.26		
Forsterite olivine (Dunite)	 0.79 0.82 0.77	 1	mean, 2-12 kb, $\varrho^* = 3.288$ to 40 kb, $\varrho^* = 3.364$ $\varrho^* = 3.324$
Galena	1.96 1.869 1.96	 7.43	 to 30 kb
Halite	4.26 1.187	51 12.4	
Hematite	0.6		
Hornblende	1.1 -1.2		
Ilmenite	0.56		

* ϱ = density

TABLE 14 (continued)

Mineral	a mb^{-1}	b mb^{-2}	Remarks
Jadeite	0.74-1.11		
Magnetite	0.547	0.82	
Mica			
Phlogopite	2.34	18.2	$\varrho = 2.877$
Muscovite	1.2	–	
Microcline	1.92	13	$\varrho = 2.557$
Orthoclase	2.123	14.5	
linear, a axis	1.013	4.8	
linear, b axis	0.559	4.9	
linear, c axis	0.468	1.3	
linear, y axis	1.097	6.9	$\perp$ b and c
Pyrite	0.680	0.87	
Quartz, α	2.706	24.0	
linear, c axis	0.718	6.2	
linear, $\perp$ c axis	0.995	7.6	
Quartz, β	1.776		600 °C
Rutile	0.483	0.92	
linear, c axis	0.105	0.24	
linear, $\perp$ c axis	0.190	0.24	
Soda niter	3.92	39	
Topaz, Japan	0.611	1.1	$\varrho = 3.544$ (mean)
linear, a axis	0.2176	0.24	
linear, b axis	0.1504	0.24	
linear, c axis	0.2429	0.24	
Tourmaline	0.816	1.95	black, $\varrho = 3.091$
linear, c axis	0.486	1.17	
linear, $\perp$ c axis	0.165	0.24	
Wurtzite	1.36		
Zircon	0.86		

TABLE 15

Compressibility of some minerals at different pressures

$(\triangle v/v_o$ or $\triangle l/l_o)$ (after CLARK, 1966)

Mineral	ϱ (kg/cm^2)						
	5,000	10,000	15,000	20,000	25,000	30,000	40,000
Apatite							
linear, ∥ c	.00121	.00240	.00362	.00484	.00603	.00716	
linear, ⊥ c	.00203	.00398	.00592	.00779	.00963	.01139	
volume	.00525	.01032	.01539	.02028	.02508	.02964	
Barite							
linear, a	.00247	.00478	.00696	.00895	.01076	.01243	
linear, b	.00323	.00634	.00934	.01222	.01495	.01748	
linear, c	.00251	.00498	.00740	.00973	.01204	.01427	
volume	.00819	.01601	.02351	.03058	.03728	.04354	
Beryl							
linear, ∥ c	.00100	.00199	.00299	.00398	.00496	.00585	
linear, ⊥ c	.00077	.00155	.00242	.00320	.00404	.00486	
volume	.00254	.00509	.00781	.01033	.01297	.01550	
Calcite							
volume		.0134		.0725		.0887	.1019
volume	.0066	.0130	.0192	.077	.082	.087	.096
Celestite							
linear, a	.00296	.00588	.00861	.01134	.01377	.01623	
linear, b	.00215	.00418	.00612	.00795	.00973	.01151	
linear, c	.00215	.00423	.00629	.00825	.01017	.01198	
volume	.00725	.01423	.02088	.02729	.03330	.03920	
Cobaltite							
volume	.00353	.00702	.01087	.01444	.01753	.02050	
Corundum							
linear, 90 °C	.00060	.00119	.00179	.00238	.00297	.00355	
linear, 10 °C	.00059	.00118	.00179	.00239	.00300	.00360	
volume	.00178	.00355	.00537	.00704	.00891	.01067	
Diopside							
volume $\varrho = 3.260$		.0088		.0169		.0245	.0318
Fluorite							
volume	.00585	.01147	.01695	.02213	.02735	.03234	
Galena							
volume	.00936	.01792	.02602	.03363	.04074		
Garnet							
volume	.00272	.00538	.00807	.01071	.01333	.01588	
$\varrho = 4.090$		.0071		.0138		.0200	.0257
Grossularite							
volume	.00325	.00636	.00947	.01248	.01542	.01852	
$\varrho = 3.475$		.0072		.0140		.0204	.0264
Halite							
volume		.0366		.0664		.0924	.1160
Hanksite							
linear, ∥ c	.00523	.01020	.01455	.01843	.02196	.02522	
linear, ⊥ c	.00369	.00738	.01073	.01379	.01662	.01927	
volume	.01256	.02476	.03558	.04531	.05419	.06241	

TABLE 15 (continued)

Mineral	ϱ (kg/cm^2)						
	5,000	10,000	15,000	20,000	25,000	30,000	40,000
Hypersthene							
volume		.0101		.0191		.0272	.0347
$\varrho = 3.421$							
Labradorite							
volume		.0133		.0260		.0381	.0495
$\varrho = 2.681$							
Magnetite							
volume	.00286	.00564	.00845	.01118	.01394	.01670	
Mica							
linear, par. cleavage	.0013	.0028	.0042	.0058	.0070	.0082	
volume	.0049	.0092	.0165	.0227	.0311	.0353	
Olivine							
volume		.0079		.0136		.0231	.0304
$\varrho = 3.364$							
Orthoclase, volume							
Madagascar $\varrho = 2.568$		.0171		.0333		.0488	.0634
Spain $\varrho = 2.556$		.0175		.0335		.0484	.0625
Periclase							
volume	.00290	.00571	.00859	.01139	.01410	.01674	
Pyrite							
volume	.00328	.00645	.00963	.01282	.01575	.01869	
Quartz							
linear, ∥ c	.00334	.00642	.00920	.01170	.01406	.01622	
linear, ⊥ c	.00480	.00909	.01308	.01688	.02056	.02411	
volume	.01289	.02440	.03495	.04478	.05418	.06308	
volume		.0236		.0441		.0625	.0792
volume	*)				.054	.061	.074
Spodumene							
linear, a	.00088	.00176	.00265	.00353	.00442	.00529	
linear, b	.00128	.00252	.00375	.00495	.00610	.00719	
linear, c	.00095	.00189	.00283	.00375	.00466	.00556	
linear	.00114	.00225	.00338	.00448	.00557	.00663	
volume	.00337	.00663	.00993	.01312	.01624	.01925	
Topaz							
linear, a	.00106	.00210	.00317	.00421	.00525	.00622	
linear, b	.00072	.00144	.00218	.00291	.00366	.00448	
linear, c	.00108	.00217	.00331	.00443	.00555	.00658	
volume	.00285	.00570	.00863	.01150	.01439	.01718	
Tourmaline							
linear, ∥ c	.00245	.00484	.00715	.00937	.01147	.01336	
linear, ⊥ c	.00094	.00185	.00273	.00359	.00444	.00525	
volume	.00431	.00851	.01257	.01647	.02023	.02368	

*) measured to 100,000 kg/cm^2

TABLE 16
Compressibility of rocks
(after CLARK, 1966)

A — Fluid pressure, K from change of length   C — Fluid pressure, K from V_p, V_s
B — Axial compression, K from E and v   D — Fluid pressure, K from volume change

Rock	ϱ	p bars	K mb^{-1} enclosed	K mb^{-1} unenclos.	Note
Albitite, Sylmar, Pa.	2.615	4,000	1.52		C
			1.93		A
Andesite, Salida, Colo.	2.618	34	2.20		C
(glassy?)		345	2.24		
		1,030	2.08		
Anorthosite					
New Glasgow, Ont.		600		1.74	B
	2.708	4,000	1.22		C
Stillwater Complex, Mont.	2.770	4,000	1.15		C
(*see also* diabase, gabbro, norite)					
Basalt, altered, Chaffee, Colo.	2.586	500	2.26		C
		5,000	1.96		
Basalt	2.901	1	2.19		A
		500	1.49		
		5,000	1.25		
Basalt, Scotch Plains, N.J.	2.911	2,000	2.42		D
(20 per cent glass)		10,000	1.68		
Basalt, diabasic, altered	2.924	1	1.59		A
		10,000	1.31		
Bronzitite					
Stillwater Complex, Mont.	3.279	4,000	.97		C
		4,000	.96		A
Bushveld Complex,	3.288	4,000	.94		C
Transvaal		4,000	.89		A
Diabase					
Vinalhaven, Me.	2.96	0	1.71	1.46	A
		120	1.59		
		720	1.26		
	2.962	4,000	1.17		A

V_p = compressional wave velocity
V_s = shear wave velocity

TABLE 16 (continued)

Rock	ϱ	p bars	K mb^{-1} enclosed	unenclos.	Note
Sudbury, Ont.	3.002	2,000		1.37	D
(Murray Mine)		10,000		1.25	
		600		1.36	B
Palisade, Granton, N.J.	2.975	2,000		1.54	D
		10,000		1.30	
Frederick, Md.	3.020	1	1.25		A
		1,000	1.28		
		5,000	1.22		
		9,000	1.17		
Frederick, Md.	3.033	2,000		1.23	D
		10,000		1.07	
Frederick, Md.	3.012	4,000	1.22		C
		4,000	1.16		A
Whin Sill, England	2.937	2,000		1.70	D
(quartz dolerite)		10,000		1.26	
Diorite, Salem, Mass.	3.025	34	1.57		C
		345	1.42		
		1,030	1.32		
Diorite porphyry	2.792	1	6.69		A
Ural Mountains		500	2.36		
		5,000	1.19		
Dolomite, Bethlehem, Pa.	2.82	0	3.71	1.19	A
		120	2.54	1.19	
		600	1.48	1.19	
Dolomite, Blair,	2.849	1	1.23		A
Martinsburg, W.Va.		1,000	1.20		
		5,000	1.14		
		9,000	1.09		
Dolomite, Webatuck, N.Y.	2.867	1	19.		A
		500	1.36		
		1,000	1.19		
		5,000	1.05		
		9,000	1.00		

TABLE 16 (continued)

Rock	ϱ	p bars	K mb^{-1} enclosed	unenclos.	Note
Dunite					
Balsam Gap, N.C.	3.27	0		1.12	A
		120		1.09	
		600		.95	
Balsam Gap, N.C.	3.288	7,000		.79	D
Balsam Gap, N.C.	3.267	4,000	.80		C
		4,000	.83		A
Twin Sisters, Wash.	3.312	4,000	.80		C
Gabbro					
San Marcos, Cal.	2.993	500	1.11		C
		5,000	1.02		
Duluth, Minn.	2.885	500	1.24		C
(bytownite)		5,000	1.15		
	2.933	500	1.24		C
(hornblende)		5,000	1.17		
New Glasgow, Ont.	3.106	2,000	1.34		D
(olivine)		10,000	1.13		
		600		1.52 (mean)	B
Mellen, Wis.	2.931	4,000	1.05		C
		4,000	1.14		A
Granite					
Quincy, Mass.	2.59?	0	7.56	1.92	A
		120	4.02	1.85	
		600	2.53	1.67	
		axial 600		3.6-3.2	B
	2.629	500	1.81		C
		5,000	1.54		
	2.621	4,000	1.72		C
		4,000	1.92		A
Rockport, Mass.	2.63	0	9.17	1.95	A
		120	5.04	1.87	
		600	2.66		
	2.624	4,000	1.62		C
		4,000	1.85		A

TABLE 16 (continued)

Rock	ϱ	p bars	K mb^{-1} enclosed	unenclos.	Note
Westerly, R.I.		1		1.95	
		600		3.3	B
	2.628	34	2.31		C
		345	1.84		
		1,030	1.72		
	2.615	500	1.99		C
		5,000	1.78		
	2.616	2,000		2.12	D
		10,000		1.82	
		4,000	1.82		C
	2.646	1	8.3		A
		500	2.89		
		1,000	2.46		
		2,000	2.16		
		5,000	1.99		
		9,000	1.87		
Bear Mt., Texas	2.610	34	2.11		C
		345	1.79		
		1,030	1.68		
Washington, D.C. (granodiorite)	2.739	2,000		2.23	D
		10,000		1.82	
Stone Mt., Georgia	2.633	2,000		2.06	D
		10,000		1.80	
	2.631	1	15.6		A
		500	3.49		
		1,000	2.48		
		2,000	2.12		
		5,000	1.91		
		9,000	1.76		
Woodbury, Vt.	2.634	500	1.84		C
		5,000	1.75		
Texas, "pink"	2.636	500	1.55		C
		5,000	1.38?		
Texas, "gray"	2.609	500	2.05		C
		5,000	1.74		
Barriefield, Ont.	2.672	500	1.45		C
		5,000	1,35?		

TABLE 16 (continued)

Rock	ϱ	p bars	K mb⁻¹ enclosed	K mb⁻¹ unenclos.	Note
Karelia, U.S.S.R.	2.641	0	4.23		A
porosity = 0.4 %		500	1.82		
		5,000	1.07?		
Limestone					
Nazareth, Pa.	2.69	0	2.92	2.47	A
(Carbonaceous)		120	2.75	2.45	
		160	2.35	2.41	
Solenhofen, Bavaria	2.602	6,000		1.36	A
		(mean to			
		12 Kb)			
	2.602	5,000		1.29	6°C A
		(mean to		1.42	100°C
		10 Kb)		1.63	270°C
				1.71	476°C
	2.656	1	1.53		C
		500	1.54		
		5,000	1.49		
Oak Hall Quarry, Pa.	2.712	1	1.35		A
		1,000	1.34		
		5,000	1.31		
		9,000	1.28		
Marble	2.71	1	18.0	1.39	A
Vermont		120	3.31	1.38	
		600	1.50	1.26	
Gunnison Co., Colo.	2.708	7,000	1.38	1.40	D
Danby, Vt.	2.704	4,000	1.28		C
	2.698	500	1.35		C
		5,000	1.21		
	2.712	1	8.7		A
		500	1.69		
		1,000	1.45		
		5,000	1.31		
		9,000	1.27		
Norite					
Sudbury, Ont.	2.85	0	3.15	1.65	A
		120	2.24	1.63	
		600	1.65	1.57	

TABLE 16 (continued)

Rock	ϱ	p bars	K mb^{-1} enclosed	unenclos.	Note
Elizabethtown, N.Y.		34	1.35		C
	3.057	345	1.18		
		1,030	1.10		
French Creek, Pa.		0	5.90	1.40	A
(gabbro)	3.05	120	4.11	1.34	
		600	1.66	1.26	
	3.054	4,000	1.11		C
		4,000	1.13		A
Pipestone (catlinite)	2.840			1.29	A
		(mean to 10 Kb)			
Quartzite, Montana	2.647	4,000	2.18		C
Quartzitic sandstone,		1	5.87	2.67	A
Bethlehem, Pa.		120	4.28	2.65	
		600	3.09	2.60	
Quartzite, Cheshire,	2.643	1	7.6		A
Rutland, Vt.		500	3.04		
		1,000	2.74		
		5,000	2.48		
		9,000	2.26		
Sandstone, Caplen dome,	2.543	500	3.5		C
Texas		5,000	2.33		
"Serpentine"					
(Talc schist)	2.875	2,000		1.79	D
Alberene, Va.		10,000		1.36	
Syenite, Augite, Ont.	2.780	4,000	1.21?		C
		4,000	1.69		A
Talc, Hewitt, N.C.	2.751	1		1.86	A
Tonalite	2.763	4,000	1,49		C
Trachyte	2.712	550	2.12		C
		5,000	1.80		

TABLE 17a

Compressibility of common minerals of low symmetry

(after BRACE, 1965b)

Mineral	Linear Compressibility			$k =$ $k_1 + k_2 + k_3$
	k_1	k_2	k_3	
Calcite	0.273	0.273	0.822	1.368
Quartz (low)	0.982	0.982	0.684	2.638
Olivine, FO_{92}	0.187	0.359	0.230	0.776
Muscovite	0.346	0.346	1.639	2.331
Biotite	0.368	0.368	1.694	2.430
Orthoclase	1.097	0.559	0.468	2.124
Augite	0.255	0.510	0.344	1.108
Hornblende	0.616	0.367	0.205	1.188
Microcline	1.109	0.290	0.613	2.012
Albite	0.978	0.408	0.498	1.884
Labradorite	0.587	0.332	0.421	1.340

Directions related to crystal axes according to NYE (1957). For monoclinic system, OX_3 parallel to crystal c axis, OX_2 parallel to b and OX_1 perpendicular to bc plane.

TABLE 17b

Linear compressibility $(Mb)^{-1}$ as a function of pressure and direction

(after BRACE, 1965b)

Rock	Direction	Pressure, kb					
		0	1	3	5	7	9
Marble, Yule, Colorado	PR	1.12	0.380	0.370	0.377	0.380	0.378
	Q	2.78	0.613	0.607	0.593	0.575	0.564
	$B^1)$	5.02	1.373	1.347	1.347	1.335	1.320
Chlorite schist, Chester, Vermont	1	0.388	0.385	0.377	0.369	0.362	0.354
	2	0.409	0.406	0.399	0.392	0.386	0.378
	3	0.466	0.459	0.445	0.431	0.417	0.402
	B	1.263	1.250	1.220	1.192	1.165	1.134
	$\parallel$	0.890	0.740	0.610	0.578	0.542	
Phyllite, East Hill, Moretown, Vermont	$\perp$	11.4	1.431	1.101	0.960	0.916	
	$B^2)$	13.2	2.911	2.321	2.116	2.000	
	$\parallel$	0.742	0.786	0.787	0.756	0.724	
Micaceous quartzite, Coniston, Ontario	Y	1.268	0.967	0.861	0.797	0.735	
	$B^3)$	3.278	2.720	2.509	2.350	2.194	
	$\parallel$	0.582	0.581	0.587	0.603	0.598	0.590
Mica schist, Gassetts, Vermont	$\perp$	1.053	1.051	0.997	0.887	0.803	0.742
	$B^2)$	2.217	2.213	2.171	2.093	1.999	1.922
	$\parallel$	0.500	0.500	0.499	0.485	0.471	0.463
Red slate	$\perp$	1.020	1.020	0.941	0.887	0.849	0.817
	$B^2)$	2.020	2.020	1.939	1.857	1.791	1.743
	$\parallel$	0.572	0.571	0.586	0.547	0.523	0.502
Black slate	$\perp$	1.460	1.381	1.229	0.974	0.827	0.781
	$B^2)$	2.604	2.523	2.401	2.068	1.873	1.785

$^1) B = 2PR + Q.$ $^2) B = 2\parallel + \perp.$ $^3) B = 2Y + \parallel.$

Compressibility of rocks is dependent upon the compressibility of individual grains, pores and cracks. Cracks in hand specimens and other discontinuities in rock mass (such as joints) are the major factors contributing to the compressibility of rocks.

Compressibility is greater for more porous rocks although grain size plays its role and may be an over-riding factor in certain cases. Brace (1965a) found that the volumetric compressibility of Stone Mountain granite (porosity 0.3%) to be almost twice that of Westerly granite (porosity 1.1%) though the porosity of Westerly granite is almost 4 times that of Stone Mountain granite. The average grain diameter in the two cases is 2.50 mm and 0.75 mm respectively indicating that a part of the compressibility is due to grain boundaries. For bedded and banded rocks, compressibility difference in the two directions is higher than in unbanded rocks.

Linear compressibility measurements on a number of rocks with preferred orientations conducted by Brace (1965a, b) have shown that the values are dependent upon the orientation of the grain and pressure and are different in different directions (Table 17). The influence of pressure on the volume compressibility of some rocks is given in Fig. 6-78. The values at

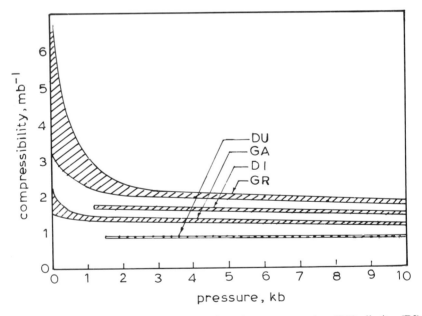

Fig. 6–78: Volume compressibility as a function of pressure; granites (GR), diorites (DI), gabbros (GA) and dunites (DU) fall within the shaded areas. Below 3 kb pressure the compressibility is affected by cracks
(after Adams and Williamson, 1923).

lower pressure (1 atm) are strongly influenced by cracks (ADAMS and WIL-
LIAMSON, 1923; BIRCH and CLARK, 1940). At low pressures, compressibility
may be many times the value at 3-4 kilobars. Above 2 to 3 kbars, most
cracks are closed and their effect is eliminated but not the effect of pores.

The influence of pores on the compressibility of a material has been considered by WALSH and BRACE (1966a).

For spherical pores

$$\frac{1}{K} - \frac{1}{K_0} = \frac{1}{K_0}\left(\frac{1-\nu}{1-2\nu}\right)\left(\frac{2\pi\bar{a}^3}{\bar{v}}\right) \tag{6.98}$$

where K = compressibility of rock with pores
         K_0 = compressibility of rock without pores
         ν = POISSON's ratio
         $\bar{a}$ = average diameter of pores
         $\bar{v}$ = porosity

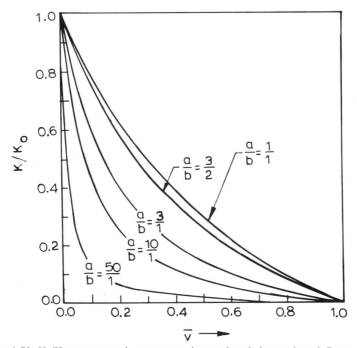

Fig. 6-79: K_o/K versus porosity, pore-pore interactions being neglected. Pores are oblate
spheroids with semi-major axis to semi-minor axis ratios of a/b
(after WARREN, 1973).

WARREN (1973) has analysed the influence of the spheroidal pores with different (a/b) ratios between the two axes. The results are given in Fig. 6-79.

The above discussion does not take into account the influence of pore-pore interaction. The influence of pore interaction is to increase (K_o/K) ratio for the same porosity and its influence increases with total porosity (WARREN, 1969; 1973) but for most practical purposes, it may not be significant.

6.10. Shock Hugoniot of Rocks

When a material is subjected to shock loading (e.g. in buried explosions) the pressure and density function for the material obtained in this way is called the Hugoniot or shock equation of state and differs from the isothermal or adiabatic equation of state. Hugoniot represents the locus of those thermo-dynamic states which may be attained in the medium starting from the same initial stage of the medium. On the pressure-density diagram the adiabatic and the isothermal are less steep than a Hugoniot which passes through the same initial state.

Besides the direct determination by subjecting the specimen to shock pressure using explosives, the other technique commonly used is that of summation, e.g.

$$P(\eta) = \sum_{i=1}^{N} f_i\, P_i\, (\eta) \tag{6.99}$$

where $P =$ shock pressure

$\eta = \varrho/\varrho_0$ is the shock compression or ratio of density behind to density ahead of the shock wave

$f =$ molecular fraction

The summation is carried out over the N elements which comprise the rock and is performed at constant compression η. By performing the summation at several values of compression, the Hugoniot of the rock in question is obtained over a pressure range, generally from 0.1 to 1.0 megabars.

The information required for synthesis includes the knowledge of Hugoniot of each constituent and the chemical analysis of the rock. The Hugoniot of chemical elements constituting the rock is obtained from experimental data (RICE, McQUEEN and WALSH, 1958). When no data for a particular element is available, substitution procedure is followed where the element of a similar atomic number is substituted (ex. silicon for aluminium) and it is assumed that the Hugoniot $P(\eta)$ of the substituting element is the same

as that of the substituted element. Eq. 6.100 is used in evaluating Eq. 6.99
for the pressure quantities $P_i(\eta)$:

$$P_i(\eta) = \frac{\varrho_{0i}\, C_i^2 \eta\, (\eta - 1)}{[\eta\, (1 - S_i) + S_i]^2} \qquad (6.100)$$

where C and S are constants of the element.

The Eq. 6.100 is obtained from the shock conservation equations

$$P = \varrho_0\, V_s\, u, \; \varrho_0\, V_s = \varrho\, (V_s - u)$$
and
$$V_s = C + Su$$

i.e. the shock velocity V_s is a linear function of particle velocity u.

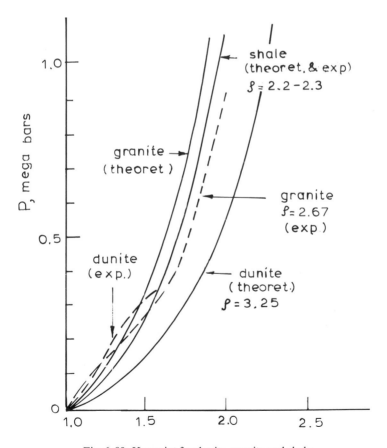

Fig. 6–80: Hugoniot for dunite, granite and shale
(after CHABAI, 1962).

CHABAI (1962) has used the summation technique for a number of rock types and the results obtained are quite in agreement with the experimental values (Figs. 6-80, 6-81 and 6-82). The various constants for Hugoniot for different rocks are given in Table 18.

The above method however fails if the rock contains pores and cracks filled with fluids. Shock wave passing through such a medium compresses the fluid and the shock disturbance in each of these components must coalesce in a macroscopic fashion to yield the wave velocities observed.

LYAKHOV (1959) suggested that Hugoniot in this case be determined by a summation of compression at constant pressure, i.e.

$$\eta (P) = \Sigma_i \, \alpha_i \, \eta_i (P) \tag{6.101}$$

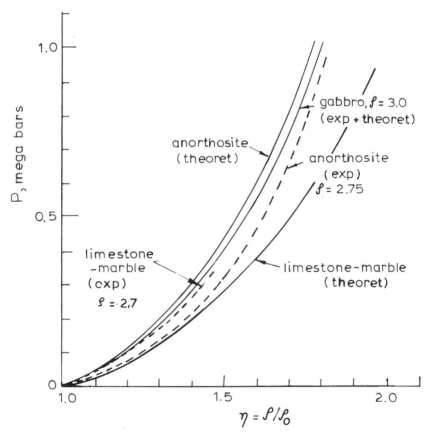

Fig. 6–81: Hugoniot for limestone-marble, gabbro and anorthosite
(after CHABAI, 1962).

where

$$a_i = \text{percentage by volume}$$
$$\eta_i(P) = \text{Hugoniot of the gas, or liquid or solid component}$$

The use of this method gives results very close to the experimental values provided the percentage of the various components are accurately known and their material variability is small. Fig. 6-83 gives the Hugoniot for volcanic tuff which clearly indicates the influence of saturation on the Hugoniot curve.

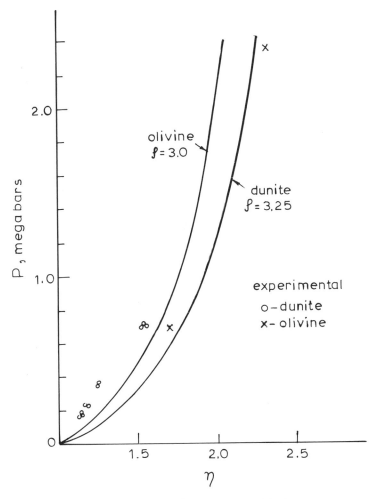

Fig. 6–82: Hugoniot for olivine and dunite
(after CHABAI, 1962).

TABLE 18
Constants for Hugoniot of nonporous[1] rocks
(after CHABAI, 1962)

Rock material	Density (g/cm^3)	Synthesis		Experiment[3]	
		C^2)	S	C^2)	S
Basalt	2.67	3.7	1.41	2.6	1.60
Gabbro	2.98	3.5	1.32	3.5	1.32
Shale	2.00	3.6	1.34	3.6	1.34
Limestone	2.50	3.4	1.27	3.4	1.27
Granite	2.67	3.7	1.34	4.6	1.00
Anorthosite	2.75	3.8	1.33	3.0	1.47
Dunite	3.30	3.1	1.31	6.3	0.65
Olivine	3.00	3.7	1.41	5.0	1.14
Halite	2.16	3.0	1.28	3.5	1.33
Dolomite	2.84	3.3	1.30		
Rhyolite	2.35	3.9	1.45		
Sandstone	2.00	4.3	1.34		
Claystone	1.00	5.8	1.32		
Diorite	2.84	3.7	1.38		
Syenite	2.76	3.7	1.38		
Diabase	2.97	3.6	1.41		
Pyroxenite	3.23	3.4	1.38		
Lherzolite	3.23	3.3	1.39		
Dacite	2.60	3.9	1.37		
Andesite	2.60	3.8	1.38		
Trachyte	2.60	3.7	1.37		
Phonolite	2.60	3.1	1.36		
Grandiorite	2.72	3.8	1.36		
Nepheline syenite	2.61	3.7	1.38		
Quartz diorite	2.81	3.7	1.38		

[1]) Nonporous here is used in the sense that no attempt was made to account for porosity in the synthesis.
[2]) Dimensions of C are mm/μsec.
[3]) Experimental values listed for C and S were not determined by least squares fit and consequently are not precise.

When shock velocity and particle velocity data are plotted together, the relationship so obtained is non-linear (GRADY, MURRI and MAHRER, 1976) at high pressures. The Hugoniot sound velocity is higher than bulk sound velocity which is most probably due to the occurrence of a partial or complete transformation during shock compression to a higher density and less compressible phase.

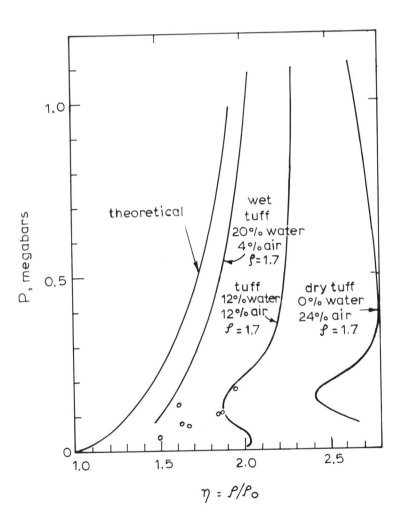

Fig. 6–83: Hugoniot for volcanic tuff
(after CHABAI, 1962).

6.11. Dilatancy in Rocks

The rock in uniaxial or triaxial compression undergoes volumetric changes which are positive (compression) in the earlier stages and negative (expansion) in the later stages of loading. This volumetric expansion or dilatancy has long been observed in granular mediums where it is associated with relative movement of grains or groups of grains causing irregularly packed interspaces and is a geometrical necessity for the deformation to occur (LOGANI, 1973) and is dominant only in densely packed materials. It has also been observed in several different ductile and brittle materials. Dynamic (wave velocity) tests by SCHOCK et al. (1973), THILL (1972) and GUPTA (1973) have also shown the occurrence of dilatancy in rocks. Direct field evidence for dilation in the regional strains field across San Andreas fault, near Parkfield, California has been reported by CHERRY and SAVAGE (1972). Indirect field evidence for dilation involving a reduction in the ratio of compressional to shear wave velocity prior to an earthquake has been reported by SEMENOV (1969), AGGARWAL et al. (1973) and ROBINSON et al. (1974).

COOK (1970) has shown that dilatancy is not a superficial phenomenon due to anomalous surface deformations but a pervasive property. Tests on hollow cylinders showed that both the internal hole and the specimen diameter increased with deformation.

Tests conducted by BRIDGMAN (1949) on soapstone and calcite marble (both are rather ductile rocks and possess low porosity) showed that at zero confining pressures, the rocks were dilatant. HANDIN et al. (1963) measured changes in porosity (increase in porosity means dilatancy) of several rocks at various effective confining pressures and found an increase in porosity at low effective confining pressures while decrease in porosity at high effective confining pressures (Fig. 6-84).

The net change in porosity is a function of the strain the specimens have undergone, the confining pressure and the initial porosity. High porosity rocks show compression at higher effective confining pressure up to a very large strain whereas low porosity rocks show dilatancy even at these pressures after a comparatively smaller deformation (Compare Berea sandstone with Hasmark dolomite – Fig. 6-84).

When volumetric changes are plotted against stress, the relationship is not linear (Fig. 6-85). The initial concavity in axial strains has been conclusively proved to be due to closing of cracks (BRACE, PAULDING and SCHOLZ, 1966; BIENIAWSKI, 1967).

The linearity portion of axial strain is due to linear elastic deformation of

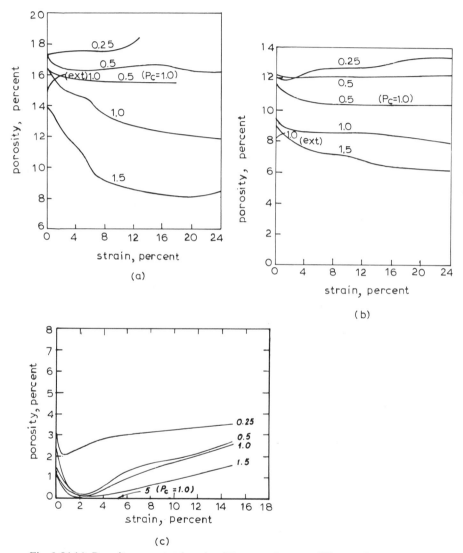

Fig. 6–84 (a): Porosity versus total strain of Berea sandstone at different effective confining pressures P_c (k bars) at 24 °C. All at $P_c = 2$ kilo bars except curve denoted $P_c = 1$. All in compression except curve marked ext (for extension).

(b): Porosity versus total strain of Marianna limestone at different effective confining pressures (kilo bars) at 24 °C. All at $P_c = 2$ kilobars except curve denoted $P_c = 1$. All in compression except curve marked ext (for extension).

(c): Porosity versus total strain of Hasmark dolomite at different effective confining pressures (kilo bars) at 24 °C. All in compression at 2 kilobars confining pressure except curve denoted

$$P_c = 1$$

(after HANDIN et al, 1963).

rocks. In this region, however, the volumetric curve (also the lateral strain curve) departs from linearity at point A. At this point crack initiation and stable crack propagation starts. The shape of the volumetric and lateral strain curves changes rapidly until a reversal in the gradient of the volumetric strain curve takes place (point B). Point A corresponds to about $1/3$rd the uniaxial compressive strength of rock. The observed values of this for various rocks are given in Table 19. The ratio of K_1 and K_2 to compressive strength (Table 19) depends upon the h/d ratio of the specimen and is lower for lower h/d ratio. Its value increases with increase in h/d ratio and indications are that for very slender specimens, K_2 may approach unity. Results on sandstone are given in Fig. 6-86.

The points A and B (Fig. 6-85) also correspond to specific energy release due to crack propagation and the crack propagation velocity in the rock. A general mechanism of rock deformation and fracture is given in Fig. 6-87 which correlates the various phenomena occurring during the course of deformation.

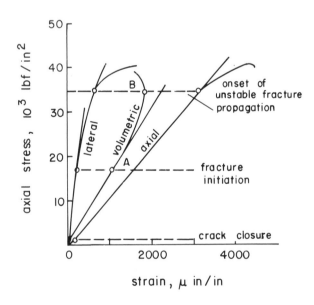

Fig. 6–85: Relationship between axial stress and axial, lateral, and volumetric strain, for quartzite in uniaxial compression tested in a conventional loading machine (after BIENIAWSKI, 1967).

The influence of confining pressure on the dilatancy is to raise the point of reversal *B* (Fig. 6-85) and may also result in greater dilation before failure for some rocks (RUMMEL, 1972; BROWN anid SWANSON, 1970; SWANSON and BROWN, 1972).

With extremely high pressure and with high porosity rocks, the tendency for dilatancy is suppressed (SWANSON and BROWN, 1972) and for very porous rocks the effect may be reversed, i.e. decrease in volume can accompany failure (EDMOND and PATERSON, 1972). Tests on a number of rocks and other materials with porosity ranging from 0.5 % (sodium chloride) to 25 % (graphite) showed negative dilatancy at very high pressure. This negative

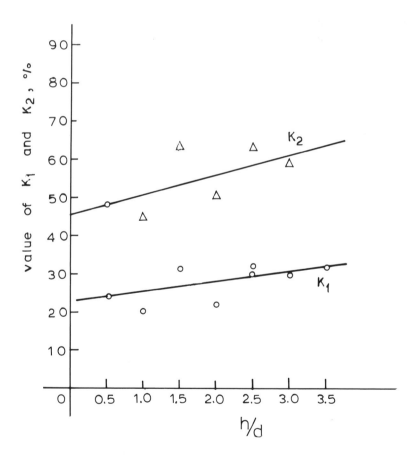

Fig. 6–86: Influence of h/d ratio on value of K_1 and K_2
(data from BORDIA, 1972).

TABLE 19
Stable and unstable crack propagation in rocks

Rock type	Stress conditions	Stable crack growth. Stress at fracture initiation as % of fracture strength, K_1	Stress at unstable crack growth as % of fracture strength, K_2	Reference
Norite	uniaxial compression	35.	73.	Bieniawski (1967)
	uniaxial tension	94.5	96.5	Bieniawski (1967)
Quartzite	uniaxial compression	40.	89.	Bieniawski (1967)
	$\sigma_1/\sigma_3 = 38.1$	25.	75.	
Granite	uniaxial compression	0.30	64.	Brace, Paulding & Scholz (1966)
	uniaxial compression (fast loading)	50.	78.	
	$\sigma_3 = 1\,kb$	–	74.	dito.
Marble	uniaxial	0.45	–	dito.
	$\sigma_3 = 0.25$	0.65	71.	
	$\sigma_3 = 0.49$	0.50	–	
Apatite	$\sigma_3 = 0$	0.45	–	dito.
	$\sigma_3 = 0.81$	0.72	–	
	$\sigma_3 = 2.38$	0.61	0.90	
	$\sigma_3 = 3.20$	0.68	–	
Sandstone	uniaxial compression (h/d = 2.5)	32.	64.	Bordia (1972)
Sandstone	uniaxial	50.	76.	Crouch (1970)
Westerly Granite	uniaxial compression	48.	70.	Zoback & Byerlee (1975)
	$\sigma_3 = 500$ bars	32.	60.	
	$\sigma_3 = 2000$ bars	31.	68.	
Indiana Limestone	uniaxial compression	30.	73.	
Porphyrite Tonalite	uniaxial compression (strain rate 3×10^{-4})	50.	67.	Perkins, Green and Friedman (1970)
	(strain rate 6×10^{-1})	50.	70.	

dilatancy was far greater than the elastic response of the materials. This can be attributed to the changes in the pore structure or grain arrangement or to the development of internal cracking. Tests on changes in pore radii of certain sandstones before and after compression of specimens support this concept (MARMORSHTEYN and MEKLER, 1973).

There is, however, no general correlation between the brittle ductile transition and any particular aspect of the volume change behaviour such as change from positive dilatancy to negative dilatancy or absence of non-elastic volume changes.

A marked dilatancy of specimens occurs on release of pressure (PATERSON, 1963; EDMOND and PATERSON, 1972). A total dilation of 6, 9 and 12 percent has been measured for Carrara marble, lithographic limestone and Gosford sandstone respectively during the pressure release from 6-8 kb. The total length increase for these rocks was 4, 4.5 and 4 percent. The initial porosities of the rocks were 1.1, 5.9 and 13 percent respectively. It indicates that pressure release effect in more compact materials (limestone and marble) is predominantly an increase in length whereas the dilatancy in sandstone is nearly isotropic. The reason for this behaviour is the anisotropy of the grains (PATERSON, 1963). In sandstone the dilatancy on pressure release occurs mainly due to the general loosening of the granular structure under the influence of contact stresses set up during deformation or high pressure and differ from those resulting simply from application of hydrostatic pressure.

Materials that have an adequate number of slip planes for homogeneous deformation will not show any marked dilatancy at high confining pressures, but other materials will show dilatancy. Dilatancy is inevitable in cataclastic deformation where an increase in volume is required to accommodate the relative movement of grains or their fragments. Also in the presence of restricted cleavage planes additional deformation is accommodated partially by kinking and at high temperatures diffusion (McLEAN, 1966). At pressures that are not greatly in excess of the brittle-ductile transition pressure, especially when temperature is low, cataclastic deformation plays an important role (PATERSON, 1969).

The lateral component of the dilatancy (lateral strain) increases with increase in axial strain and as fracture approaches, the rate of increase accelerates rapidly. The results of tests on Wombeyan marble at different confining pressures are given in Fig. 6-88. The maximum slope of the lateral component of the dilatancy – axial strain curve is greater when the failure is brittle than when it is ductile. Norite, sandstone and quartzite show similar behaviour (CROUCH, 1970). Tests on limestone from St-Marc-des-Carrieres, Quebec [porosity $< 1 \%$, $\sigma_c = 78$ MPa (11,700 lbf/in^2)] showed ratio of volumetric strain to axial strain on failure dropping from 4.22 to 0.63 as the confining pressure was raised from 13 to 500 bars (190 to 7500 lbf/in²) (LADANYI and DON, 1970).

The higher rate of loading decreases dilatancy (Fig. 6-89) and the strain at the point of onset of unstable fracture propagation (point *B*) is smaller for higher strain rates. The ratio of stress at unstable crack growth as a percentage of fracture strength seems to be unchanged.

The influence of cyclic loading on the dilatancy is quite marked. The stress at which dilatancy starts (Fig. 6-90) is reduced with each successive cycling. This has been confirmed in uniaxial and triaxial tests in a variety of rocks (SWANSON and BROWN, 1972; SCHOLZ and KRANZ, 1974; HAIMSON, 1974; ZOBACK and BYERLEE, 1975). Each cycle results in an increased amount of

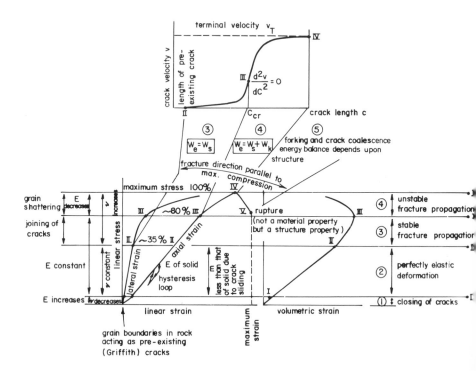

Fig. 6–87: Mechanism of brittle fracture of rock in multiaxial compression (after BIENIAWSKI, 1967).

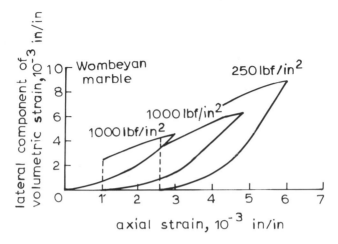

Fig. 6–88: Lateral volumetric strain-axial strain curves for repeated loading of Wombeyan marble at different confining pressures
(after CROUCH, 1970).

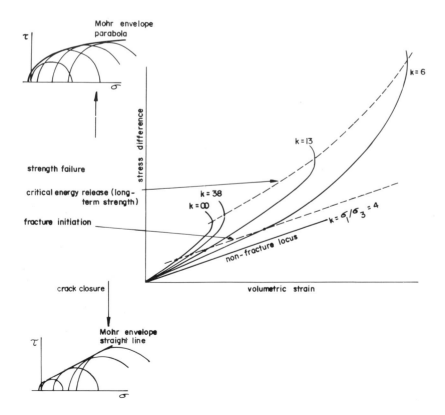

dilatancy and at the end of cycle there is a net gain in the volume of the
specimen. Net gain in crack porosity measurements for Westerly granite
for each cycle are given in Table 20. These values have an uncertainty of
20% and sensitivity of 50×10^{-6}.

The volumetric changes are associated with acoustic emissions and large
acoustic emission counts have been noted near the last loading cycle prior
to failure (HAIMSON and KIM, 1975).

Dilatancy with time has a similar shape (Fig. 6-91) to the axial strain creep
curve which proves that dilatancy and creep is the result of microfracturing
and collapse of structure.

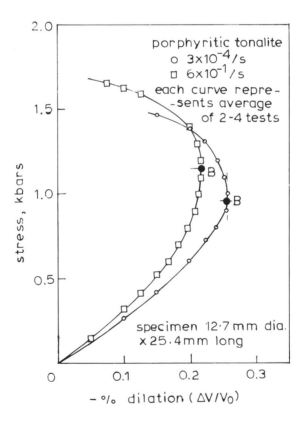

Fig. 6–89: Volume change as a function of stress for specimens loaded parallel to bulk
core axis
(after PERKINS, GREEN and FRIEDMAN, 1970).

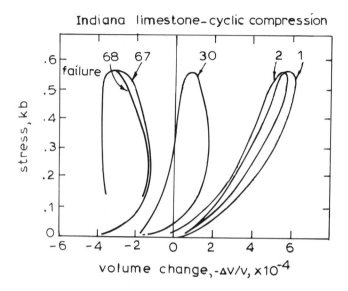

Fig. 6–90: Typical stress-volumetric strain behaviour in cyclic uniaxial compression—
Indiana limestone
(after HAIMSON, 1974).

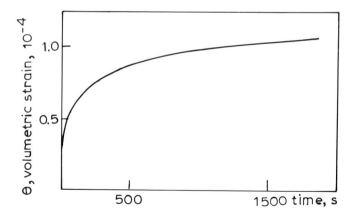

Fig. 6–91: Volumetric strain versus time during a creep test on Westerly granite at room
temperature and pressure, and a compressive load of 2.03 kb
(after SCHOLZ, 1970).

TABLE 20
Cyclic tests on Westerly granite in uniaxial compression
(after BRACE, PAULDING & SCHOLZ, 1966)

Cycle	Total porosity x 10^{-6}	Gain in porosity x 10^{-6}
0	1580	0
1	1560	$-$ 20
2	1620	$+$ 40
3	1710	$+$ 130
4	2050	$+$ 470
5	2770	$+$ 1190

6.12. Summary and Conclusions

The preparation of specimens for the determination of its deformation under different loading conditions is as important as for the determination of its strength. The various requirements along with the type of test are described in Vol. I. The loading arrangement and deformation measuring equipment vary depending upon test conditions and are partly discussed in Vol. I and in this chapter.

Most of the loading machines available meet the loading requirements. Normally a 1 MN capacity universal testing machine will be quite suitable for most of the test conditions.

Electrical resistance strain gauges are the most commonly used strain measuring equipment and are suitable under most of the test conditions. LVDT s can be used where space is not a problem. Mechanical dial gauges and optical gauges can be used for uniaxial compression and tension tests for both lateral and axial deformation measurements. The gauges are difficult to set up, but require no sophisticated reading equipment.

While mounting deformation gauges on test specimens, points of high strain gradients should be avoided. It is advisable to measure deformation at more than one point and take an average value for calculations.

The moduli in compression and tension for rocks are not the same and their ratio may vary from 1–10 or more depending upon the rock type. This results in difficulties in interpreting the results of tests where the stresses change from compression to tension (e.g. bending tests) and appropriate care should be taken in their analysis. Where more than one elastic relationship holds good and can be used in the calculation, it is suggested to weigh each of these relationships in the light of observed stress-strain behaviour. While reporting the modulus of elasticity, the secant value or tangent value at 50 % fracture strength of specimen should be calculated and mentioned along with other results. The method of measurement of strain such as gauges mounted on the specimen or between platens, test technique, (e.g. bending or Brazilian etc.) should be stated.

The variation in modulus value with stress is due to temporary or permanent changes occurring in the structure of the specimens and hence the loading and unloading path followed by the specimens may be quite different.

Modulus values of homogeneous rocks are not influenced by specimen size, but for non-homogeneous specimens, the (h/d) ratio and specimen size play an important role. The drop in value with increase in (h/d) ratio may be considerable (a factor of 2–3). The post-failure stiffness of the specimens is greatly affected by increase in (h/d) ratio and increases rapidly with slender specimens. When brush platens are used, the post-failure modulus is not influenced by changes in (h/d) ratio.

Platen conditions, rate of loading and lateral pressure affect the stress-strain curve, but not to a great extent. The modulus values calculated will vary only by 5–10 % between extreme conditions.

The influence of temperature on the stress-strain curve becomes marked particularly when higher lateral pressure is in existence. The ductility of rocks increases with increase in temperature and pressure and otherwise brittle rocks show ductile behaviour. The intermediate principal stress exerts a decisive influence on the ductility, and strain before faulting decreases with increase in intermediate principal stress.

Cracks and pores play an important role in the modulus values calculated from tests. The low value of deformation modulus is due to closure of cracks at the initial stage of loading. The modulus value obtained on unloading represents to a greater measure the modulus of the intact material.

The ratio between the unloading and loading modulus is a measure of the degree of fracturing of specimens.

Modulus anisotropy is the result of micro and macro fabric of the rock. Rocks containing spherical grains or randomly distributed grains have low or no modulus anisotropy. Elongated shapes result in modulus anisotropy in rocks and the ratio may vary between 1–2.5.

Presence of joints and bedding planes greatly influence modulus values. Modulus of deformation for rocks stressed parallel to the bedding planes is greater than when stressed at right angle to their bedding planes. The normal stiffness of joints being usually much lower than rock substance, the modulus values when loaded at right angle to joint planes are greatly reduced.

POISSON's ratio of rocks measured from stress-strain curves is very much dependent upon the stress level and is greatly influenced by the opening or closing of cracks. Measured values at different stress levels may vary from 0.1 to 1.0 or even more. Under such circumstances, samples should preferably be loaded to a stress level lower than that at which any permanent changes occur in rocks (e.g. lower than point A in Fig. 6-85) and POISSON's ratio be calculated using incremental values. In other cases, it is suggested to subject specimens to lateral pressures to about 20–40 % of their strength and then load axially and measure lateral and axial strains at about 50 % of their strength under these test conditions.

Compressibility of rocks is a function of crack closure, pore deformation and material compression. Most of the cracks get closed at 1 kb hydrostatic pressure, but pore collapse takes place at much higher pressures.

Rock dilatation is an all pervasive property of hard rocks and the stress value at which it starts is a measure of the permanent changes occurring in rock structure. This stress is many times referred to as the time-dependent strength of rock.

References to Chapter 6

1. ADAMS, F. D. and NICOLSON, J. T.: An experimental investigation into the flow of marble. Phil. Trans. Royal Soc. London, Series A, Vol. 195, 1901, pp. 363–401.
2. ADAMS. L. H. and WILLIAMSON, E. D.: On the compressibility of minerals and rocks. J. Franklin Inst., Vol. 195, 1923, pp. 475–529.
3. ADLER, L.: Evaluating double elasticity in drill cores under flexure. Int. J. Rock Mech. Min. Sci., Vol. 7, 1970, pp. 357–370.
4. AGGARWAL, Y. P., SYKES, L. R., ARMBRUSTER, J. and SBAR, M. L.:Premonitory changes in seismic velocities and prediction of earthquakes. Nature, Vol. 241, Jan. 12, 1973, pp. 101–104.
5. ALEXANDROV, K., RYSHORA, T. V., BELIKOV, B. P. and SHABANOVA, L. A.: Anisotropy of elastic properties of rock. Int. Geol. Rev., Vol. 11, 1969, pp. 539–574.
6. ARAMBURU, J. A.: La Angostura dam underground power house: Prediction and measurement of displacements during excavations. Proc. 3rd Cong. Int. Soc. Rock Mech., Denver, 1974, Vol. 2, Part B, pp. 1231–1241.
7. ARTHUR, J. R. F. and MENZIES, B. K.: Inherent anisotropy in a sand. Geotechnique, Vol. 22, 1972, pp. 115–128.
8. A. S. T. M.: Standard method of test for elastic moduli of rock core specimens in uniaxial compression. A. S. T. M. Year Book, 1974, pp. 400–403.

9. ATKINSON, J. H.: Anisotropic elastic deformations in laboratory tests on undisturbed London clay. Geotechnique, Vol. 25, 1975, pp. 357–374.

10. ATTEWELL, P. B.: Triaxial anisotropy of wave velocity and elastic moduli in slate and their axial concordance with fabric and tectonic symmetry. Int. J. Rock Mech. Min. Sci., Vol. 7, 1970, pp. 193–207.

11. BAMFORD, W. E.: Anisotropy, and the natural variability of rock properties. Proc. Symp. Rock Mech., Univ. Sydney, 1969, pp. 1–10.

12. BARLA, G. and GOFFI, L.: Direct tensile testing of anisotropic rocks. Proc. 3rd Cong. Int. Soc. Rock Mech., Denver, 1974, Vol. 2, Part A, pp. 93–98.

13. BAULE, H. and MUELLER, E.: Messung elastischer Eigenschaften von Gesteinen. (Measurements of the elastic properties of rocks). In German. Handbuch der Physik, Springer, Vol. 47, 1956.

14. BHASKARAN, R.: Variability in strength and deformation characteristics of anisotropic clays. Proc. Symp. Recent Developments Analysis Soil Behaviour and Application to Geotechnical Structures, Univ. N. S. W., 1975.

15. BIENIAWSKI, Z. T.: Mechanism of brittle fracture of rock. Part I – Theory of the fracture process; Part II – Experimental studies; Part III – Fracture in tension and under long-term loading. Int. J. Rock Mech. Min. Sci., Vol. 4, 1967, pp. 365–430.

16. BIENIAWSKI, Z. T.: Time-dependent behaviour of fractured rock. Rock Mech., Vol. 2, No. 3, Sept., 1970, pp. 123–137.

17. BIRCH, F.: Compressibility; Elastic constants. In Handbook of Physical Constants (Edited by S. P. Clark). Geol. Soc. Am. Mem. 97, 1966, pp. 97–173.

18. BIRCH, F. and CLARK, H.: The thermal conductivity of rocks and its dependence upon temperature and composition. Am. J. Sci., Vol. 238, 1940, pp. 613–635.

19. BLAIR, B. E.: Physical properties of mine rock. Part 3. U.S.B.M.R.I. 5130, 1955, 69p.

20. BORDIA, S. K.: Complete stress – volumetric strain equation for brittle rock up to strength failure. Int. J. Rock Mech. Min. Sci., Vol. 9, 1972, pp. 17–24.

21. BRACE, W. F.: Effect of pressure on electric – resistance strain gauges. Exp. Mech., Vol. 4, 1964, pp. 212–216.

22. BRACE, W. F.: Some new measurements of linear compressibility of rocks. J. Geophy. Res., Vol. 70, 1965a, pp. 391–398.

23. BRACE, W. F.: Relation of elastic properties of rocks to fabric. J. Geophys. Res., Vol. 70, 1965b, pp. 5657–5667.

24. BRACE, W. F.: Micro-mechanics in rock systems. In Structure, Solid Mechanics and Engineering Design (Editor – M. TEENI). New York, Wiley, 1971, pp. 187–204.

25. BRACE, W. F. and JONES, A. H.: Comparison of uniaxial deformation in shock and static loading of three rocks. J. Geophys. Res., Vol. 76, No. 20, July 10, 1971, pp. 4913–4921.

26. BRACE, W. F., PAULDING, B. W. and SCHOLZ, C.: Dilatancy in the fracture of crystalline rocks. J. Geophys. Res., Vol. 71, 1966, pp. 3939–3953.

27. BRACE, W. F. and RILEY, D. K.: Static uniaxial deformation of 15 rocks to 30 kb. Int. J. Rock. Mech. Min. Sci., Vol. 9, 1972, pp. 271–288.

28. BRADY, B. T.: A mechanical equation of state for brittle rock. Part I – The prefailure behaviour of brittle rock. Int. J. Rock Mech. Min. Sci., Vol. 7, 1970, pp. 385–421.

29. BRADY, B. T.: A mechanical equation of state for brittle rock. Part II – The prefailure initiation behaviour of brittle rock. Int. J. Rock Mech. Min. Sci. & Geomech. Abstr., Vol. 10, 1973, pp. 291–309.

30. BRAGG, W. L. and FRYE, J. F.: A dynamical model of a crystal structure. Part 1. Proc. Roy. Soc. London, Series A, Vol. 190, 1947, pp. 474–481.

31. BRIDGMAN, P. W.: Volume changes in the plastic stages of simple compression. J. Appl. Phys., Vol. 20, 1949, pp. 1241–1251.

32. BROWN, W. S. and SWANSON, S. R.: Constitutive relations for rocks based on experimental measurement of stress-strain and fracture properties. Univ. Utah, Dept. Mech. Eng., Feb., 1970.

33. BUDIANSKY, B. and O'CONNELL, R. J.: Elastic moduli of a cracked solid. Int. J. Solids Structures, Vol. 12, 1976, pp. 81–97.

34. BUSCHING, H. W., GOETZ, W. H. and HARP, M. E.: Stress-deformation behaviour of anisotropic bituminous mixtures. Proc. Assoc. Asphalt Paving Tech., Vol. 36, 1967, pp. 632–671.

35. CHABAI, A. J.: Synthesis of shock Hugoniots for rock materials. Proc. 5th Symp. Rock Mech., Minneapolis, Minn., 1962, pp. 347–366.

36. CHENEVERT, M. E. and GATLIN, C.: Mechanical anisotropies of laminated sedimentary rocks. Soc. Pet. Eng. J., Vol. 5, 1965, pp. 67–77.

37. CHERRY, J. T. and SAVAGE, J. C.: Rock dilatancy and strain accumulation near Parkfield, California. Bull. Seism. Soc. Am., Vol. 62, 1972, pp. 1343–1347.

38. CHRISTENSEN, N. I. and RAMANANANTOANDRO, R.: Elastic moduli and anisotropy of dunite to 10 kilobars. J. Geophy. Res., Vol. 76, 1971, pp. 4003–4010.

39. CLARK, S. P. (Editor): Handbook of physical constants. Geol. Soc. Am. Mem. 97, 1966, 587 p.

40. COOK, N. G. W.: The failure of rock. Int. J. Rock Mech. Min. Sci., Vol. 2, 1965, pp. 389–403.

41. COOK, N. G. W.: An experiment proving dilatancy is a pervasive volumetric property of brittle rock loaded to failure. Rock Mech., Vol. 2, 1970, pp. 181–188.

42. CORDING, E. J.: The stability during the construction of three large underground openings in rock. Ph. D. Thesis, Univ. Illinois, Urbana, 1967.

43. CROUCH, S. L.: Experimental determination of volumetric strains in failed rock. Int. J. Rock Mech. Min. Sci., Vol. 7, 1970, pp. 589–603.

44. CROUCH, S. L.: A note on post-failure stress-strain path dependence in norite. Int. J. Rock Mech. Min. Sci., Vol. 9, 1972, pp. 197–204.

45. DALLY, J. W. and RILEY, W. F.: Experimental stress analysis. New York, McGraw-Hill, 1965, 520 p.

46. DAVIES, J. D. and STAGG, K. G.: Splitting tests on rock specimens. Proc. 2nd Cong. Int. Soc. Rock Mech., Belgrade, 1970, Vol. 2, pp. 343–349.

47. DAYRE, M., DESSENNE, J. L. and WACK, B.: Local and mean changes of the density of chalk samples tested under triaxial conditions. Proc. 2nd Cong. Int. Soc. Rock Mech., Belgrade, 1970, Vol. 1, pp. 373–381.

48. DEAN, M. III.: Techniques for protecting and waterproofing resistance wire strain gauges. David Taylor Model Basin, Rep. 797, Oct., 1957.

49. DE BEER, E., DELMER, A. and WALLAYS, M.: Essais de charge en galene et en surface avec plaques de grandes dimensions. Proc. Int. Symp. Rock Mech., Madrid, 1968, pp. 13–33.

50. DE BEER, E. E., GRAULICH, J. M. and WALLAYS, M.: In situ testing for determining the modulus of compressibility of rock mass consisting of shales. Proc. 3rd Cong. Int. Soc. Rock Mech., Denver, 1974, Vol. 2, Part B, pp. 645–653.

51. DEKLOTZ, E. J., BROWN, J. W. and STEMLER, O. A.: Anisotropy of schistose gneiss. Proc. 1st Cong. Int. Soc. Rock. Mech., Lisbon, 1966a, Vol. 1, pp. 465–470.

52. DEKLOTZ, E. J. and HECK, W. J.: Tests for strength characteristics of rock, Piledriver project. U. S. Army Corps Eng., Mo. River Div., Omaha, N. B., MRD Lab. 64/474, 1965.

53. DEKLOTZ, E. J., HECK, W. J. and NEFF, T. L.: Tests for strength characteristics of a schistose gneiss: First Interim Rep., Preliminary tests. U. S. Army Corps Eng., Mo. River Div., Omaha, N. B., MRD Lab. 64/126, 1965.

54. DEKLOTZ, E. J., HECK, W. J. and NEFF, T. L.: Strength parameters of selected inter-mediate quality rocks: First Interim Rep. U. S. Army Corps Eng., Mo. River Div., Omaha, N. B., MRD Lab. 64/493, 1966 b .

55. DIETZE, W.: The state of further exploration of rock mechanics in the Mansfeld copper-slate mining. Proc. Int. Strata Control Cong., Leipzig, 1958, pp. 173–190.

56. DONATH, F. A.: Some information squeezed out of rock. Am. Sci., Vol. 58, 1970, pp. 54–72.

57. DOUGLASS, P. M. and VOIGHT, B.: Anisotropy of granites: A reflection of microscopic fabric. Geotechnique, Vol. 19, 1969, pp. 376–398.

58. DREYER, W.: The science of rock mechanics – Part 1 – The strength properties of rocks. Clausthal-Zellerfeld, Trans Tech Publications, 1972, 501 p.

59. DUNCAN, J. M. and CHANG, C. Y.: Nonlinear analysis of stress and strain in soils. J. Soil Mech. Found. Div., Am. Soc. Civ. Eng., Vol. 96, No. SM 5, Sept., 1970, pp. 1629–1653.

60. DUVALL, W. I.: The effect of anisotropy on the determination of dynamic elastic constants of rock. Trans. Soc. Min. Eng. AIME, Vol. 232, No. 4, Dec., 1965, pp. 309–316.

61. EDMOND, J. M. and PATERSON, M. S.: Volume changes during the deformation of rocks at high pressures. Int. J. Rock Mech. Min. Sci., Vol. 9, 1972, pp. 161–182.

62. EVANS, I. and POMEROY, C. D.: The strength, fracture and workability of coal. London, Pergamon Press, 1966, 277 p.

63. EVDOKIMOV, P. D. and SAPEGIN, D. D.: Stability, shear sliding resistance, and defor-mation of rock foundations. Translated from Russian. Jerusalem, Israel Program for Scientific Translations, 1967, 147 p.

64. EVERLING, G.: Gesteinmechanische Untersuchungen und Grundlagen zur Ermittlung des Gebirgsdruckes aus Bohrlochverformung. (Rock mechanics investigations into determination of rock pressure from drill hole deformations.) In German. Glückauf, Vol. 96, 1960, pp. 390–409.

65. FAIRHURST, C.: Laboratory measurement of some physical properties of rock. Proc. 4th Symp. Rock Mech., Univ. Park, Penn., 1961, pp. 105–118.

66. FUNG, P. K.: Instrumentation for measuring load and deformation under high pressure. Exp. Mech., Vol. 15, 1975, pp. 61–66.

67. GERRARD, C. M.: Background to mathematical modelling in geomechanics – the role of fabric and stress history. Proc. Symp. Computational Methods in Soil and Rock Mech., Karlsruhe, 1975.

68. GERRARD, C. M., DAVIS, E. H. and WARDLE, L. J.: Estimation of the settlement of cross-anisotropic deposits using isotropic theory. Univ. Sydney, School of Civil Eng., Res. Rep. No. R – 191, 1972, 35 p.

69. GIBSON, R. E.: The analytical method in soil mechanics. Geotechnique, Vol. 24, No. 2, 1974, pp. 114–140.

70. GONANO, L. P.: Stress gradient and size effect phenomenon in brittle materials. Ph. D. Thesis, James Cook Univ., Townsville, 1974.

71. GRADY, D. E., MURRI, W. J. and MAHRER, K. D.: Shock compression of dolomite. J. Geophys. Res., Vol. 81, 1976, pp. 889–893.

72. GRAY, K. E.: Some rock mechanics aspects of petroleum engineering. Proc. 9th Symp. Rock Mech., Golden, Colorado, 1967, pp. 405–433.

73. GREEN, S. J., LEASIA, J. D., PERKINS, R. D. and JONES, A. H.: Triaxial stress behaviour of Solenhofen limestone and Westerly granite at high strain rates. Unpublished Report. 1970.

74. GREEN, S. J. and PERKINS, R. D.: Uniaxial compression tests at strain rates from 10^{-4} to 10^4/sec on three geologic materials. Rep. DASA – 2199, 1969, 44 p.

75. GRESZCZUK, L. B.: Effect of foreign inclusions on properties of solids. J. Eng. Mech. Div., Am. Soc. Civ. Eng., 1966, pp. 63–78.

76. GRIGGS, D. T.: Deformation of rocks under high confining pressures. I. Experiments at room temperature. J. Geol., Vol. 44, No. 5, July–Aug., 1936, pp. 541–577.

77. GRIGGS, D. T., TURNER, F. J. and HEARD, H. C.: Deformation of rocks at 500° to 800 °C. Geol. Soc. Am. Mem. 79, 1960, pp. 30–104.

78. GUPTA, I. N.: Dilatancy and premonitory variations of P, S travel times. Bull. Seism. Soc. Am., Vol. 63, 1973, pp. 1157–1161.

79. GUSTKIEWICZ, J.: Uniaxial compression testing of brittle rock specimens with special consideration given to bending moment effects. Int. J. Rock Mech. Min. Sci. & Geomech. Abstr., Vol. 12, No. 1, Jan., 1975a, pp. 13–25.

80. GUSTKIEWICZ, J.: Strain fluctuations in heterogeneous rocks. Int. J. Rock Mech. Min. Sci. & Geomech. Abstr., Vol. 12, Nos. 5/6, June, 1975b, pp. 181–189.

81. HAIMSON, B. C.: Mechanical behaviour of rock under cyclic loading. Proc. 3rd Cong. Int. Soc. Rock Mech., Denver, 1974, Vol. 1, pp. 373–378.

82. HAIMSON, B. C. and KIM, K.: Acoustic emission and fatigue mechanism in rocks. Proc. First Conf. Acoustic Emission in Geologic Structures and Materials, Univ. Park, Pa., 1975, Trans Tech Publications, Clausthal-Zellerfeld, 1977.

83. HANDIN, J., FRIEDMAN, M., LOGAN, J. M., PATTISON, L. J. and SWOLFS, H. S.: Experimental folding of rocks under confining pressure: Buckling of single – layer rock beams. In Flow and Fracture of Rocks. Geophysical Monograph 16, Am. Geophys. Union, Washington, D. C., 1972, pp. 1–28.

84. HANDIN, J. and HAGER, R. V.: Experimental deformation of sedimentary rocks under confining pressure: Tests at room temperature on dry samples. Bull. Am. Assoc. Pet. Geol., Vol. 41, 1957, pp. 1–50.

85. HANDIN, J. and HAGER, R. V.: Experimental deformation of sedimentary rocks under confining pressure: Tests at high temperature. Bull. Am. Assoc. Pet. Geol., Vol. 42, 1958, pp. 2892–2934.

86. HANDIN, J., HAGER, R. V., FRIEDMAN, M. and FEATHER, J. N.: Experimental deformation of sedimentary rocks under confining pressure: Pore pressure tests. Bull. Am. Assoc. Pet. Geol., Vol. 47, No. 5, May, 1963, pp. 717–755.

87. HASHIN, Z.: The elastic moduli of heterogeneous materials. J. Appl. Mech., Vol. 29, 1962, pp. 143–150.

88. HAWKES, I. and MELLOR, M.: Uniaxial testing in rock mechanics laboratories. Eng. Geol., Vol. 4, 1970, pp. 177–285.

89. HEARD, H. C.: Transition from brittle fracture to ductile flow in Solenhofen limestone as a function of temperature, confining pressure, and interstitial fluid pressure. Geol. Soc. Am. Mem. 79, 1960, pp. 193–226.

90. HEARD, H. C.: Experimental deformation of rocks and the problem of extrapolation to nature. NSF Adv. Sci. Semin. Rock Mech., Boston College, 1967, Terr. Sci. Lab., Air Force Cambridge Res. Lab., Bedford, Mass., pp. 439–507.

91. HEARD, H. C.: Steady-state flow in polycrystalline halite at pressure of 2 kilobars. In Flow and Fracture of Rocks. Geophysical Monograph 16, Am. Geophy. Union, Washington, D. C., 1972, pp. 191–209.

92. HETENYI, M. (Editor): Handbook of experimental stress analysis. New York, Wiley, 1960, 1077 p.

93. HILL, R.: The elastic behaviour of a crystalline aggregate. Proc. Phys. Soc., Vol. 65, Series A, 1952, pp. 349–354.

94. HOBBS, D. W.: Stress-strain-time behaviour of a number of coal measure rocks. Int. J. Rock Mech. Min. Sci., Vol. 7, 1970, pp. 149–170.

95. HOFFMANN, H.: Investigations into carbonic rocks under triaxial pressure for the purpose of rock stress computation. Proc. Int. Strata Control Cong., Leipzig, 1958.

96. HONDROS, G.: The evaluation of Poisson's ratio and the modulus of materials of a low tensile resistance by the Brazilian (indirect tensile) test with particular reference to concrete. Aust. J. Appl. Sci., Vol. 10, 1959, pp. 243–268.

97. HORIBE, T. and KOBAYASHI, R.: Physical and mechanical properties of coal measure rocks under triaxial pressure. Proc. 3rd Int. Conf. Strata Control, Paris, 1960, pp. 175–186.

98. HORIBE, T. and KOBAYASHI, R.: On the mechanical behaviour of rocks under various loading rates. In Japanese. J. Soc. Mat. Sci., Japan, Vol. 14, 1965.

99. HOSHINO, K. and KOIDE, H.: Process of deformation of the sedimentary rocks. Proc. 2nd Cong. Int. Soc. Rock Mech., Belgrade, 1970, Vol. 1, pp. 353–359.

100. HUDSON, J. A., BROWN, E. T. and FAIRHURST, C.: Shape of the complete stress-strain curve for rock. Proc. 13th Symp. Rock Mech., Urbana, Illinois, 1971, pp. 773–795.

100a. HUDSON, J. A., CROUCH, S. L. and FAIRHURST, C.: Soft, stiff and servo-controlled testing machines: A review with reference to rock failure. Eng. Geol. 'Vol. 6, 1972' pp. 155–189.

101. HUGHES, D. S. and MAURETTE, C. K.: Variation of elastic moduli of igneous rocks with pressure and temperature. Geophysics, Vol. 21, No. 2, 1956, pp. 23–31.

102. JAEGER, J. C.: Behaviour of closely jointed rock. Proc. 11th Symp. Rock. Mech., Berkeley, California, 1969, pp. 57–68.

103. JOHN, M.: The influence of loading rate on mechanical properties and fracture processes of rock. Rep. S. African C. S. I. R., MEG 1115, 1972, 28 p.

104. JONES, A. H. and FROULA, N. H.: Uniaxial strain behaviour of four geologic materials to 50 kilobars. Rep. DASA – 2209, 1969, 62 p.

105. KAARSBERG, E. A.: Elastic studies of isotropic and anisotropic rock samples. Trans A. I. M. E., Vol. 241, 1968, pp. 470–475.

106. KHACHIKIAN, G. G.: Determination of the elastic characteristics of stratified anisotropic rock for pressure tunnel designs. Gidrotechnicheskoe Stroiltelstvo, Vol. 9, 1966, pp. 26–28. (Translation by U. S. Bureau of Reclamation, Denver, Colorado, 1967).

107. KING, M. S.: Ultrasonic compressional and shear-wave velocities of confined rock samples. Proc. 5th Can. Rock Mech. Symp., Toronto, 1968, pp. 127–156.

108. KO, H. Y. and GERSTLE, K. H.: Constitutive relation of coal. Proc. 14th Symp. Rock Mech., Univ. Park, Penn., 1972, pp. 157–188.

109. KO, K. C. and HAAS, C. J.: The effective modulus of rock as a composite material. Int. J. Rock Mech. Min. Sci., Vol. 9, 1972, pp. 531–541.

110. KULHAWY, F. H.: Stress deformation properties of rock and rock discontinuities. Eng. Geol., Vol. 9, 1975, pp. 327–350.

111. KULHAWY, F. H. and DUNCAN, J. M.: Stresses and movements in Oroville dam. J. Soil Mech. Found. Div., Am. Soc. Civ. Eng., Vol. 98, SM 7, July, 1972, pp. 653–665.

112. KUMAZAWA, M.: The elastic constant of polycrystalline rocks and nonelastic behaviour inherent to them. J. Geophys. Res., Vol. 74, 1969, pp. 5311–5320.

113. LADANYI, B. and DON, N.: Study of strains in rock associated with brittle failure. Proc. 6th Can. Rock Mech. Symp., Montreal, 1970, pp. 49–64.

114. LAFEBER, D. and WILLOUGHBY, D. R.: Fabric symmetry and mechanical anisotropy in natural soils. Proc. 1st Australia-New Zealand Conf. Geomechanics, Melbourne, 1971, Vol. 1, pp. 165–174.

115. LAMA, R. D.: Elasticity and strength of coal seams in situ and an attempt to determine the energy in pressure bursting of roadsides. D. Sc. Tech. Thesis, Faculty of Mining, Academy of Min. & Metall., Cracow, Poland, 1966.

116. LAMA, R. D.: Effect of non homogeneities and discontinuities on deformational behaviour and strength of rocks. Metals and Minerals Review, Vol. VIII, No. 8, 1969, pp. 3–10.

117. LAMA, R. D.: In situ and laboratory strength of coal. Proc. 12th Symp. Rock Mech., Rolla, Missouri, 1970, pp. 265–300.
118. LAMA, R. D.: Creep of fractured rock. SFB–77, Jahresbericht 1973, Inst. Soil Mech. and Rock Mech., Univ. Karlsruhe, Karlsruhe.
119. LAMA, R. D.: Unpublished. 1975.
120. LAMA, R. D.: Unpublished. 1976.
121. LAMA, R. D. and GONANO, L. P.: Size effect considerations in the assessment of mechanical properties of rock masses. Proc. 2nd Symp. Rock. Mech., Dhanbad, 1976.
122. LAMB, E. H.: A small roller extensometer. Eng., London, Vol. 113, 1922.
123. LAUFFER, H. and SEEBER, G.: Design and control of linings of pressure tunnels and shafts, based on measurements of the deformability of the rock. Trans. 7th Cong. Large Dams, Rome, 1961, pp. 679–709.
124. LEEMAN, E. R. and GROBBELAAR, C.: A compressometer for obtaining stress-strain curves of rock specimens up to fracture. J. Sci. Inst., Vol. 34, 1957a, pp. 279–280.
125. LEEMAN, E. R. and GROBBELAAR, C.: A lateral extensometer for the determination of Poisson's ratio of rock. J. Sci. Inst., Vol. 34, 1957b, pp. 503–505.
126. LEKHNITSKII, S. G.: Stress distribution close to a horizontal working of elliptical shape in a transversely isotropic mass with inclined planes of isotropy. Mech. Solids, Vol. 1, No. 2, 1966, pp. 35–41.
127. LEKHNITSKII, S. G.: Anisotropic plates. Translated from the 2nd Russian edition. New York, Gordon and Breach, 1968.
128. LEPPER, H. A.: Compression tests on oriented specimens of Yule marble. Am. J. Sci., Vol. 247, 1949, pp. 570–574.
129. LNEC: Esludo das propriedades mecanicas da rocha de fundacao da barragem da aquieira. LNEC, Lisbon, Feb., 1964.
130. LOGANI, K.: Dilatancy model for the failure of rocks. Ph. D. Thesis, Iowa State Univ., Dept. Civil Eng., 1973, 212 p.
131. LÖGTERS, G. and VOORT, H.: In situ determination of the deformational behaviour of a cubical rock mass sample under triaxial load. Rock Mech., Vol. 6, 1974, pp. 65–79.
132. LOVE, A. E. H.: A treatise on the mathematical theory of elasticity. Cambridge, Cambridge University Press, 1927, 643 p.
133. LOZINSKA-STEPIEN, H.: Deformations during compression of the Cretaceous clay limestones and marls in an interval of loads from 0 to the boundary of proportionality in the light of laboratory examinations. Proc. 1st Cong. Int. Soc. Rock Mech., Lisbon, 1966, Vol. 1, pp. 381–384.
134. LYAKHOV, G. M.: Shock waves in multi-component media. Izvest. Akad. Nauk U.S.S.R., Otdel Teknicheskikh Nauk, Mekhanika i Mashionostroyeniye, Moscow, No. 1, Jan., 1959.
135. MARMORSHTEYN, L. M. and MEKLER, Y. B.: Influence of the deformation time upon the physical properties of sandstone reservoir. Akad. Sci. U.S.S.R., Physics of the Solid Earth, No. 5, 1973, pp. 345–347.
136. MASURE, P.: Behaviour of rocks with two-dimensional discontinuous anisotropy. Proc. 2nd Cong. Int. Soc. Rock Mech., Belgrade, 1970, Vol. 1, pp. 197–207.
137. MCLAMORE, R. T.: Strength-deformation characteristics of anisotropic sedimentary rocks. Thesis, Univ. Texas, Austin, 1966.
138. MCLEAN, D.: The physics of high temperature creep in metals. Reports on Progress in Physics, Vol. 29, 1966, pp. 1–33.
139. MESRI, G. and GIBALA R.: Engineering properties of a Pennsylvanian shale. Proc. 13th Symp. Rock Mech., Urbana, Illinois, 1971, pp. 57–75.
140. MILLER, F. E. and DOERINGSFELD, H. A.: Mechanics of materials. Scranton, Pa., International Textbook, 1955.

141. MOGI, K.: Experimental study of deformation and fracture of marble. Bull. Earth-quake Res. Inst., Tokyo Univ., Vol. 37, Part 1, March, 1959, pp. 155–170.

142. MOGI, K.: Deformation and fracture of rocks under confining pressure. Part 1. Compression tests on dry rock sample. Bull. Earthquake Res. Inst., Tokyo Univ., Vol. 42, Part 3, 1964, pp. 491–514.

143. MOGI, K.: Deformation and fracture of rocks under confining pressure. Part 2. Elasticity and plasticity of some rocks. Bull. Earthquake Res. Inst., Tokyo Univ., Vol. 43, Part 2, 1965, pp. 349–379.

144. MOGI, K.: Fracture and flow of rocks under high triaxial compression. J. Geophys. Res., Vol. 76, No. 5, Feb. 10, 1971, pp. 1255–1269.

145. MOGI, K.: Fracture and flow of rocks. Tectonophysics, Vol. 13, 1972, pp. 541–568.

146. MORGAN, J. R. and GERRARD, C. M.: Anisotropy and nonlinearity in sand properties. Proc. Eighth Conf. Soil. Mech. Found. Engng., Moscow, Vol. 1, No. 2, 1973, pp. 287–292.

147. NISHIHARA M.: Stress-strain relation of rocks. Doshisha Eng. Rev., Vol. 8, No. 2, Aug., 1957, pp. 13–55.

148. NISHIMATSU, Y.: The torsion test and elastic constants of the orthotropic rock substance. Proc. 2nd Cong. Int. Soc. Rock Mech., Belgrade, 1970, Vol. 1, pp. 479–484.

149. NUR, A. and SIMMONS, G.: Stress-induced velocity anisotropy in rock: An experimental study. J. Geophys. Res., Vol. 74, 1969, pp. 6667–6674.

150. NYE, J. F.: Physical properties of crystals. Oxford, Clarendon Press, 1957, 322 p.

151. OBERT, L. and DUVALL, W. I.: Rock mechanics and the design of structures in rock. New York, Wiley, 1967, 650 p.

152. PANOV, S. I., SAPEGIN, D. D. and KHRAPKOV, A. A.: Some specific features of deformability of rock masses adjoining a gallery. Proc. 2nd Cong. Int. Soc. Rock Mech., Belgrade , 1970, Vol. 1, pp. 485–490.

153. PATERSON, M. S.: Experimental deformation and faulting in Wombeyan marble. Geol. Soc. Am. Bull., Vol. 69, 1958, pp. 465–476.

154. PATERSON, M. S.: Secondary changes in length with pressure in experimentally deformed rocks. Proc. Roy. Soc. London, Series A, Vol. 271, 1963, pp. 57–87.

155. PATERSON, M. S.: The ductility of rocks. In Physics of Strength and Plasticity. (Editor – ALI S. ARGON). Cambridge, M.I.T. Press, 1969, pp. 377–392.

156. PAUL, B.: Prediction of elastic constants of multiphase materials. Trans. A.I.M.E., Vol. 218, 1960, pp. 36–41.

157. PAULDING, B. W.: Crack growth during brittle fracture in compression. Ph. D. Thesis, M.I.T., Cambridge, Mass., 1965, 214 p.

158. PENG, S. S.: A note on the fracture propagation and time-dependent behaviour of rocks in uniaxial tension. Int. J. Rock Mech. Min. Sci. & Geomech. Abstr., Vol. 12, 1975, pp. 125–127.

159. PENG, S. and PODNIEKS, E. R.: Relaxation and the behaviour of failed rock. Int. J. Rock Mech. Min. Sci., Vol. 9, 1972, pp. 699–712.

160. PERKINS, R. D., GREEN, S. J. and FRIEDMAN, M.: Uniaxial stress behaviour of porphyritic tonalite at strain rates to 10^3/ second. Int. J. Rock Mech. Min. Sci., Vol. 7, 1970, pp. 527–535.

161. PERRY, C. C. and LISSNER, H. R.: The strain gage primer. New York, McGraw-Hill, 1962, 332 p.

162. PETERSEN, C. F., MURRI, W. J. and COWPERTHWAITE, M.: Hugoniot and release-adiabat measurements for selected geologic materials. J. Geophys. Res., Vol. 75, 1970, pp. 2063–2071.

163. PHILLIPS, D. W.: Tectonics of mining. Sheffield Univ. Min. Mag., 1948a.

164. PHILLIPS, D. W.: Tectonics of mining. Coll. Eng., Vol. 25, 1948b, pp. 199–202, 206; 278–282; 312–316; 349–352.

165. PINTO, J. L.: Deformability of schistous rocks. Proc. 2nd Cong. Int. Soc. Rock Mech., Belgrade, 1970, Vol. 1, pp. 491–496.

166. PRICE, N. J.: A study of the time-strain behaviour of coalmeasure rocks. Int. J. Rock Mech. Min. Sci., Vol. 1, 1964, pp. 277–303.

167. PROTODYAKONOV, M. M.: Methods of studying the strength of rocks, used in the U.S.S.R. Proc. Int. Symp. Min. Res., Rolla, Missouri, 1961, Vol. 2, pp. 649–668.

168. RALEIGH, C. B. and PATERSON, M. S.: Experimental deformation of serpentinite and its tectonic implications. J. Geophys. Res., Vol. 70, 1965, pp. 3965–3985.

169. REUSS, A.: Berechnung der Fließgrenze von Mischkristallen auf Grund der Plastizitätsbedingung für Feinkristalle. Z. Angew. Math. u. Mech., Vol. 9,1929, pp. 49, 58.

170. RICE, J. R.: On the stability of dilatant hardening for saturated rock masses. J. Geophys. Res., Vol. 80, 1975, pp. 1531–1536.

171. RICE, M. H., McQUEEN, R. G. and WALSH, J. M.: Compression of solids by strong shock waves. (Ed. F. SEITZ and D. TRUMBULL). Vol. 6, New York, Academic Press, 1958.

172. RICKETTS, T. E. and GOLDSMITH, W.: Wave propagation in an anisotropic half-space. Int. J. Rock Mech. Min. Sci., Vol. 9, 1972, pp. 493–512.

173. ROBINSON, R., WESSON, R. L. and ELLSWORTH, W. L.: Variation of P-wave velocity before the Bear Valley, California earthquake of 24 February, 1972. Science, Vol. 104, 1974, pp. 1281–1283.

174. ROCHA, M. and SILVA, J. N. DA: A new method for the determination of deformability in rock masses. Proc. 2nd Cong. Int. Soc. Rock Mech., Belgrade, 1970, Vol. 1, pp. 423–437.

175. RODRIGUES, F. P.: Anisotropy of granites. Proc. 1st Cong. Int. Soc. Rock Mech., Lisbon, 1966, Vol. 1, pp. 721–731.

175a. RODRIGUES, F. P. Anisotropy of rocks: Most probable surfaces of the ultimate stresses and of the moduli of elasticity. Proc. 2nd Cong. Int. Soc. Rock Mech., Belgrade, 1970, Vol. 1, pp. 133–142.

176. ROGATKINA, Zh. E.: Effect of anisotropy of clay soils on their physico-mechanical properties. Soil Mech. Found. Eng., No. 1, 1967, pp. 23–26.

177. RUMMEL, F.: Dilation von Ruhrsandstein und Marmorproben bei konstanten Manteldrucken bis 1 kb. SFB-77, Jahresbericht, 1972, Institute Soil Mech. and Rock Mech., Univ. Karlsruhe, Karlsruhe.

178. RUSCH, H.: Researches toward a general flexural theory for structural concrete. Proc. Am. Conc. Inst., Vol. 57, July, 1960, pp. 1–28.

179. SAADA, A. S. and OU, C.: Strain-stress relations and failure of anisotropic clays. J. Soil Mech. Found. Div., Am. Soc. Civ. Eng., Vol. 99, No. SM 12, 1973, pp. 1091–1111.

180. SALUSTOWICZ, A.: Zarys mechaniki gorotworu. Katowice, Wydawnictwo ,,Slask", 1965.

181. SAUCIER, K. L.: Properties of Cedar City tonalite. U.S. Army Corps Eng., Waterways Exp. Station, Vicksburg, Miss., Misc. Paper C–69–9, 1969.

182. SCHMIDT, H.: Statische Probleme des Tunnel- und Druckstollenbaues. Berlin, 1926.

183. SCHOCK, R. N., HEARD, H. C. and STEPHENS, D. R.: Stress-strain behaviour of a granodiorite and two graywackes on compression to 20 kilobars. J. Geophys. Res., Vol. 78, 1973, pp. 5922–5941.

184. SCHWARTZ, A. E.: Failure of rock in the triaxial shear test. Proc. 6th Symp. Rock Mech., Rolla, Missouri, 1964, pp. 109–151.

185. SCHOLZ, C.: The role of micro-fracturing in rock deformation. Proc. 2nd Cong. Int. Soc. Rock Mech., Belgrade, 1970, Vol. 1, pp. 323–327.

186. SCHOLZ, C. H. and KRANZ, R.: Note on dilatancy recovery. J. Geophys. Res., Vol. 79, 1974, pp. 2132–2135.

187. SEEBER, G.: Ten years use of TIWAG radial jack. Proc. 2nd Cong. Int. Soc. Rock Mech., Beldrade, 1970, Vol. 1, pp. 439–448.

188. SELDENRATH, I. T. and GRAMBERG, J.: Stress-strain relations and breakage of rocks. Proc. Conf. Mech. Prop. Non-metallic Brittle Materials, London, 1958, pp. 79–105.

189. SEMENOV, A. H.: Variation in the travel time of transverse and longitudinal waves before violent earthquakes. Invest. Earth Phys., No. 4, 1969, pp. 72–77.

190. SERAFIM, J. L.: Rock mechanics considerations in the design of concrete dams. Proc. Int. Conf. State of Stress in the Earth's Crust, Santa Monica, California, 1963, pp. 611–645.

191. SERAFIM, J. L. and LOPES, J. J. B.: Insitu shear tests and triaxial tests of foundation rocks of concrete dams. Proc. 5th Int. Conf. Soil Mech. Found. Eng., Paris, 1961, Vol. 1, pp. 533–539.

192. SIMMONS, G.: Single crystal elastic constants and calculated aggregate properties. J. Grad. Res. Centre, Vol. 34, 1965, pp. 1–269.

192a. SIMONS, N. E.: The stress path method of settlement analysis applied to London clay. Proc. Roscoe Memorial Symp., Cambridge Uni., 1971, pp. 241–252.

193. SINCLAIR, S. R. and BROOKER, E. W.: The shear strength of Edmonton shale. Proc. Geotechnical Conf. Shear Strength Properties of Natural Soils and Rocks, Oslo, 1967, Vol. 1, pp. 295–299.

194. SKEMPTON, A. W., SCHUSTER, R. L. and PETLEY, D. J.: Joints and fissures in the London clay at Wraysbury and Edgeware. Geotechnique, Vol. 19, 1969, pp. 205–217.

195. SKEMPTON, A. W. and HUTCHINSON, J.: Stability of natural slopes and embankment foundations. Proc. 7th Int. Conf. Soil Mech. Found. Eng., Mexico, 1969, State of Art Volume, pp. 291–340.

196. STEPANOV, V. and BATUGIN, S.: Assessing the effect of the anisotropy of rocks on the accuracy of stress determination by the relief method. Sov. Min. Sci., Vol. 3, 1967, pp. 312–315.

197. STORIKOVA, M. F.: Anisotropy of the elastic properties of rocks of the Sakhalin Islands. Akad. Nauk. U.S.S.R., Izvest. Geological Series, No. 8, 1965, pp. 61–74.

198. STOWE, R. L.: Strength and deformation properties of granite, basalt, limestone and tuff at various loading rates. U.S. Army Corps Eng., Waterways Exp. Station, Vicksburg, Miss., Misc. Paper C–69–1, 1969.

199. STOWE, R. L. and AINSWORTH, D. L.: Effect of rate of loading on strength and Young's modulus of elasticity of rock. Proc. 10th Symp. Rock Mech., Univ. Texas, Austin, Texas, 1968, pp. 3–34.

200. SWANSON, S. R. and BROWN, W. S.: The influence of state of stress on the stress-strain behaviour of rocks. J. Basic Engineers, Trans. A.S.M.E., Vol. 94, March, 1972, pp. 238–242.

201. SYN-DZAO-MIN: Laboratory tests on the strength of model pillars on coal and the influence of size and shape. In Polish. D. Sc. Tech. Thesis, Faculty of Mining, Academy of Min. & Metall., Cracow, Poland, 1965.

202. THILL, R. E.: Acoustic methods for monitoring failure in rock. Proc. 14th Symp. Rock Mech., Univ. Park, Penn., 1972, pp. 649–687.

202a. THILL, R. E., Willard, R. J. and Bur, T. R.: Correlation of longitudinal velocity variation with rock fabric. J. Geophys. Res., 1969, Vol. 74, No. 20, pp. 4897–4909.

203. TREMMEL, E. and WIDMANN, R.: Deformation properties of gneiss. Proc. 2nd Cong. Int. Soc. Rock Mech., Belgrade, 1970, Vol. 1, pp. 567–575.

204. VOIGHT, W.: Lehrbuch der Kristallphysik. Leipzig, Teubner, 1910.

205. VON KARMAN, Th.: Festigkeitsversuche unter allseitigem Druck. Z. Ver. dt. Ing., Vol. 55, 1911, pp. 1749–1757.

206. WALSH, J. B.: The effect of cracks on the uniaxial elastic compression of rocks. J. Geophys. Res., Vol. 70, 1965a, pp. 399–411.

207. WALSH, J. B.: The effect of cracks in rocks on Poisson's ratio. J. Geophys. Res., Vol. 70, 1965b, pp. 5249–5257.
208. WALSH, J. B.: The effect of cracks on the compressibility of rock. J. Geophys. Res., Vol. 70, 1965c, pp. 381–389.
209. WALSH, J. B. and BRACE, W. F.: Cracks and pores in rocks. Proc. 1st Cong. Int. Soc. Rock Mech., Lisbon, 1966a, Vol. 1, pp. 643–646.
210. WALSH, J. B. and BRACE, W. F.: Elasticity of rock: A review of some recent theoretical studies. Rock Mech. Eng. Geol., Vol. 4, 1966b, pp. 283–297.
211. WARD, W. H., SAMUELS, S. G. and BUTLER, M. E.: Further studies of the properties of London clay. Geotechnique, Vol. 9, 1959, pp. 33–58.
212. WARD, W. H., MARSLAND, A. and SAMUELS, S. G. Properties of the London clay at the Ashford Common shaft. Geotechnique, Vol. 15, 1965, pp. 321–344.
213. WARREN, N.: Elastic constants versus porosity for a highly porous ceramic, perlite. J. Geophys. Res., Vol. 74, 1969, pp. 713–719.
214. WARREN, N.: Theoretical calculation of the compressibility of porous media. J. Geophys. Res., Vol. 78, 1973, pp. 352–362.
215. WATSTEIN, D.: Effect of straining rate on the compressive strength and elastic properties of concrete. J. Am. Conc. Inst., Vol. 24, 1953, pp. 729–744.
216. WAWERSIK, W. R.: Technique and apparatus for strain measurements on rock in constant confining pressure experiments. Rock Mech., Vol. 7, No. 4, Nov., 1975, pp. 231–241.
217. WAWERSIK, W. R. and BRACE, W. F.: Post failure behaviour of a granite and diabase. Rock Mech., Vol. 3, 1971, pp. 61–65.
218. WAWERSIK, W. R. and FAIRHURST, C.: A study of brittle rock fracture in laboratory compression experiments. Int. J. Rock Mech. Min. Sci., Vol. 7, No. 5, Sept., 1970, pp. 561–575.
219. WIEBENGA, W. A., MANN, P. E. and DOODLEY, J. C.: Meadowbank dam site; Seismic determination of rock constants. Bureau of Mineral Resources (Australia) Record No. 1964/114, 1964.
220. WILHELM, N.: An inductive axial strain measurement device for fatigue investigations at notch root. Exp. Mech., Vol. 15, No. 12, 1975, pp. 19N – 20N.
221. WILLARD, R. J. and McWILLIAMS, J. R.: Effect of loading rate on transgranular-intergranular fracture in Charcoal gray granite. Int. J. Rock Mech. Min. Sci., Vol. 6, 1969, pp. 415–421.
222. WUERKER, R. G.: Influence of stress rate and other factors on strength and elastic properties of rocks. Proc. 3rd Symp. Rock Mech., Golden, Colo., 1959, pp. 3–31.
223. ZISMAN, W. A.: A comparison of the statically and seismologically determined elastic constants of rocks. Proc. Nat. Acad. Sci., Vol. 19, 1933, pp. 680–686.
224. ZNANSKI, J.: Sklonności skal do tapania. Archiwum Gornictwa i Huntictwa, Tom 2, 1954.
225. ZOBACK, M. D. and BYERLEE, J. D.: The effect of cyclic differential stress on dilatancy in Westerly granite under uniaxial and triaxial conditions. J. Geophys. Res., Vol. 80, 1975, pp. 1526–1530.

Uncited References to Chapter 6

1. ANDERSON, O. L., SCHREIBER, C., LIEBERMANN, R. L. and SOGA, N.: Some elastic constants data on minerals relevant to geophysics. Rev. Geophys. Space Phys., Vol. 6, 1968, pp. 491–524.
2. BERRY, P., CREA, G., MARTINO, D. and RIBACHI, R.: Influence of fabric on the deformability of anisotropic rock. 3rd Cong. Int. Soc. Rock Mech., Denver, Colorado, Vol. II–A, 1974, pp. 105–110.

3. BRADY, B. T.: The non-linear mechanical behaviour of brittle rock, Part I and Part II. Int. J. Rock Mech. Min. Sci., Part I, Vol. 6, 1969, pp. 211–225, Part II, Vol. 6, 1969, pp. 301–310.

4. BRADY, B. T.: The effect of confining pressure on the elastic-stress distribution in a radially end-constrained circular cylinder. Int. J. Rock Mech. Min. Sci., Vol. 8, 1971, pp. 153–64.

5. BRADY, B. T., HOOKER, V. E. and AGAPITO, J. F.: Laboratory and insitu mechanical behaviour studies of fractured oil shale pillars. Rock Mech., Vol. 7, No. 2, 1975, pp. 101–120.

6. BURSHTEIN, L. S.: Deformation characteristics and fracture mechanics of rock during tension tests by the crushing method. Soviet Min. Sci., Vol. 6, 1974, pp. 37–40.

7. BYERLEE, J. D.: Brittle ductile transition in rocks. J. Geophys. Res., Vol. 73, 1968, pp. 4741–4750.

8. DONATH, F. A. and TOBING, D. G.: Deformational mode in experimentally deformed rock. Geol. Soc. Amer. Bull., Vol. 82, No, 6, 1971, pp. 1441–1462.

9. DREYER, W.: Gebirgsmechanik in Salz. Ferdinand Enke, Stuttgart, 1974, 905 pp.

10. FARRAN, J. and PERAMI, R.: Microfissuration, deformation et compressibilité des roches sour charges triaxiales. 3rd Cong. Int. Soc. Rock Mech., Denver, Colorado, Vol. II–A, 1974, pp. 138–143.

11. FARZIN, M. H., KRIZEK, R. J. and COROTIS, R. B.: Evaluation of modulus and Poisson's ratio from triaxial tests. Transp. Res. Rec. N537, 1975, pp. 69–80.

12. FELDMAN, C.: New controlled temperature chamber for use with tensile testing machines. Am. Soc. Mech. Engrs. Paper No. 67-WA-DE-IS for Meeting, Nov. 1927, 12 p.

13. HAVARD, D. G. and TOPPER, T. H.: New equipment for cyclic biaxial testing. Exp. Mechanics, Vol. 9, No. 12, 1969, pp. 550–557.

14. HEARMON, R. F. S.: An introduction to applied anisotropic elasticity. Oxford Univ. Press, 1961.

15. HOLMES, A. M. C.: Continuous servo-controlled alignment of specimens in material testing. Exp. Mech., Vol. 15, No. 9, 1975, pp. 358–64.

16. HOLUBEC, I. and FINN, P. J.: A lateral deformation transducer for triaxial testing. Canad. Geotech. J., Vol. 6, 1969, pp. 353–356.

17. HOUPERT, R.: Mechanical behaviour of crystalline rocks of quasi isotropic structure. Ph. D. Thesis, Univ. Nancy, 1973, 166 p.

18. ISMAIL, I. A. and MURREL, S. A.: Dilatancy and the strength of rocks containing pore water under undrained conditions. Geophys. J. R. Astr. Soc., Vol. 44, No. 1, 1976, pp. 107–134.

19. IVISON, J.M. and BLANCHARD, J.: A simple telemetering system for strain-measurements. Strain, Vol. 5, No. 3, 1969, pp. 160–162.

20. JONES, F. O.: A laboratory study of the effects of confining pressure on fracture, flow and storage capacity in carbonate rocks. J. Petrol. Tech., Vol. 27, Jan., 1975, pp. 21–22.

21. KAZUO, H. and KOIDE, H.: Process of deformation of sedimentary rock. 2nd Int. Cong. Belgrade, Paper 2–13, 1970, pp. 353–359.

22. KLINK, S. A.: POISSON's ratio variations in concrete. Experimental Mechanics, Vol. 15, April, 1975, pp. 138–141.

23. KULAKOV, G. I.: Simultaneous measurement of the components of the stress tensors and the determination of the elastic constants of a rock. Soviet Mining Sci., Vol. 10, 1975, pp. 659–63.

24. LAFEBER, D. and WILLOUGHBY, D.: Fabric symmetry and mechanical anisotropy in natural soils. 1st Aust. New Zealand Conf. on Geomech., Melbourne, 1971, pp. 165–174.

25. MOGI, K.: Pressure dependence of rock strength and transition from brittle fracture to ductile flow. Bull. Earthquake Res. Inst. Tokyo Univ., Vol. 44, 1966, pp. 215–232.

26. MORGENSTERN, N. R. and TAMULY-PHUKAN, A. L.: Non-linear stress-strain relationship for a homogeneous sandstone. Int. J. Rock Mech. Min. Sci., Vol. 6, 1968, pp. 127–142.

27. MORLAND, L. W.: Elastic anisotropy of regularly jointed media. Rock Mech., Vol. 8, 1975, pp. 35–53.

28. MOSHER, S., BERGER, R. L. and ANDERSON, D. E.: Fracturing characteristics of two granites. Rock Mech., Vol. 7, 1975, pp. 167–176.

29. MUTINANSKY, J. M. and SINGH, M. M.: A statistical study of relationships between rock properties. 9th Symp. Rock Mech., 1967, pp. 161–177.

30. NISHIMATSU, Y.: The torsion test and the elastic constants of the orthotropic rock substance. 2nd Cong. Int. Soc. Rock Mech., Belgrade, 1970, Pap. I–28.

31. ONODERA, T. F. and YOSHINAKA, R.: Weathering and its relation to mechanical properties of granite. Proc. 3rd Cong. Int. Soc. Rock Mech., Denver, Vol. II–A, 1974, pp. 71–78.

32. PAINTING, A. L.: A study of the end effects in specimen cores under compression tests, with a view to the elimination of these tests. J.S. Afr. Inst. Min. Metal., Vol. 7, 1975, pp. 333–339.

33. PATERSON, M. S.: Effect of pressure on stress-strain properties of materials. Geol. J. R. Astron., Vol. 14, 1967, pp. 13–17.

34. REMO, J. L. and JOHNSON, A. A.: A preliminary study of the ductile-brittle transition under impact condition in material from a octahedrite. J. Geophys. Res., Vol. 80, No. 26, 1975, pp. 3744–46.

35. RICE, J. R.: On the stability of dilatant hardening for saturated rock masses. J. Geophys. Res., Vol. 80, 1975, pp. 1531–36.

36. ROMERO, S. U. and GOMEZ, B. B.: Brittle and plastic failure of rocks. 2nd Cong. Int. Soc. Rock Mech., Belgrade, Vol. II, 1970, Pap. III–20.

37. SAMANTA, S. K.: On the limit of plastic deformation in compression of circular cylinders. Int. J. Fracture, Vol. 11, 1975, pp. 301–313.

38. SIMMONS, G. and BRACE, W. F.: Comparison of static and dynamic measurements of compressibility of rock. J. Geophys. Res., Vol. 70, 1965, pp. 5649–5656.

39. SMART, P.: Exponential stress-strain curves, Civ. Eng., Aug, 1975, p. 39.

40. STEPHENS, D. R., LILLEY, E. M. and LOUIS, H.: Pressure volume equation of state of consolidated and fractured rocks to 40 kb. Int. J. Rock Mech. Min. Sci., Vol. 7, 1970, pp. 257–296.

41. STUART, W. D. and DIETRICH, J. D.: Continuum theory of rock dilatancy, 3rd Cong. Int. Soc. Rock Mech., Denver, Colorado, Vol. II–A, 1974, pp. 530–534.

42. TAYLOR, R. K. and SPEARS, D. A.: The break-down of British coal measure rocks. Int. J. Rock Mech. Min. Sci., Vol. 7, 1970, pp. 481–501.

43. YOSHINAKA, R.: Mechanical properties of soft rock. Rock Mechanics in Japan, Vol. 1, 1970, pp. 35–37.

44. ZASLAVSKY, V. I., USOLTSEV, Yu. K. and ALEKSANDROV, K. S.: Elastic properties of some rock forming minerals. Invest. Physics of Solid Earth, Academy of Sci., U.S.S.R., Vol. 10, 1974, pp. 835–837.

45. NELSON, H. G., TETELMAN, A. S. and WILLIAMS, D. P.: The kinetic and dynamic aspects of corrosion fatigue in a gaseous hydrogen environment. National Assn. of Corrosion Eng. NACE 2. Corrosion Fatigue Conferences, Storrs, 1971, pp. 359–365.

CHAPTER 7

Dynamic Elastic Constants of Rock

7.1. Introduction

In Chapter 6, the elastic constants of rock have been considered from the point of view of its reaction to static stresses. However, a rock may be subject to transient dynamic loading and the way it reacts to the dynamic stresses is important.

Dynamic elastic constants can be calculated from the elastic wave velocities measurements. The resonance and ultrasonic pulse methods are used to determine the elastic wave velocities in laboratory. Seismic wave propagation method is used for field determinations. These methods are discussed in detail.

The ability of a rock to withstand high dynamic stresses without failure has wide importance and its determination is discussed in this chapter.

7.2. Elastic Waves

There are two basic types of elastic waves: body waves which travel through the interior of the rock body, and surface waves which can only travel along the surface of the rock. Body waves can be subdivided into two modes: compression or primary (P) waves and shear or secondary (S) waves. P-waves induce longitudinal oscillatory particle motions similar in many ways to simple harmonic vibrations and when they impinge on a free boundary in any direction other than head-on, one of the resultant effects of the displacement is the induction of S-waves in which the particles move in a transverse direction without compressing the material. P-waves, of course, travel in any direction in a material which resists compression, but since S-waves depend upon the ability of the transmitting material to resist changes in shape, they can only exist in solid.

The velocity equations of these waves are as follows:

$$V_p = \left[\frac{E(1-\nu)}{\varrho(1+\nu)(1-2\nu)}\right]^{1/2} \tag{7.1}$$

$$V_s = \left[\frac{G}{\varrho}\right]^{1/2} = \left[\frac{E}{2\varrho(1+\nu)}\right]^{1/2} \tag{7.2}$$

where

V_p = velocity of compression waves, m/s or in/s
V_s = velocity of shear waves, m/s or in/s
E = dynamic modulus of elasticity, Pa or lbf/in²
G = dynamic modulus of rigidity, Pa or lbf/in²
ν = POISSON's ratio, and
ϱ = density, kg/m³ or lbs²/in⁴.

7.3. Methods of Determining Dynamic Elastic Constants in Laboratory

Two methods are usually used to determine the dynamic elastic constants; (i) the resonance method and (ii) the ultrasonic pulse method. The wave velocities most commonly determined for these purposes are the longitudinal and torsional wave velocities by the resonance method or the compression and shear wave velocities by the pulse method. It is assumed, as a first approximation for mathematical simplicity, that rock is homogeneous, isotropic and elastic. Theoretically, in such a material, determinations of elastic constants by either the resonance or pulse method should yield equivalent results.

7.3.1. Resonance Method

The method of inducing resonance and determining the velocity of propagation in cylindrical or prismatic bars has been used extensively to study the elasticity of rocks. Natural resonances are excited in the specimen in the longitudinal, flexural (transverse) or torsional modes. The longitudinal and flexural vibrations are functions of modulus of elasticity, whereas the torsional vibration is related to the modulus of rigidity of the material.

The resonance method is based on a standing-wave phenomenon. When the specimen is undergoing longitudinal or torsional vibration, the length of the specimen contains an integral number n of half-wave lengths, or

$$l = \frac{n\lambda}{2} \qquad (7.3)$$

where l = length of specimen and
λ = wave-length of vibration

The velocity of the wave, V, is then expressed by

$$V = \lambda f = \frac{2\, lf}{n} \qquad (7.4)$$

where

f = resonant frequency of any mode of vibration.

However, this simple formula does not apply to flexural resonance since the nodes are not at the quarter points, as simple standing waves would require.

7.3.1.1. Longitudinal vibration

If the specimen has the shape of a cylindrical rod or square bar, modulus of elasticity can be determined from the longitudinal resonances of the specimen. The equation which relates the longitudinal resonant frequency and modulus of elasticity is in the form

$$E = \frac{1}{U}\left[\frac{2lf_1}{n}\right]^2 \varrho \qquad (7.5)$$

where

E = dynamic modulus of elasticity, Pa or lbf/in²
l = length of specimen, m or in
f_1 = longitudinal resonant frequency, Hz or cycles/s
n = number of mode of vibration, being one for the fundamental mode
ϱ = density of the specimen, kg/m³ or lb.s²/in⁴ and
U = correction factor which involves the shape and size of the specimen, wave-length and POISSON's ratio.

In the longitudinal resonance vibration of cylindrical rods, wave-length λ of standing waves is expressed by Eq. 7.3, and U for the case where $d/\lambda \ll 1$ is in the form (RAYLEIGH, 1945)

$$U \simeq 1 - \left(\frac{\pi n\, vd}{\sqrt{8\, l}}\right)^2 \qquad (7.6)$$

where

d = diameter of the specimen.

For specimens with a square cross-section, d in Eq. 7.6 is replaced by $2b/\sqrt{3}$ where b is the length of the side.

An approximate equation for bars with a rectangular cross-section has not been derived, and the use of the flexural vibration is therefore recommended.

For an infinitely thin cylindrical bar, $d/\lambda < 0.1$, U can generally be ignored and the modulus of elasticity is conveniently obtained by determining the fundamental longitudinal frequency using the relationship

$$E = \varrho\, V_0^2 \qquad (7.7)$$

where

V_0 = longitudinal bar velocity for an infinitely thin rod.

Comparison of Eq. 7.7 with Eq. 7.1 shows that the longitudinal bar velocity and longitudinal, free-medium velocity are related by

$$V_p = V_0 \left[\frac{(1-\nu)}{(1+\nu)(1-2\nu)} \right]^{1/2} \qquad (7.8)$$

7.3.1.2. Flexural vibration

Generally speaking, it is easier to excite the flexural vibration than the longitudinal vibration, especially for thin specimens. For this reason, the flexural vibration is more practical and important for determining modulus of elasticity.

The general equation which relates modulus of elasticity and the flexural resonant frequency f_f is (TIMOSHENKO and GOODIER, 1951)

$$E = \left[\frac{2\pi\, l^2 f_f}{K\, m^2} \right]^2 \varrho\, T \qquad (7.9)$$

where

K = radius of gyration of the cross-section about the axis perpendicular to the plane of vibration

m = constant depending on the mode of vibration   and

T = shape factor, which depends upon the shape, size and POISSON's ratio of the specimen and the mode of vibration.

The value of K is known to be $d/4$ for a circular cross-section and $a/\sqrt{12}$ for a rectangular cross-section with the dimension a in the direction of vibration. The value of m is 4.7300 for the fundamental flexural vibration, 7.8532 for the first overtone and 10.9956 for the second overtone.

Cylindrical rods: From the Eq. 7.9 the relationship between modulus of elasticity and the fundamental flexural vibration for cylindrical rods is

$$E = 1.261886 \left[\frac{l^2 f_t}{d}\right]^2 \varrho \, T_1 \tag{7.10}$$

and for the first overtone it is

$$E = 0.1660703 \left[\frac{l^2 f_t}{d}\right]^2 \varrho \, T_2 \tag{7.11}$$

Correction factors T_1 and T_2 are given in Tables 21 and 22.

Rectangular bars: For a rectangular cross-section, the equation to be used is of the form (SCHREIBER, ANDERSON and SOGA, 1973)

$$E = 0.94642 \left[\frac{l^2 f_t}{t}\right]^2 \varrho \, T_3 \tag{7.12}$$

where

$t =$ the dimension of the cross-section parallel to the direction of vibration   and

$$T_3 = 1 + 6.585 \, (1 + 0.0752 \nu + 0.8109 \nu^2 \, (t/l)^2 - 0.868 \, (t/l)^4$$
$$- \frac{8.340 \, (1 + 0.2023 \nu + 2.173 \nu^2) \, (t/l)^4}{1 + 6.338 \, (1 + 0.14081 \nu + 1.536 \nu^2) \, (t/l)^2}$$

The values of T_3 for different values of ν and (t/l) are given in Table 23.

7.3.1.3. Torsional vibration

The general equation which relates the modulus of rigidity G and the torsional resonant frequency f_t is

$$G = \frac{4 \varrho \, R \, l^2 f_t^2}{n^2} \tag{7.13}$$

TABLE 21

Correction factor for the fundamental mode of flexural vibration of cylinders T_1 as a function of the diameter-to-length ratio d/l and POISSON's ration v

(after SCHREIBER, ANDERSON and SOGA, 1973)

Diameter-to-length ratio d/l	POISSON's ratio v				
	0.00	0.05	0.10	0.15	0.20
0.00	1.000000	1.000000	1.000000	1.000000	1.000000
0.02	1.001954	1.001979	1.002004	1.002029	1.002053
0.04	1.007804	1.007905	1.008004	1.008102	1.008199
0.06	1.017522	1.017748	1.017968	1.018186	1.018405
0.08	1.031064	1.031461	1.031848	1.032233	1.032618
0.10	1.048367	1.048983	1.049580	1.050174	1.050765
0.12	1.069357	1.070225	1.071077	1.071920	1.072753
0.14	1.093950	1.095111	1.096256	1.097378	1.098495
0.16	1.122061	1.123544	1.125007	1.126452	1.127884
0.18	1.153592	1.155430	1.157245	1.159039	1.160817
0.20	1.188458	1.190676	1.192869	1.195038	1.197191
0.22	1.226567	1.229189	1.231781	1.234351	1.236906
0.24	1.267868	1.270881	1.273894	1.276886	1.279865
0.26	1.312189	1.315668	1.319120	1.322555	1.325980
0.28	1.359546	1.363477	1.367381	1.371276	1.375167
0.30	1.409842	1.414233	1.418606	1.422974	1.427352
0.32	1.463014	1.467873	1.472724	1.477584	1.482465
0.34	1.519006	1.524340	1.529680	1.535043	1.540446
0.36	1.577766	1.583580	1.589417	1.595298	1.601240
0.38	1.639253	1.645548	1.651888	1.658300	1.664800
0.40	1.703422	1.710202	1.717054	1.724007	1.731082
0.42	1.770243	1.777505	1.784874	1.792382	1.800052
0.44	1.839681	1.847424	1.855317	1.863393	1.871677
0.46	1.911712	1.919934	1.928355	1.937012	1.945932
0.48	1.986311	1.995009	2.003964	2.013216	2.022795
0.50	2.063458	2.072628	2.082123	2.091985	2.102247
0.52	2.143136	2.152775	2.162814	2.173303	2.184276
0.54	2.225331	2.235431	2.246024	2.257157	2.268871
0.56	2.310030	2.320588	2.331739	2.343539	2.356026
0.58	2.397222	2.408232	2.419951	2.432439	2.445736
0.60	2.486902	2.498357	2.510654	2.523855	2.538002

TABLE 21 (continued)

Poisson's ratio v					
0.25	0.30	0.35	0.40	0.45	0.50
1.000000	1.000000	1.000000	1.000000	1.000000	1.000000
1.002077	1.002100	1.002124	1.002147	1.002170	1.002193
1.008295	1.008388	1.008482	1.008575	1.008666	1.008757
1.018619	1.018826	1.019038	1.019245	1.019450	1.019653
1.032994	1.033360	1.033733	1.034096	1.034459	1.034818
1.051344	1.051916	1.052484	1.053050	1.053610	1.054170
1.073577	1.074393	1.075202	1.076008	1.076808	1.077604
1.099599	1.100694	1.101782	1.102864	1.103941	1.105015
1.129302	1.130711	1.132113	1.133509	1.134901	1.136288
1.162585	1.164337	1.166086	1.167830	1.169569	1.171308
1.199332	1.201464	1.203591	1.205714	1.207838	1.209961
1.239449	1.241986	1.244521	1.247057	1.249596	1.252141
1.282836	1.285807	1.288779	1.291757	1.294746	1.297747
1.329403	1.332832	1.336270	1.339723	1.343194	1.346687
1.379065	1.382977	1.386911	1.390869	1.394857	1.398879
1.431747	1.436169	1.440625	1.445121	1.449661	1.454250
1.487380	1.492337	1.497345	1.502411	1.507540	1.512737
1.545901	1.551420	1.557012	1.562682	1.568439	1.574286
1.607259	1.613367	1.619572	1.625884	1.632309	1.638853
1.671405	1.678129	1.684983	1.691975	1.699113	1.706401
1.738298	1.745669	1.753207	1.760921	1.768818	1.776904
1.807904	1.815954	1.824214	1.832694	1.841401	1.850342
1.880193	1.888956	1.897979	1.907273	1.916846	1.926702
1.955140	1.964653	1.974485	1.984646	1.995143	2.005980
2.032727	2.043030	2.053718	2.064802	2.076288	2.088178
2.112937	2.124075	2.135673	2.147741	2.160285	2.173305
2.195762	2.207781	2.220345	2.233465	2.247142	2.261377
2.281194	2.294146	2.307741	2.321984	2.336876	2.352415
2.369231	2.383174	2.397866	2.413311	2.429508	2.446448
2.459873	2.474869	2.490733	2.507467	2.525064	2.543511
2.553126	2.569244	2.586362	2.604478	2.623578	2.643647

TABLE 22

Correction factor for the first overtone of flexural vibration of cylinders T_2 as a function of the diameter-to-length ratio d/l and POISSON's ratio v

(after SCHREIBER, ANDERSON and SOGA, 1973)

Diameter-to-length ratio d/l	POISSON's ratio v				
	0.00	0.05	0.10	0.15	0.20
0.00	1.000000	1.000000	1.000000	1.000000	1.000000
0.01	1.001352	1.001376	1.001399	1.001422	1.001445
0.02	1.005404	1.005499	1.005592	1.005683	1.005774
0.03	1.012144	1.012355	1.012564	1.012770	1.012973
0.04	1.021552	1.021926	1.022295	1.022660	1.023019
0.05	1.033605	1.034188	1.034760	1.035326	1.035885
0.06	1.048267	1.049105	1.049924	1.050734	1.051537
0.07	1.065511	1.066641	1.067750	1.068844	1.069926
0.08	1.085296	1.086759	1.088198	1.089613	1.091015
0.09	1.107583	1.109415	1.111219	1.112996	1.114756
0.10	1.132330	1.134568	1.136772	1.138949	1.141103
0.11	1.159496	1.162177	1.164820	1.167429	1.170012
0.12	1.189048	1.192055	1.195318	1.198394	1.201441
0.13	1.220947	1.224613	1.228230	1.231807	1.235353
0.14	1.255166	1.259371	1.263523	1.267633	1.271711
0.15	1.291674	1.296449	1.301168	1.305844	1.310487
0.16	1.330448	1.335824	1.341140	1.346414	1.351656
0.17	1.371471	1.377477	1.383422	1.389325	1.395199
0.18	1.414728	1.421393	1.427999	1.434563	1.441104
0.19	1.460212	1.467565	1.474861	1.482122	1.489364
0.20	1.507918	1.515991	1.524010	1.532000	1.539980
0.21	1.557849	1.566672	1.575448	1.584202	1.592958
0.22	1.610014	1.619619	1.629185	1.638741	1.648312
0.23	1.664427	1.674847	1.685239	1.695636	1.706063
0.24	1.721106	1.732377	1.743634	1.754912	1.766240
0.25	1.780080	1.792238	1.804400	1.816603	1.828880
0.26	1.841381	1.854465	1.867575	1.880751	1.894027
0.27	1.905048	1.919102	1.933207	1.947405	1.961734
0.28	1.971129	1.986199	2.001348	2.016625	2.032066
0.29	2.039679	2.055815	2.072063	2.088477	2.105095
0.30	2.110759	2.128016	2.145425	2.163041	2.180906
0.31	2.184442	2.202881	2.221516	2.240405	2.259595
0.32	2.260809	2.280496	2.300430	2.230672	2.341270
0.33	2.339949	2.360959	2.382273	2.403955	2.426056
0.34	2.421965	2.444380	2.467163	2.490383	2.514090
0.35	2.506968	2.530881	2.555234	2.580098	2.605527
0.36	2.595085	2.620598	2.646633	2.673262	2.700540

TABLE 22 (continued)

Poisson's ratio v					
0.25	0.30	0.35	0.40	0.45	0.50
1.000000	1.000000	1.000000	1.000000	1.000000	1.000000
1.001467	1.001489	1.001511	1.001533	1.001554	1.001576
1.005863	1.005952	1.006039	1.006126	1.006212	1.006297
1.013174	1.013373	1.013569	1.013763	1.013957	1.014149
1.023376	1.023728	1.024077	1.024421	1.024763	1.025103
1.036440	1.036987	1.037530	1.038064	1.038597	1.039128
1.052327	1.053108	1.053883	1.054653	1.055417	1.056178
1.070995	1.072054	1.073103	1.074146	1.075181	1.076211
1.092401	1.093775	1.095137	1.096491	1.097837	1.099177
1.116497	1.118223	1.119937	1.121641	1.123337	1.125025
1.143235	1.145352	1.147455	1.149547	1.151630	1.153705
1.172573	1.175115	1.177643	1.180161	1.182668	1.185170
1.204465	1.207470	1.210461	1.213439	1.216410	1.219376
1.238874	1.242376	1.245865	1.249343	1.252816	1.256285
1.275764	1.279800	1.283822	1.287838	1.291851	1.295863
1.315106	1.319710	1.324305	1.328896	1.333488	1.338084
1.356876	1.362084	1.367288	1.372493	1.377707	1.382928
1.401055	1.406905	1.412756	1.418616	1.424490	1.430382
1.447633	1.454162	1.460700	1.467255	1.473834	1.480442
1.496602	1.503850	1.511116	1.518411	1.525741	1.533111
1.547966	1.555973	1.564010	1.572090	1.580217	1.588400
1.601732	1.610540	1.619395	1.628306	1.637282	1.646330
1.657917	1.667572	1.677290	1.687084	1.696961	1.706930
1.716543	1.727092	1.737726	1.748455	1.759290	1.770238
1.777642	1.789136	1.800738	1.812460	1.824312	1.836303
1.841254	1.853748	1.866375	1.879151	1.892085	1.905185
1.907428	1.920979	1.934694	1.948588	1.962672	1.976953
1.976222	1.990892	2.005761	2.020844	2.036152	2.051692
2.047703	2.063560	2.079655	2.096003	2.112614	2.129495
2.121950	2.139068	2.156467	2.174161	2.192161	2.210474
2.199055	2.217513	2.236300	2.255430	2.274913	2.294753
2.279119	2.299006	2.319274	2.339937	1.261003	2.382477
2.362262	2.383673	2.405523	2.427824	2.450585	2.473806
2.448613	2.471654	2.495197	2.519253	2.543828	2.568923
2.538324	2.563111	2.588468	2.614407	2.640929	2.668033
2.631559	2.658220	2.685528	2.713488	2.742103	2.771366
2.728506	2.757184	2.786589	2.816726	2.847593	2.879180

TABLE 23
Shape factor T_3 as a function of t/l and v
(after SCHREIBER, ANDERSON and SOGA, 1973)

t/l	Poisson's ratio v								
	0.05	0.10	0.15	0.20	0.25	0.30	0.35	0.40	0.45
0.00000	1.00000	1.00000	1.00000	1.00000	1.00000	1.00000	1.00000	1.00000	1.00000
0.00500	1.00016	1.00016	1.00016	1.00017	1.00017	1.00018	1.00018	1.00019	1.00019
0.01000	1.00066	1.00066	1.00067	1.00068	1.00070	1.00072	1.00074	1.00076	1.00078
0.01500	1.00148	1.00150	1.00152	1.00155	1.00158	1.00162	1.00166	1.00171	1.00177
0.02000	1.00264	1.00267	1.00271	1.00275	1.00281	1.00288	1.00296	1.00305	1.00315
0.02500	1.00413	1.00417	1.00423	1.00430	1.00439	1.00450	1.00462	1.00476	1.00492
0.03000	1.00595	1.00601	1.00609	1.00619	1.00632	1.00648	1.00666	1.00686	1.00708
0.03500	1.00809	1.00817	1.00829	1.00843	1.00861	1.00882	1.00906	1.00933	1.00964
0.04000	1.01057	1.01067	1.01082	1.01101	1.01124	1.01151	1.01182	1.01218	1.01258
0.04500	1.01337	1.01350	1.01368	1.01392	1.01421	1.01456	1.01496	1.01541	1.01592
0.05000	1.01650	1.01666	1.01688	1.01718	1.01754	1.01796	1.01845	1.01901	1.01963
0.05500	1.01995	1.02014	1.02041	1.02077	1.02120	1.02172	1.02231	1.02298	1.02374
0.06000	1.02372	1.02395	1.02428	1.02470	1.02521	1.02582	1.02653	1.02733	1.02822
0.06500	1.02781	1.02808	1.02847	1.02896	1.02956	1.03028	1.03110	1.03204	1.03309
0.07000	1.03223	1.03254	1.03298	1.03355	1.03425	1.03508	1.03604	1.03712	1.03834
0.07500	1.03696	1.03732	1.03783	1.03848	1.03928	1.04023	1.04132	1.04257	1.04396
0.08000	1.04201	1.04242	1.04299	1.04374	1.04464	1.04572	1.04696	1.04838	1.04995
0.08500	1.04738	1.04784	1.04848	1.04932	1.05034	1.05155	1.05295	1.05454	1.05632
0.09000	1.05306	1.05357	1.05429	1.05522	1.05637	1.05772	1.05929	1.06107	1.06306
0.09500	1.05905	1.05961	1.06042	1.06145	1.06273	1.06423	1.06597	1.06795	1.07016
0.10000	1.06534	1.06597	1.06686	1.06800	1.06941	1.07107	1.07300	1.07518	1.07763
0.10500	1.07195	1.07264	1.07361	1.07487	1.07642	1.07825	1.08036	1.08275	1.08545
0.11000	1.07886	1.07961	1.08068	1.08206	1.08375	1.08575	1.08807	1.09069	1.09364
0.11500	1.08607	1.08689	1.08805	1.08956	1.09140	1.09358	1.09611	1.09897	1.10217
0.12000	1.09358	1.09447	1.09574	1.09737	1.09937	1.10174	1.10448	1.10759	1.11107
0.12500	1.10189	1.10236	1.10372	1.10549	1.10765	1.11021	1.11318	1.11654	1.12031
0.13000	1.10949	1.11054	1.11201	1.11391	1.11625	1.11901	1.12221	1.12583	1.12989
0.13500	1.11789	1.11901	1.12060	1.12264	1.12515	1.12812	1.13156	1.13546	1.13983
0.14000	1.12658	1.12778	1.12948	1.13167	1.13436	1.13755	1.14123	1.14542	1.15010
0.14500	1.13556	1.13685	1.13866	1.14100	1.14388	1.14729	1.15123	1.15570	1.16071
0.15000	1.14483	1.14620	1.14813	1.15063	1.15370	1.15733	1.16154	1.16631	1.17166
0.15500	1.15438	1.15584	1.15789	1.16055	1.16382	1.16769	1.17216	1.17724	1.18294
0.16000	1.16421	1.16576	1.16794	1.17077	1.17424	1.17835	1.18310	1.18850	1.19455
0.16500	1.17422	1.17596	1.17828	1.18127	1.18495	1.18930	1.19434	1.20007	1.20648
0.17000	1.18471	1.18644	1.18890	1.19207	1.19595	1.20056	1.20590	1.21196	1.21875
0.17500	1.19537	1.19721	1.19979	1.20314	1.20725	1.21212	1.21775	1.22416	1.23133
0.18000	1.20631	1.20824	1.21097	1.21450	1.21883	1.22397	1.22991	1.23667	1.24424
0.18500	1.21752	1.21955	1.22243	1.22614	1.23070	1.23611	1.24237	1.24949	1.25746
0.19000	1.22899	1.23113	1.23415	1.23806	1.24286	1.24854	1.25513	1.26261	1.27100
0.19500	1.24074	1.24298	1.24615	1.25025	1.25529	1.26126	1.26818	1.27604	1.28486
0.20000	1.25274	1.25510	1.25842	1.26272	1.26800	1.27427	1.28152	1.28977	1.29902
0.20500	1.26501	1.26748	1.27546	1.27546	1.28100	1.28756	1.29516	1.30380	1.31349
0.21000	1.27754	1.28012	1.28376	1.28848	1.29426	1.30113	1.30908	1.31813	1.32828
0.21500	1.29033	1.29302	1.29683	1.30175	1.30780	1.31498	1.32329	1.33275	1.34336
0.22000	1.30338	1.30619	1.31016	1.31530	1.32161	1.32911	1.33779	1.34767	1.35875
0.22500	1.31667	1.31960	1.32375	1.32911	1.33569	1.34351	1.35257	1.36288	1.37445
0.23000	1.33623	1.33328	1.33759	1.34318	1.35004	1.35819	1.36763	1.37838	1.39044
0.23500	1.34403	1.34721	1.35170	1.35751	1.36465	1.37314	1.38297	1.39416	1.40673
0.24000	1.35808	1.36138	1.36605	1.37210	1.37953	1.38836	1.39859	1.41024	1.42332
0.24500	1.37238	1.37581	1.38067	1.38695	1.39467	1.40385	1.41448	1.42660	1.44020

TABLE 23 (continued)

t/l	Poisson's ratio ν								
	0.05	0.10	0.15	0.20	0.25	0.30	0.35	0.40	0.45
0.25000	1.38692	1.39049	1.39553	1.40205	1.41007	1.41960	1.43065	1.44324	1.45738
0.25500	1.40171	1.40541	1.41064	1.41741	1.42573	1.43562	1.44710	1.46016	1.47485
0.26000	1.41674	1.42057	1.42600	1.43302	1.44165	1.45191	1.46381	1.47737	1.49261
0.26500	1.43201	1.43598	1.44160	1.44888	1.45782	1.46846	1.48079	1.49485	1.51066
0.27000	1.44753	1.45164	1.45745	1.46498	1.47425	1.48526	1.49805	1.51262	1.52900
0.27500	1.46327	1.46753	1.47354	1.48134	1.49093	1.50233	1.51557	1.53066	1.54762
0.28000	1.47926	1.48366	1.48988	1.49794	1.50786	1.51966	1.53335	1.54897	1.56653
0.28500	1.49548	1.50002	1.50645	1.51479	1.52504	1.53724	1.55141	1.56756	1.58573
0.29000	1.51193	1.51662	1.52327	1.53188	1.54248	1.55508	1.56972	1.58642	1.60521
0.29500	1.52862	1.53346	1.54032	1.54921	1.56015	1.57317	1.58830	1.60555	1.62497
0.30000	1.54553	1.55053	1.55761	1.56678	1.57808	1.59152	1.60714	1.62496	1.64501
0.30500	1.56268	1.56783	1.57513	1.58460	1.59625	1.61012	1.62624	1.64463	1.66533
0.31000	1.58005	1.58536	1.59289	1.60265	1.61466	1.62897	1.64559	1.66457	1.68594
0.31500	1.59765	1.60312	1.61088	1.62094	1.63332	1.64807	1.66521	1.68478	1.70682
0.32000	1.61547	1.62111	1.62910	1.63946	1.65222	1.66742	1.68508	1.70526	1.72797
0.32500	1.63352	1.63933	1.64755	1.65822	1.67136	1.68701	1.70521	1.72600	1.74941
0.33000	1.65179	1.65777	1.66623	1.67721	1.69074	1.70685	1.72559	1.74700	1.77112
0.33500	1.67029	1.67643	1.68514	1.69644	1.71036	1.72694	1.74623	1.76827	1.79310
0.34000	1.68900	1.69532	1.70427	1.71589	1.73021	1.74727	1.76712	1.78980	1.81535
0.34500	1.70794	1.71443	1.72364	1.73558	1.75030	1.76785	1.78826	1.81159	1.83789
0.35000	1.72709	1.73376	1.74322	1.75550	1.77063	1.78867	1.80965	1.83365	1.86069
0.35500	1.74646	1.75331	1.76303	1.77564	1.79119	1.80972	1.83130	1.85596	1.88376
0.36000	1.76605	1.77309	1.78306	1.79601	1.81198	1.83102	1.85319	1.87853	1.90711
0.36500	1.78585	1.79307	1.80331	1.81661	1.83301	1.85256	1.87533	1.90136	1.93072
0.37000	1.80587	1.81328	1.82379	1.83743	1.85427	1.87434	1.89771	1.92445	1.95460
0.37500	1.82610	1.83370	1.84448	1.85848	1.87575	1.89635	1.92035	1.94779	1.97875
0.38000	1.84654	1.85433	1.86539	1.87975	1.89747	1.91861	1.94322	1.97139	2.00317
0.38500	1.86719	1.87518	1.88652	1.90124	1.91941	1.94109	1.96635	1.99524	2.02785
0.39000	1.88805	1.89624	1.90786	1.92296	1.94159	1.96381	1.98971	2.01935	2.05279
0.39500	1.90912	1.91752	1.92942	1.94489	1.96399	1.98677	2.01322	2.04371	2.07800
0.40000	1.93040	1.93900	1.95120	1.96704	1.98661	2.00996	2.03717	2.06832	2.10348
0.40500	1.95189	1.96069	1.97318	1.98942	2.00946	2.03338	2.06126	2.09318	2.12921
0.41000	1.97358	1.98259	1.99538	2.01201	2.03253	2.05703	2.08559	2.11829	2.15521
0.41500	1.99548	2.00470	2.01779	2.03481	2.05582	2.08091	2.11016	2.14365	2.18147
0.42000	2.01758	2.02702	2.04042	2.05783	2.07934	2.10502	2.13497	2.16926	2.20799
0.42500	2.03988	2.04954	2.06325	2.08107	2.10308	2.12936	2.16001	2.19512	2.23477
0.43000	2.06239	2.07227	2.08628	2.10451	2.12703	2.15393	2.18529	2.22122	2.26180
0.43500	2.08510	2.09519	2.10953	2.12817	2.15121	2.17872	2.21081	2.24757	2.28910
0.44000	2.10800	2.11833	2.13298	2.15205	2.17560	2.20374	2.23656	2.27416	2.31665
0.44500	2.13110	2.14166	2.15664	2.17613	2.20021	2.22898	2.26254	2.30100	2.34445
0.45000	2.15441	2.16519	2.18050	2.20042	2.22503	2.25444	2.28876	2.32808	2.37251
0.45500	2.17791	2.18892	2.20457	2.22492	2.25007	2.28013	2.31520	2.35540	2.40082
0.46000	2.20160	2.21285	2.22883	2.24963	2.27533	2.30604	2.34188	2.38296	2.42939
0.46500	2.22549	2.23698	2.25330	2.27454	2.30079	2.33217	2.36879	2.41076	2.45821
0.47000	2.24957	2.26130	2.27797	2.29966	2.32647	2.35852	2.39592	2.43880	2.48727
0.47500	2.27384	2.28582	2.30284	2.32489	2.35236	2.38508	2.42328	2.46708	2.51659
0.48000	2.29831	2.31054	2.32790	2.35051	2.37845	2.41187	2.45087	2.49560	2.54616
0.48500	2.32296	2.33544	2.35317	2.37623	2.40476	2.43887	2.47869	2.52435	2.57597
0.49000	2.34781	2.36054	2.37862	2.40216	2.43127	2.46608	2.50673	2.55333	2.60603
0.49500	2.37284	2.38583	2.40428	2.42829	2.45799	2.49351	2.53499	2.58255	2.63634
0.50000	2.39806	2.41130	2.43012	2.45462	2.48492	2.52115	2.56347	2.61200	2.66689

where

R = a shape factor depending upon the shape of the cross-section of the specimen and

n = an integer which is unity for the fundamental mode, two for the first overtone, etc.

This equation applies to rods where the cross-section is circular, square or rectangular – the geometry being taken account of by the shape factor R. For a cylinder, $R = 1$. If the cross-section is rectangular with sides a and b $(a > b, s = a/b)$, $1/R$ is given by (SCHREIBER, ANDERSON and SOGA, 1973)

$$\frac{1}{R} = \frac{3}{4(1+s^2)}\left[\frac{16}{3} - \frac{2}{b}\left(\frac{4}{\pi}\right)^5 \sum_{p=0}^{\infty} \frac{\tanh(2p+1)\pi s}{(2p+1)^5 2}\right] \qquad (7.14)$$

When $s = 1$, the cross-section is square, and $1/R$ becomes 0.8435.

When the length of the bar becomes short in comparison with its cross-sectional dimensions, the above equation does not hold explicitly. For a short bar with square cross-section, the following correction must be made:

$$\frac{R'}{R} = 1 + 0.00851 \left(\frac{nb}{l}\right)^2 \qquad (7.15)$$

where

R' = new shape factor.

For an infinite bar with rectangular cross-section, the following equation may be used (where $b > a$) (TIMOSHENKO and GOODIER, 1951)

$$R = \frac{1 + (b/a)^2}{4 - 2.521\,(a/b)\left(1 - \dfrac{1.991}{e^{\pi b/a} + 1}\right)} \qquad (7.16)$$

7.3.1.4. Calculation of modulus of elasticity from flexural resonant frequency

Since the equation which relates modulus of elasticity with flexural resonant frequency requires a knowledge of POISSON's ratio (ν), one starts the calculation by assuming a reasonable value for that ratio, say 0.25. From the modulus of rigidity calculated from Eq. 7.13, and the tentative value of modulus of elasticity (using $\nu = 0.25$), POISSON's ratio is calculated from equation

$$\nu = \frac{E}{2G} - 1 \qquad (7.17)$$

If this value disagrees with the assumed value, then the new value of POISSON's ratio is used to recalculate E until the values of both POISSON's ratio and modulus of elasticity each converge.

7.3.1.5. Measuring system

The system consists of (1) an exciter circuit (2) a pickup circuit and (3) specimen support. An outline of the apparatus used by the U.S. Bureau of Mines (LEWIS and TANDANAND, 1974) is given in Fig. 7-1. Mechanical vibrations are induced in the specimen and cause it to vibrate as a whole in one of its natural frequency modes as the sine wave oscillator and exciter are tuned to a resonant frequency of the rock. Mechanical vibrations of the specimen are converted to electrical signals by the pickup, amplified, and monitored on the Y-axis of the oscilloscope. The output of the oscillator is fed directly to the X-axis of the oscilloscope for comparison with the signal received by the pickup; frequency is monitored by the frequency counter. A transformer is used for matching the impedance between the oscillator and exciter. A function generator with a frequency range from 0.008 Hz to 1 MHz and sine wave output of 30 V peak-to-peak maximum into a 600-ohm load supplies the input signal to the exciter. The exciter used is an electromagnetic record cutting head (tweeter type i.e. directional speaker

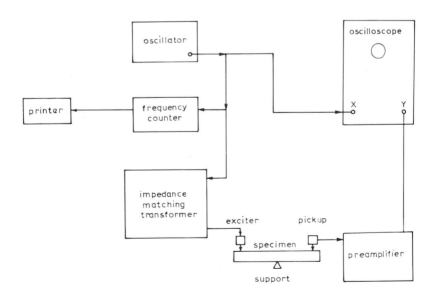

Fig. 7-1: Resonant velocity measurement system of U.S. Bureau of Mines
(after LEWIS and TANDANAND, 1974).

can also be used through air-coupling), whereas the pickup is a crystal phonograph cartridge. The resonant frequencies can be identified by observing both the amplitude and phase of the Lissajous figures on the oscilloscope. The frequency resonance is displayed on an electric counter and recorded on tape by a digital printer.

Frequency can be scanned continuously either by manual tuning or by voltage tuning in an automatic scan adaptation of the basic resonance system (Fig. 7-2). In the automatic frequency scan adaptation, the voltage control generator synchronises the frequency of the voltage tuned oscillator with the chart motion of the recorder. A switching circuit permits frequencies to be scanned continuously and automatically through any of the four frequency ranges: 3,000 to 9,525 Hz, 7,250 to 13,775 Hz, 12,500 to 19,025 Hz and 13,750 to 20,275 Hz. The a.c. signal of the pickup is converted to a representative d.c. signal by the a.c. to d.c. converter, amplified and recorded by a strip chart recorder. In the automated system, resonant frequencies are determined from the peak amplitudes of the resonance curves.

Specimen support, which provides the coupling of (1) energy from the exciter to the specimen and (2) detecting the vibrations of the specimen with the

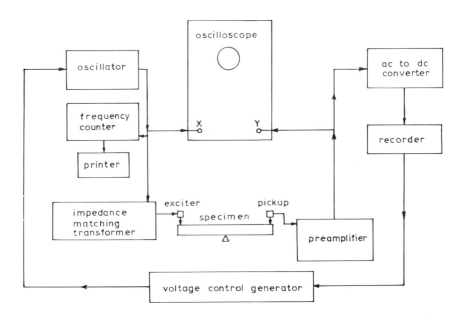

Fig. 7-2: Automatic frequency scan resonance system of U.S. Bureau of Mines
(after LEWIS and TANDANAND, 1974).

pickup, is a crucial element in determining the elastic constants of a rod or cylinder. Since a mechanical restraint of any sort applied to a specimen may affect its natural resonant frequencies, thus causing differing values of elastic constants for the specimen to be realised, one must try to minimise it. A widely adopted method is to support the specimen on knife edges of foam rubber at the nodal points and excite it through flexible wires extending from the transducer or directly with an air column by means of a tweeter as shown in Fig. 7-3. The nodes of the vibrating specimen are the positions of zero displacement in the direction of vibration. This depends upon the mode of vibration, as given in Table 24. However, best results have been observed on suspending the specimen from threads, one thread attached directly to the exciter and the other to the pickup. To obtain accurate results, the position of the threads should be close to the nodes. When the position of the thread is far from the nodes, a frequency shift of a few cycles may occur. The thread may be cotton, nylon, glass fibre, or even thin wire of a refractory metal for use at elevated temperatures. If the specimen is suspended in the manner shown in Fig. 7-3b, one can obtain both torsional vibration and flexural vibration measurements.

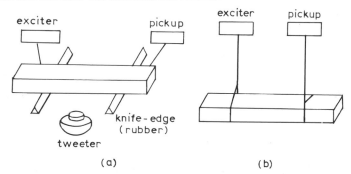

Fig. 7-3: Illustration of methods of coupling acoustic energy
(after SCHREIBER, ANDERSON and SOGA, 1973).

TABLE 24

Position of the nodes expressed as a function of the length of the specimen
(after SCHREIBER, ANDERSON and SOGA, 1973)

Mode of vibration	Type of vibration		
	longitudinal	flexural	torsional
Fundamental	0.500	0.224 0.776	0.500
First overtone (2nd mode)	0.250 0.750	0.132 0.500 0.868	0.250 0.750

7.3.1.6. Identification of the vibrating mode

As described previously, there are three types of vibrations normally excited, namely, longitudinal, flexural and torsional. As the scanning frequency approaches the resonant frequency of the specimen, a Lissajous pattern appears on the oscilloscope. The frequency at which maximum signal amplitude is observed is the resonant frequency. Since the Lissajous pattern, whose shape is controlled by the phase shift of the electronic signal, is an exact reflection of the mechanical vibration, it is possible to determine whether the motion of the specimen at the position of the exciter and the pickup is in the same phase or opposite. When the Lissajous pattern is adjusted so that the direction of the diagonal is down to the left for the same phase, there will be no pattern at the nodes and a down-to-the-right direction will occur for the·opposite phase, as shown in Fig. 7-4. This search can be done by either moving the position of the pickup or exciter separately or simultaneously, or by traversing the side of the specimen with the pickup, without changing any other condition. By counting the number of nodes along the bar, one can determine the mode of vibration. Torsional modes give the change in pattern along the cross-sectional face of the bar, as shown in Fig. 7-4. Except for some specimens which have an unusually low shear modulus, the fundamental torsional resonant frequency for thin rods or bars is higher than the fundamental flexural resonant frequency. The fundamental flexural vibration has two nodes. The fundamental torsional vibration has only one node at the centre. By suspending the specimen as shown in Fig. 7-3b, the longitudinal vibration is almost impossible to excite. Therefore, if one observes a resonant condition which has a node at the centre, after observing the resonant condition with two nodes, one can anticipate that it is the torsional vibration.

An alternative approach is to identify the mode of vibration from the position of the nodes. As shown in Table 24, torsional and longitudinal vibra-

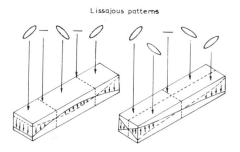

Fig. 7-4: The resonant vibrations in a rectangular bar
(after SCHREIBER, ANDERSON and SOGA, 1973).

tions give the same nodal position: the fundamental having a node at the centre and the first overtone at 0.25 l and 0.75 l from one end of the specimen. Flexural vibrations give slightly different nodal positions, which are 0.224 l and 0.776 l for the fundamental mode.

The placement of exciter and pickup for the three types of vibrations is given in Fig. 7-5. This figure also gives the shapes of Lissajous patterns (as well as voltmeter readings) when pickup is placed at various positions on the specimen.

7.3.1.7. Measurements at high temperatures

Resonance method can be used for measuring the elastic constants of solids as a function of temperature. This is because the exciter and pickup can easily be located outside the furnace, so that only the specimen is exposed to high temperatures, and also that both flexural and torsional resonant frequencies can be determined. LEWIS and TANDANARD (1974) applied this system to high-temperature work on rocks.

The following care must be taken in the measurement: (1) The temperature of the specimen should be as uniform as possible from one end to the other, preferably within 2 °C (3.8 °F); (2) The support wires, if they are of refractory metals such as nichrome, kanthal, or tungsten, might exhibit their own resonant condition near the resonant frequency of the specimen. Therefore, it is advisable to use glass threads instead of metal wires to as high a temperature as can be reached (possibly 700 to 800 °C (1292 to 1472 °F), using fused silica threads) in order to cross-check the results. Asbestos strings were used by LEWIS and TANDANARD (1974).

The elastic constants at high temperatures can be calculated from a knowledge of the temperature dependence of the resonant frequencies and the thermal expansion of the solid.

7.3.1.8. Some other methods of employing resonance

Other methods may be classified as (1) piezoelectric effect, (2) electromagnetic effect, and (3) electrostatic effect depending upon the choice of excitation and detection of the resonant condition. In the case of rocks, it is rather difficult to set a specimen in resonance by these techniques without applying a conductive film or attaching transducers. Consequently, a correction must be made in order to obtain the absolute value of elastic constants of the specimen.

Piezoelectric effect: A piezoelectric transducer is cemented to the rod or bar and this composite system is made to resonate. When a thin cylindrical specimen is cemented to a cylindrical transducer of equal cross-section, the

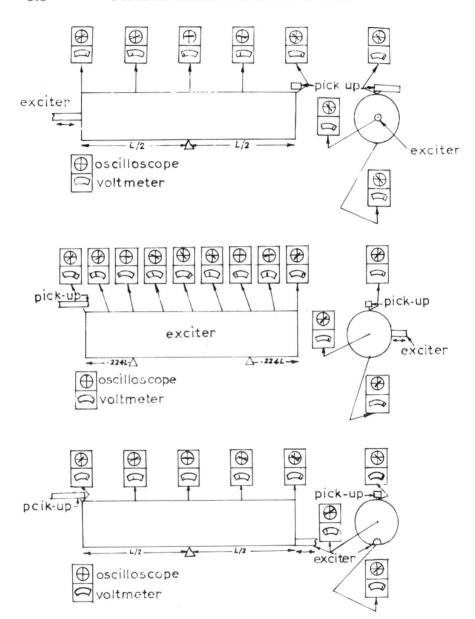

Fig. 7-5: The placement of exciter and pickup for the three types of vibrations.
(a) Longitudinal resonant frequency
(b) Flexural resonant frequency
(c) Torsional resonant frequency
(after SUTHERLAND, 1962).

mechanical resonant frequencies of the composite cylinder are given by
(SCHREIBER, ANDERSON and SOGA, 1973)

$$V_1\, \varrho_1 \tan\left(\frac{2\,\pi f l_1}{V_1}\right) + V_2\, \varrho_2 \tan\left(\frac{2\,\pi f l_2}{V_2}\right) = 0 \qquad (7.18)$$

where ϱ_1 and ϱ_2, l_1 and l_2 = densities and lengths of the specimen and
transducer, respectively

V_1 = longitudinal or torsional velocity of the specimen in the
direction of the cylinder axis

V_2 = longitudinal or torsional velocity of the transducer   and

f = longitudinal or torsional resonant frequency of the composite cylinder.

Since $V_i = 2 l_i\, f_i$, the above equation can be expressed as

$$M_1 f_1 \tan\left(\frac{\pi f}{f_1}\right) + M_2 f_2 \tan\left(\frac{\pi f}{f_2}\right) = 0 \qquad (7.19)$$

where M_1 and M_2 are the mass of the specimen and transducer. The resonant frequency f_1 of the specimen, therefore, can be calculated from M_1,
M_2, f_2 and f.

The general arrangement for observing the resonance of the composite
oscillator is shown in Fig. 7-6. When the frequency of the applied voltage

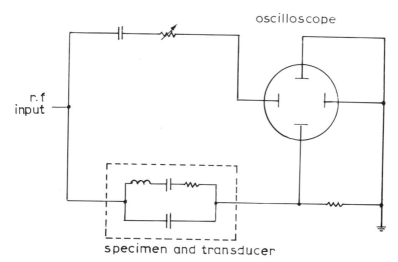

Fig. 7-6: Simple circuit for the composite oscillator
(after SCHREIBER, ANDERSON and SOGA, 1973).

approaches a frequency of free vibration of the composite oscillator, the electrical characteristic of the oscillator may be regarded as a fixed capacity shunted by an inductance, capacitance, and resistance in series. Since the actual frequency of free vibration is identical with the resonant frequency of the series branch alone, it can be deduced from the observed electric behaviour of the oscillator. The output current will be a maximum at the resonant frequency of the composite f. Quartz transducers used for the composite oscillator are shown in Fig. 7-7. The cement may be a thin layer of shellac.

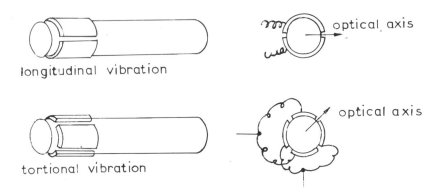

longitudinal vibration

tortional vibration

optical axis

optical axis

Fig. 7-7: Quartz transducer for the composite oscillator
(after SCHREIBER, ANDERSON and SOGA, 1973).

TERRY (1957) regarded the cement as a third cylinder and estimated the error in the calculated value of the elastic wave velocity V_1 due to neglecting the cement. The error ΔV is given by

$$\Delta V = \frac{M_3}{M_1}\left[\cos^2 \theta_1 + \left(\frac{\varrho_1 V_1}{\varrho_3 V_3}\right)^2 \sin^2 \theta_1\right]\left[\frac{2\theta_1}{2\theta_1 - \sin 2\theta_1}\right] \tag{7.20}$$

where $\theta_i = 2\pi f/f_i$ and the subscripts 1 and 3 refer to the specimen and the cement, respectively. If the composite rod is vibrated in its fundamental mode and the specimen and transducer have been selected to have approximately equal natural frequencies, so that $f_1 \simeq f_2 \simeq 2f$, the error due to the cement can be expressed as

$$\Delta V \simeq -\frac{M_3}{M_1}\left(\frac{\varrho_1 V_1}{\varrho_3 V_3}\right)^2 \tag{7.21}$$

Since $\varrho V^2 = C$ and $M/\varrho = \pi r^2 l$ for a cylinder, the above equation becomes

$$\varDelta V \simeq \frac{l_3\,C_1}{l_1\,C_3} \qquad (7.22)$$

where C is the appropriate elastic modulus.

EVANS and POMEROY (1966) described the use of a nickel cylinder in determining the dynamic elastic constants of coal. The coal specimen, in the form of a small prism or cylinder, is cemented end-to-end to a nickel cylinder which is selected so as to have approximately the same natural mechanical frequency as the specimen. The nickel rod is magnetostrictively excited to mechanical vibration by placing it axially in a small coil in which an alternating current is flowing. The mechanical vibrations are transmitted to the coal specimen and the mechanical resonance vibrations of the composite rod are detected by a second small coil placed about the nickel rod.

Depending on the polarisation of the nickel transducer, either longitudinal or torsional vibrations may be produced, so that the moduli of elasticity and the moduli of rigidity of the coal specimens may be determined. For longitudinal vibrations, prisms of approximately square cross-section of side about 3 mm ($^1/_8$ in) were used. These were prepared from blocks of coal by means of a high speed carborundum slitting wheel. For torsional vibrations, cylinders of coal, 9 mm ($^3/_8$ in) in diameter, were prepared in a lathe using a small grinding wheel. All the specimens used were 2 to 3 cm (1 in) in length and were resonated at frequencies between 4 kc/s and 20 kc/s. The experimental accuracy of the method is limited by the precision with which the specimens can be prepared, but is somewhat better than 1 per cent.

Electromagnetic effect: OBERT, WINDES and DUVALL (1946) utilised this effect in developing their apparatus and a schematic diagram is given in Fig. 7-8. Vibration in the specimens is produced by cementing to one end

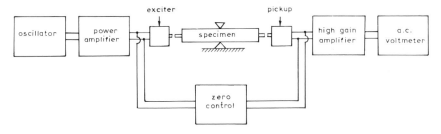

Fig. 7-8: Resonant frequency apparatus based on electromagnetic effect.
(after OBERT, WINDES and DUVALL, 1946).

of the core a soft iron pole-piece that interacted with an electromagnetic exciter unit (Fig. 7-9). The type of vibration produced – that is longitudinal or torsional – depends upon the type of exciter employed. An identical pole-piece cemented to the other end of the core interacted with an equivalent electromagnetic pickup to sense the vibration amplitude. To determine the resonant frequency the oscillator is tuned to give the maximum reading on the voltmeter.

The soft-iron pole-pieces cemented to the ends of the specimens reduce both the longitudinal and torsional frequency, thereby producing an error. This error can be corrected by using the following equations:

$$f_1 = f_1'\left(1 + \frac{M'}{M}\right) \tag{7.23}$$

$$f_t = f_t'\left(1 + k^2\frac{M'}{M}\right) \tag{7.24}$$

where f_1 = longitudinal frequency without pole-pieces
  f_1' = longitudinal frequency with pole-pieces
  f_t = torsional frequency without pole-pieces
  f_t' = torsional frequency with pole-pieces
  M' = mass of both pole pieces
  M = mass of specimen and
  k = ratio of the radius of gyration of the pole-pieces to the radius of gyration of the specimen.

Electrostatic effect: An electrostatic technique developed by DICKSON and STRAUCH (1959) and used by LeCOMTE (1963) is given below. In this method (Fig. 7-10a), a cylindrical specimen is supported at its centre, a nodal plane, between two metal electrodes which form two parallel plate condensers with the ends of the specimen. Rocks being nonconductors, the end faces of the specimen are coated with a thin film of silver paint. Static fields are created between the electrodes and the specimen by applying a high d.c. potential to the specimen. Any a.c. voltage of frequency "*f*" superimposed across one condenser produces an alternating force of the same frequency *f*. The electronic circuitry forms a regenerative loop and oscillations build up when *f* is close to the fundamental resonance frequency of the specimen. This resonance frequency is measured with a frequency counter and used to calculate the modulus.

The above method excites longitudinal vibrations in the specimen. By fixing small metal bars to the specimen and rearranging the electrodes (Fig. 7-10b), torsional vibrations can be obtained by using the same electrostatic method.

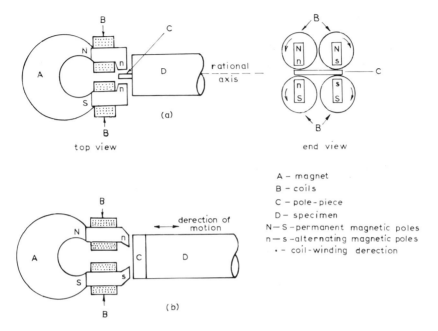

A — magnet
B — coils
C — pole-piece
D — specimen
N—S —permanent magnetic poles
n—s —alternating magnetic poles
• — coil-winding derection

Fig. 7-9: Exciter and pickup heads
(a) Torsional;   (b) Longitudinal
(after OBERT, WINDES and DUVALL, 1946).

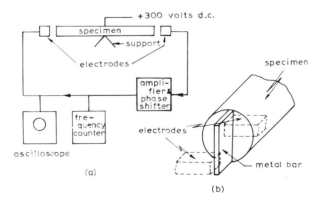

Fig. 7-10(a): Block diagram of electrostatic apparatus to produce
longitudinal vibration in a rock specimen
(b): Arrangement of electrodes to produce torsional vibration
(after LeCOMTE, 1963).

7.3.1.9. Practical limitations

Some of the practical limitations to accurate determination of resonant frequency are associated with the degree of refinement in the system. Good coupling between exciter and specimen must be achieved, and the contact pressure of a mechanically driven system can influence the results. The uniformity of a specimen as to density and geometry will determine the location of nodal points, so that clear definition of the mode of vibration observed may sometimes be difficult.

For resonance determination, the ratio of depth to width may vary widely, but requires less "shape" adjustment as a square or circular cross-section is approached. The length-to-width ratio should be kept low for avoidance of confusing overtones. Furthermore, as the specimen mass decreases, the influence of coupling between exciter, specimen and pickup becomes more critical if distortion of resonance behaviour is to be avoided.

Calculation of modulus of elasticity from resonant frequency in fundamental mode of flexural vibration demands an assumption of POISSON'S ratio. Generally, for each specimen, the modulus of elasticity is first calculated from the longitudinal resonant frequency and this value used with modulus of rigidity (as obtained from torsional resonant frequency) to calculate the value of POISSON'S ratio (Eq. 7.17). Modulus of elasticity is then computed from the transverse resonant frequency. Another method is given in section 7.3.1.4.

7.3.2. Ultrasonic* Pulse Method

In the pulse method, a mechanical impulse is imparted to the specimen and the time required for the transient pulse to traverse the length of the specimen is used to calculate the velocity of the wave by the formula

$$V = l/t \tag{7.25}$$

where

V = velocity
l = distance traversed by the wave   and
t = travel time

Variations used in elastic property determinations by ultrasonics are given in Fig. 7-11.

---

* The term *"ultrasonic"* is used here since the pulse frequencies are above the audible range.

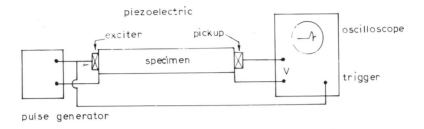

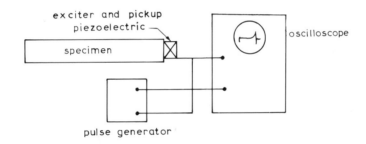

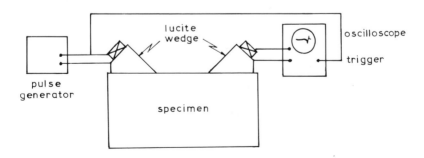

Fig. 7-11: Variations used in elastic property determinations by ultrasonics
(after SHOOK, 1963).

The elastic constants are obtained from density and the velocities, using the following expressions derived from the theory of elasticity for homogeneous, isotropic solids.

$$E = \frac{\varrho \, V_s^2 \, (3 \, V_p^2 - 4 \, V_s^2)}{V_p^2 - V_s^2} \tag{7.26}$$

$$G = \varrho \, V_s^2 \tag{7.27}$$

$$\nu = \frac{V_p^2 - 2 \, V_s^2}{2 \, (V_p^2 - V_s^2)} \tag{7.28}$$

where

E = modulus of elasticity, Pa
G = modulus of rigidity, Pa
ν = Poissons's ratio
ϱ = density, kg/m³
V_p = longitudinal infinite medium velocity, m/s   and
V_s = shear velocity, m/s.

7.3.2.1. Measuring system

The ultrasonic pulse measurement system used by the U.S. Bureau of Mines is given in Fig. 7-12. This employs separate driving and receiving transducers (i.e. pulse transmission technique). Transducer elements commonly used for longitudinal wave propagation are lead-zirconate titanate, or barium titanate ceramic discs operating in a thickness expansion mode. For shear wave transmission, a.c. cut quartz or specially cut ceramic elements are frequently used, or a longitudinal wave is mode converted with suitably cut prisms (JAMIESON and HOSKINS, 1963; YOUASH, 1970).

The specimen is placed in physical contact between two piezoelectric transducers; one acts as a driver, and the other acts as a receiver. A pulse generator supplies a short-duration electrical pulse to the driver transducer. The electrical pulse is converted into a mechanical wave or impact by the driver transducer, and this wave is transmitted to the specimen. After travelling through the test specimen, the pulse is picked up by the receiving transducer, reconverted to an electrical signal, and displayed on the screen of a cathode ray oscilloscope. The transit time required for the mechanical pulse to pass through the specimen is used to determine elastic wave velocity. Transit times are displayed on a timer and recorded by a printer or, in the automated system, plotted by an X-Y recorder (Fig. 7-13). Am-

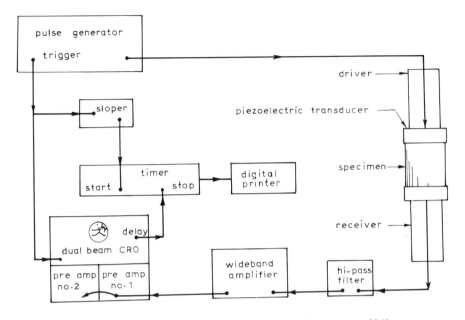

Fig. 7-12: Ultrasonic pulse measurement system of U.S. Bureau of Mines
(after Lewis and Tandanand, 1974).

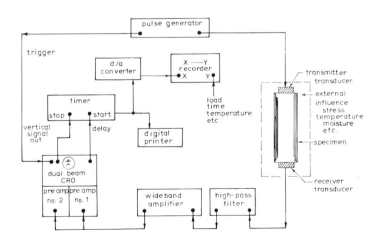

Fig. 7-13: Automated ultrasonic pulse measurement system of U.S. Bureau of Mines
(after Lewis and Tandanand, 1974).

plitude of the pulse used is typically in the range of 50 to 700 V. Pulse rise time, the time that the pulse takes to rise from 10 to 90 percent of its final amplitude, is typically about 0.2 μs or less in order to excite the pulse frequencies commonly employed in testing rock; that is, those from about 100 kHz to 1 MHz. Pulse length is set in the range from about 1.5 to 10 μs, and the pulse recurrence frequency usually is set at about 60 Hz. The couplant found most satisfactory for the transfer of longitudinal wave energy between transducer and specimen is one or two thicknesses of plastic electrician's tape placed on the coverplates of the transducers. Couplants used for shear wave propagation include Salol, Canada balsam, Lakeside 70 cement, and Nonaq stopcock grease.

The adaptation of the compression test setup for velocity determinations in controlled environments is given in Fig. 7-14.

The A.S.T.M. (1969) issued a standard method for laboratory determination of pulse velocities and ultrasonic elastic constants of rock and the system is the same as described above.

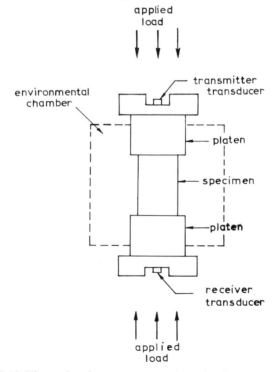

Fig. 7-14: Ultrasonic pulse measurement adaptation for compression testing (after LEWIS and TANDANAND, 1974).

7.3.2.2. Limitations

Scattering due to randomly oriented discontinuities can cause significant attenuation, so that the pulse may be considerably weakened when it reaches the receiver. Attenuation losses usually decrease with decreasing frequency, but the wave length must be held to a reasonably small value in relation to the dimensions of the test piece. The internal structure can therefore limit the application of the technique due to grain size, porosity, inclusions, etc.

The coupling between a piezoelectric transducer and the specimen is important in reducing interface losses.

Most read out systems utilise an oscillograph technique for observing the detected signal. The time-base usually available at the short times involved in sound transmission may not be of sufficient accuracy, and it is desirable to supply timing marks or compare results against a reference material.

Anisotropy of structure and of elastic parameters will influence the results. The random orientation of the grains with anisotropic elastic properties will yield average numbers for velocity of propagation. The occurrence of boundaries between materials of differing sound velocity will also give rise to reflection and mode conversion, so that backward disturbances can become serious in heterogeneous materials.

To use the equations 7.26, 7.27 and 7.28, the rock must be isotropic or possess only a slight degree of anisotropy. In order to estimate the degree of anisotropy of the rock, the compression wave velocity is measured in three orthogonal directions. These measurements are made with the same geometry, that is, all between flat surfaces or all across diameters. The equations for an isotropic medium are not applied if any of the three compression wave velocities varies by more than 2 percent from their average value (A.S.T.M., 1969). The error in E and G due to both anisotropy and experimental error will then normally not exceed 6 percent. The maximum possible error in v depends markedly upon the relative values of V_p and V_s as well as upon testing errors and anisotropy. In common rock types the error in v may be as large as or even higher than 24%. For greater anisotropy, the possible percent of error in the elastic constants would be still higher.

7.3.3. Comparison between Resonance and Ultrasonic Pulse Methods

The most notable difference between the resonance and pulse methods is the frequency at which the velocity is determined, with the resonance method operating commonly in the range from 1 to 30 kHz and the pulse method in the range from 50 to 10,000 kHz. Thus, there arises the question whether frequency dispersion in velocity might cause discrepancies in the dynamic elastic constants determined by the resonance or pulse method. Evidence from many investigations supports the view that the elastic constants are nearly independent of the frequency of excitation over a wide range of frequencies (BIRCH and BANCROFT, 1938, 1940; WYLLIE, GREGORY and GARDNER, 1958; PESELNICK and OUTERBRIDGE, 1961; SIMMONS, 1965; SIMMONS and BRACE, 1965; JOHNSON, 1970). On the other hand, some frequency dependence in the dynamic elastic constants in rocks is indicated in the studies of BRUCKSHAW and MAHANTA (1954) over the frequency range 40 to 120 Hz, and AUBERGER and RINEHART (1961) over the range 250 to 1000 kHz. ATTEWELL (1970) noted apparent frequency dependence in shear wave velocity over the frequency range from about 200 kHz, but suggested that such frequency dispersion probably was a function of the critical angle technique, rather than of the velocity itself.

Another difference between the resonance and pulse methods pointed out by LeCOMTE (1963) is that these methods differ in the stress and strain distributions that they impose in the rock. The resonance method creates a standing wave, whereas the pulse method generates a transient wave packet in the specimen. Since the stress level and strain amplitude are very low by either method, however, response of the rock in either case usually is considered elastic and the elastic constants by either method are anticipated to be unaffected by such differences in the stress and strain distribution.

POPOVIC and CVETKOVIC (1972) used both resonance and ultrasonic methods in their investigations on rocksalt for dynamic elastic constants. The modulus of elasticity obtained by the ultrasonic method ($E_{ultrasonic}$) is slightly higher than that obtained by the resonance method ($E_{resonance}$). The values are:

$$E_{ultrasonic} = 308 \text{ kgf/cm}^2$$
$$E_{resonance} = 306 \text{ kgf/cm}^2$$
$$E_{ultrasonic}/E_{resonance} = 1.11$$

THILL and PENG (1974) determined elastic constants by both the bar resonance and ultrasonic pulse methods in a large number of specimens of St. Cloud Gray granodiorite and Tennessee marble in the same test environment. They concluded that the methods do not give equivalent results in nearly isotropic rock. The amount of difference varies for each constant, with the least difference ($< 5\%$) occurring in the modulus of rigidity. The discrepancy in resonance and pulse results probably is caused by frequency dispersion effects and by difference in error sensitivity in the various formulas used to calculate the elastic constants.

For less homogeneous materials the resonance method is not as satisfactory because in a bar vibrating in either its fundamental longitudinal or torsional mode the strain reaches a maximum at the midpoint and is always zero at the ends of the bar. Hence a mechanical defect (fracture, joint, etc.) will have a greater effect if it is at or near the centre than if it is at or near either end. Also the effects of anisotropy become evident if POISSON's ratio is calculated from the longitudinal and torsional bar velocity. In fact if POISSON's ratio is calculated from Eq. 7.17, negative values or values greater than 0.5 are sometimes obtained, which on the basis of theory of elasticity is impossible. This anomaly results from the fact that in the longitudinal vibration the strain is in the axial direction, whereas in torsional vibration the shear strain is in any plane parallel to the specimen axis. Despite these possible sources of error, about one-third of rocks tested by this method give reasonable values for POISSON's ratio (OBERT and DUVALL, 1967).

The discrepancy between the two methods probably is not of great significance in E and G determinations, but may be very significant regarding ν determinations. Consequently, while either method might provide an adequate estimate of E or G, ν determination by the resonance method is not recommended because one may have great uncertainity with its determination.

For best results both the resonant and ultrasonic pulse methods should be used to determine, on the same cylindrical specimen, the propagation velocities, V_0 (bar velocity by resonance method), V_p (compression wave velocity by pulse method) and V_s (shear wave velocity by both methods). If both methods are used, two independent measurements of V_s are obtained, one from the resonant torsional frequency and one from the shear pulse travel time. These two independent shear velocity determinations should agree within the experimental error of the measurements if the rock specimen does not have any major defects near its centre section. The average value of V_s is used to calculate G by means of Eq. 7.27. The longitudinal bar velocity V_0 is used to calculate E by means of Eq. 7.7. The longitudinal pulse velocity V_p and the longitudinal bar velocity V_0 are used to calculate POISSON's ratio by means of the equation

$$\nu = \frac{1}{4}\left\{\left[\left(9 - \frac{V_0^2}{V_p^2}\right)^2 \left(1 - \frac{V_0^2}{V_p^2}\right)\right]^{1/2} - \left(1 - \frac{V_0^2}{V_p^2}\right)\right\} \qquad (7.29)$$

This equation is obtained by solving Eq. 7.1 for ν. If both V_0 and V_p are obtained by producing an axial strain along the length of the same cylindrical bar, both the velocities are related to the same elastic constant E in the axial direction. The difference between the longitudinal pulse velocity and the bar velocity is a result of the average POISSON's ratio in the plane normal to the axis of the bar. Thus even if the rock is anisotropic, the above equation will give a good average value for ν.

7.4. In Situ Test

The method of determination of dynamic elastic constants is based upon the measurement of wave propagation velocity through the rock medium. The experimental technique for in situ test is quite simple. Fig. 7-15 shows a typical experimental set up for measuring longitudinal and shear velocities.

To measure longitudinal and shear velocities of seismic waves in rock masses exposed on the surface, an explosive is detonated in a shallow drill hole. The wave motions generated on the rock surface are measured by velocity gauges (accelerometers) at several distances from the source and recorded with a tape recorder, oscillograph or an oscilloscope. Zero shot time is determined by an ionisation probe attached to the charge. Velocities are computed from the time intervals between the detonation of the explosive and the arrival of a longitudinal or shear wave at each gauge. Shear wave arrivals are most frequently recorded by transversely oriented gauges.

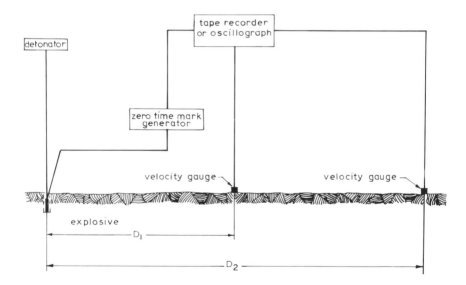

Fig. 7-15: Experimental setup for longitudinal and shear velocity measurements (after LEWIS and TANDANAND, 1974).

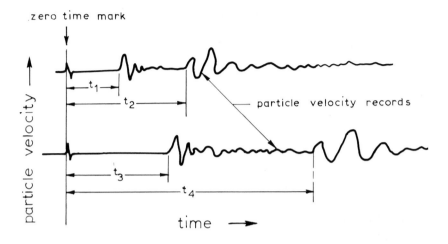

t_1 and t_3 arrival times for longitudinal wave
t_2 and t_4 arrival times for shearwave

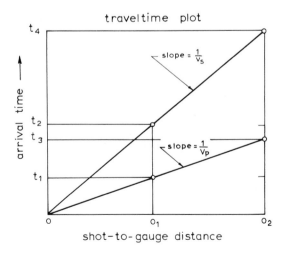

Fig. 7-16: Velocity records and computation of longitudinal
and shear velocities from traveltime plot.

Longitudinal and shear velocities are computed by dividing the shot-to-gauge distance by the time interval between the detonation of the charge and the arrival of the appropriate signal at the gauge position. The velocities can also be computed from travel time curves as shown in Fig. 7-16. Fig. 7-16 shows some typical particle velocity records and the computation of longitudinal and shear velocity from a travel time curve.

Utilising rock density and velocities determined in tests, all the elastic constants can be calculated from the standard elastic equations 7.26, 7.27 and 7.28.

While this technique is an attractive one, it is not without problems. Often it is quite difficult to generate shear waves and even more difficult to measure arrival rates. Since the velocity of shear waves is in the neighbourhood of $1/2$ of the longitudinal velocities for rock, both shear and longitudinal waves normally generated by the original longitudinal wave being reflected from a free face may mask the original shear waves. Also, the number of paths to be measured results in tedious calculations.

The method can be used to determine the depth of the various rock beds provided they show sufficient variations in wave propagation velocities.

A wave starting from the detonator position O (Fig. 7-17) may reach the receiver G_0 placed very close to the detonator directly and the detector G_1, by one of three paths; the direct path OG_1, the reflected path OBG_1, or the refracted path $OACG_1$. Similarly these various paths are available for

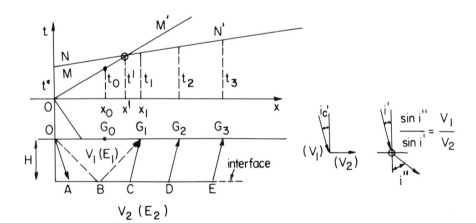

Fig. 7-17: Seismic wave propagation
(after Jaeger, 1972).

other detectors G_2, G_3 etc. If time-distance plot is made for the various detectors, the curve obtained is a straight line OMM', passing through the origin. Its slope is by definition equal to the velocity V_1. For a refracted path, the pulse travels down to the interface (between two different density beds) with the velocity V_1 and its incident is at the so-called 'critical angle φ_c'. After refraction, the impulse travels along the interface with the velocity V_2 (supposedly higher than V_1) and leaves the interface at the critical angle to reach G_1. As the distances AC, AD, AE are covered along the interface at the higher velocity V_2, it is possible that there will be a point when the wave starting from O and following the refracted path will arrive at one of the detectors G_1, G_2 or G_3, before the direct waves. If these arrival times are plotted on time-distance curve, it gives a line NN', which does not pass through the origin O. If $ON = t^*$, it can be shown that the depth H of the upper layer to the interface is given by

$$H = \frac{V_1 V_2 t^*}{2 \sqrt{V_2^2 - V_1^2}} = \frac{x^2}{2}\left(\frac{1 - \sin i'}{\cos i'}\right) = \frac{x'}{2}\sqrt{\left(\frac{V_2 - V_1}{V_2 + V_1}\right)} \quad (7.30)$$

In some cases the method does not work well; when the two layers are similar and E_1 only slightly less than E_2, or when the upper layer is covered by another thin layer with high modulus of elasticity $E_o > E_1$.

Seismic tests concern rather low stress levels by means of frequency vibration of very small amplitude and results are not comparable with jacking tests. They can, however, economically indicate variations in rock quality over a greater portion of the foundation.

BERNABINI and BORELLI (1974) evaluated the average dynamic elastic properties of a rock mass through velocity measurements between pairs of holes. Generally, the distance between receivers in the same hole and the distance between two subsequent positions of the emitter are chosen to be of the same order of magnitude (or at least equal to $1/2$–$1/3$) as the distance between the holes, in order not to have equivalent paths. The receivers, the subsequent positions of the emitter at different depths and the hypothetical rectilinear paths between emitter and receivers are shown in Fig. 7-18 for a pair of holes.

The location of holes is chosen according to the shape of the mass under examination. An example of a hole arrangement relative to the investigation of a rock mass is given in Fig. 7-19. The test begins by placing geophones in holes 2 and 3 and shooting at various depths in hole 1. The geophones are taken out of hole 2 and put in holes 4 and 5 and shots are made in hole 2 of hole 3 and so on. Each rock prism so formed (e.g. F_1–F_2–F_3; F_2–F_3–F_4–F_5; etc.) is analysed separately.

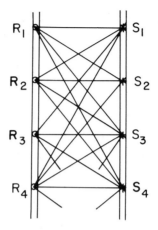

Fig. 7-18: Positions of receivers R and source S for a pair of holes
(after Bernabini and Borelli, 1974).

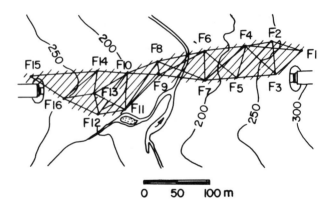

Fig. 7-19: Hole arrangement used for a survey
(after Bernabini and Borelli, 1974).

The speed values along the various emitter-receiver paths are normally scattered, and the statistical methods are employed to obtain average parameters characteristic of the rock.

Comparison of Laboratory and In Situ Results: For realistic comparison, the environmental conditions in the laboratory and in situ should be the same. Wyllie, Gregory and Gardner (1958) measured the velocities in samples from the cores taken in a West Texas well and compared these velocities with the velocities obtained in situ by a continuous velocity log. They applied a uniaxial pressure to the samples and saturated them with a brine which had the same probable propagation velocity as the mud filtrate at the depth the samples were cored, in order to simulate an in situ environment. Fig. 7-20 shows that the two measurements compared favourably.

The studies of NICHOLLS (1961) on salt and granite-gneiss, however, gave higher in situ values than laboratory values for modulus of velocity and in situ values for POISSON's ratio were more reasonable.

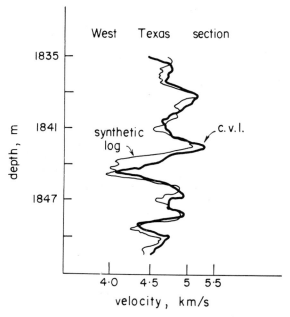

Fig. 7-20: Comparison of field continuous velocity log with synthetic laboratory log (after WYLLIE, GREGORY and GARDNER, 1958).

7.5. Comparison of Static and Dynamic Elastic Constants

Static methods give rise to a large scatter of results, but can be extended to the high strains 10^{-2} occurring in mining processes. In dynamic methods, low strains of 10^{-5} are involved with high rates of loading and the scatter is comparatively small.

Since dynamic methods usually involve low stresses, a comparison of static and dynamic values of modulus of elasticity is meaningful only if the values of the static modulus of elasticity are taken at comparable stress levels, i.e., using initial or zero stress tangent modulus.

It was found that values obtained by static techniques are lower than those obtained by dynamic methods, the difference in the constants so determined varying from 0 to 300 percent. The difference was explained as being due to the presence of fractures, cracks or cavities, with the static yielding being increased by deformation of cracks and cavities and the dynamic measurements being less influenced by these factors (ZISMAN, 1933 and IDE, 1936). The greater the degree of compactness, the more nearly static and dynamic elastic constants may agree as evidenced by the following comparisons:

	Modulus of elasticity		POISSON's ratio	
	static	dynamic	static	dynamic
	(dynes/cm²)			
Quincy granite	3.5×10^{11}	4.3×10^{11}	0.10	0.33
Sudbury norite	8.36×10^{11}	8.82×10^{11}	0.22	0.27

The Sudbury norite is more compact than the Quincy granite as indicated by the much higher value of modulus of elasticity and the closer agreement of the static and dynamic values.

The results obtained by the U.S. Bureau of Reclamation (1948) investigation on static and dynamic moduli of Davis Dam foundation rock agree with the conclusions of ZISMAN (1933) and IDE (1936). The mean value of the modulus of elasticity in dynamic measurements was found to be 10^6 lbf/in² (6895 MPa) and that in static measurements was 5×10^4 lbf/in² (345 MPa).

Similar results were obtained by DVORAK (1957). If the dynamic modulus is compared with the static modulus for a medium which is under pressure, the differences will not exceed 50 percent because the fractures are closed by the pressure.

The ratio of the dynamic to static moduli may vary between 0.85 and 2.9, according to data determined by the U.S. Bureau of Reclamation (1953). The data is given in Table 25. The discrepancy between static and dynamic values is less for rocks which have a greater modulus of elasticity.

Eight other types of rock were tested by SUTHERLAND (1962) using static and dynamic methods and he found that static values of E and G were consistently lower than dynamic values, while statically determined values of POISSON's ratio were higher. He reasoned this to be due to static tests being affected by cracks in rocks, planes of weakness, cavities, etc.

CHENEVERT (1964) and YOUASH (1970) determined elastic constants for Leuders limestone and Green River shale statically and dynamically respectively. Their results are given in Table 26. The static values of E and v are consistently lower than dynamic values.

TABLE 25

Static and dynamic elastic constants of rocks

(after U.S. Bureau of Reclamation, 1953)

Rock name	E_{static}	$E_{dynamic}$	$\dfrac{E_{dynamic}}{E_{static}}$	ν_{static}	$\nu_{dynamic}$
	$(lbf/in^2 \times 10^6)$				
Chalcedonic limestone	8.0	6.8	0.85	0.18	0.25
Limestone	9.7	10.3	1.06	0.25	0.28
Oolitic limestone	6.6	7.8	1.18	0.18	0.21
Quartzose shale	2.4	3.2	1.33	0.08	–
Monzonite porphyry	6.0	8.2	1.36	0.18	0.21
Quartz diorite	3.1	4.4	1.42	0.05	0.19
Stylolitic limestone	5.6	8.2	1.46	0.11	0.27
Biotite schist	5.8	8.6	1.48	0.01	0.16
Limestone	2.4	4.1	1.70	0.18	0.20
Limestone	4.9	7.6	1.86	0.17	0.31
Siltstone	1.9	3.9	2.05	0.05	0.08
Subgraywacke	1.8	3.8	2.11	0.03	0.19
Sericite schist	1.1	2.6	2.36	−0.02	0.44
Subgraywacke	1.6	3.8	2.37	0.02	0.06
Quartzose phyllite	1.1	2.7	2.45	−0.03	–
Calcareous shale	2.3	3.6	2.56	0.02	–
Subgraywacke	1.4	3.6	2.57	0.02	0.29
Granite (slightly altered)	0.8	2.2	2.75	0.04	0.10
Graphitic phyllite	1.4	3.9	2.78	–	–
Subgraywacke	1.3	3.8	2.90	0.05	0.08

TABLE 26

**Comparison of the values of modulus of elasticity and
POISSON's ratio determined by CHENEVERT (1964) and
YOUASH (1970) statically and dynamically respectively**

Rock type	Orientation of cores with respect to bedding	Modulus of elasticity $\times 10^6$ lbf/in^2		POISSON'S ratio	
		Static	Dynamic	Static	Dynamic
Leuders limestone	normal	3.50	4.84	0.21	0.22
Leuders limestone	parallel	3.60	4.84	0.21	0.22
Green River shale	normal	4.30	5.81	0.18	0.22
Green River shale	parallel	5.10	6.17	0.17	0.27

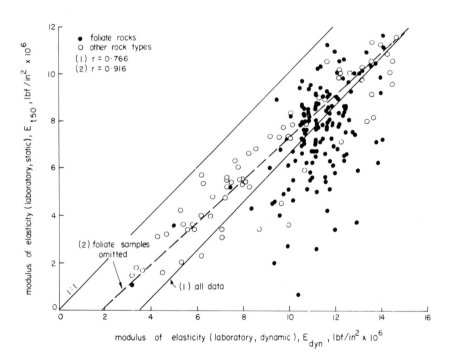

Fig. 7-21: Comparison of laboratory static and dynamic moduli
(after COON, 1968).

TABLE 27
Static and dynamic elastic properties of some rocks
(after RZHEVSKY and NOVIK, 1971)

Rock	Static modulus of elasticity E_{st} x 10^{-5}, kgf/cm^2	Dynamic modulus of elasticity E_{dyn} x 10^{-5}, kgf/cm^2	Ratio $\dfrac{E_{dyn}}{E_{st}}$	Bulk modulus K x 10^{-5}, kgf/cm^2	Rigidity or shear modulus G x 10^{-5}, kgf/cm^2
Sandstone with chalcedonic cement	7.30	7.78	1.07	2.30	3.20
Equigranular dolomite	5.05	5.31	1.00	5.21	1.88
Limestone	1.88	2.42	1.29	4.35	0.66
Calcareous dolomite	3.49	4.72	1.35	3.70	1.30
Fine-grained detrital limestone	4.77	5.71	1.20	4.60	1.80
Granite	6.60	7.10	1.08	4.73	2.60
Gabbro	7.10	7.50	1.06	5.70	3.60
Dunite	14.90	16.40	1.03	10.73	5.90
Syenite	7.40	8.10	1.10	5.40	2.90

TABLE 28
Comparison of static and dynamic results
(after RAMANA and VENKATANARAYANA, 1973)

Sample No.	Rock type	Density (g/cm^3)	Porosity (%)	Static moduli				Dynamic constants			
				E_s	G_s	K_s	v_s	E_d	G_d	K_d	v_d
				(10^{11} dyn/cm^2)				(10^{11} dyn/cm^2)			
88	Tremolite schist	3.011	−6−	8.96	4.03	3.83	0.11	09.27	3.60	7.23	0.29
61	Uralite basalt	3.062	−9−	7.85	3.66	3.74	0.15	10.47	4.10	8.80	0.28
13	Hornblende schist	3.011	−−	9.82	3.85	7.44	0.28	10.42	4.05	8.09	0.29
26	Uralite diabase	3.162	−3−	9.10	3.63	6.11	0.25	08.20	3.10	7.76	0.32
30	Dolerite	3.000	−−	8.20	3.63	3.69	0.13	09.19	3.55	7.45	0.29
38	Dolerite	3.136	−4−	9.39	3.91	5.22	0.20	10.93	4.36	7.37	0.25
95	Actinolite schist	3.499	−6−	7.79	3.02	6.18	0.29	14.86	5.91	7.11	0.26

E − modulus of elasticity
G − modulus of rigidity
K − bulk modulus
v − POISSON's ratio

Fig. 7-21 shows the relationship between static and dynamic moduli for intact specimens (COON, 1968). The static modulus was calculated by the slope of the tangent of the stress-strain curve at 50% of the compressive strength. The correlation coefficient for all data is 0.766.

An examination of the data shows that the degree of correlation is lowered by the foliate rock specimens which form a near vertical band with a dynamic modulus of from 10 to 14 $\times$ 10^6 lbf/in^2 (68,950 to 96,530 MPa). The poor relationship is probably caused by variations in the foliation which have a significant effect on static modulus but produce only minor changes in the dynamic modulus. The dashed correlation line was determined by considering only the nonfoliate rock specimens. The correlation coefficient for these specimens is 0.916. The laboratory dynamic moduli are generally 1 to 6 $\times$ 10^6 lbf/in^2 (6,895 to 41,370 MPa) higher than corresponding static moduli.

Table 27 gives static and dynamic elastic properties of some rocks (RZHEVSKY and NOVIK, 1971). The ratio of the dynamic to static moduli varies between 1.00 and 1.35.

The results of static and dynamic tests on Kolar rocks, conducted by RAMANA and VENKATANARAYANA (1973) are given in Table 28. From these results, it appears that the dynamic values of E and v are generally higher than the respective static values, but not abnormally high.

In situ tests also indicate that the elastic moduli observed from jacking tests and pressure chamber tests are definitely lower than the moduli calculated from seismic velocities. The variation of the ratio $E_{seismic}/E_{static}$ is probably due to differences in rock quality from site to site. The static elastic modulus ranged from one-sixteenth to one-third of the seismic values. For the same sites, which were grouted to increase the rock quality, the static elastic moduli ranged from 0.14 to 0.39 the value of the seismic values (CLARK, 1965).

The results obtained by COON (1968) are shown in Figs. 7-22 and 7-23. The scatter noted in both graphs may be due to the variable quality of the tests reported and to the relative size of the areas tested.

7.6. Parameters Affecting Propagation Velocity of Waves in Rocks

The parameters affecting propagation velocity of waves in rocks (hence dynamic elastic constants) are: rock type, texture, density, porosity, anisotropy, stress, water content, and temperature.

7.6.1. Rock Type

Values of propagation velocity for some rocks and minerals are given in Tables 29, 30 and 31. (Some dynamic properties are also given in Tables 30 and 31.) Generally it can be said that velocities are higher for more dense and compact rocks, less high for less dense and compact rocks.

For some rocks, the range of variation in velocity is larger than for the others. For example, in limestone, the velocities vary between 2 and 6 km/s. This variation is due to various textures in limestone. Granite and sandstone both contain principally silica. Because of different structures, their propagation velocities are 5 km/s and 3 km/s respectively.

7.6.2. Texture

The velocity in a rock may be related with the velocities in its various mineral components. BIRCH (1943) assumed that over any length L, reasonably long with respect to the mean crystal diameter, a pulse travels through each mineral of the rock in proportion to its amounts by volume in the rock. Thus he obtained for the travel time

$$T = \frac{L}{V} = \frac{Lx_1}{V_1} + \frac{Lx_2}{V_2} + \ldots\ldots\ldots + \frac{Lx_i}{V_i} \qquad (7.31)$$

where

T = travel time
L = path length
V = velocity in the rock
$V_1, V_2, \ldots\ldots V_i$ = velocities in minerals $1, 2, \ldots\ldots i$ and
$x_1, x_2, \ldots\ldots x_i$ = proportion by volume of minerals $1, 2, \ldots\ldots i$.

The velocity in the rock as a function of the velocities in its various components is expressed by:

$$V = \frac{1}{\sum \dfrac{x_i}{V_i}} \qquad (7.32)$$

This equation takes into account only the composition of the rock. It gives satisfactory results at high pressure, perhaps because the acoustical properties of the given boundaries are more uniform at high pressure, but this equation does not take into account such factors as grain size, or preferred orientation of crystals which may be of considerable importance.

The velocity of waves is influenced by the size of the grains constituting the rock. The velocity is greater as a rule in fine-grained rocks than in coarse-grained ones.

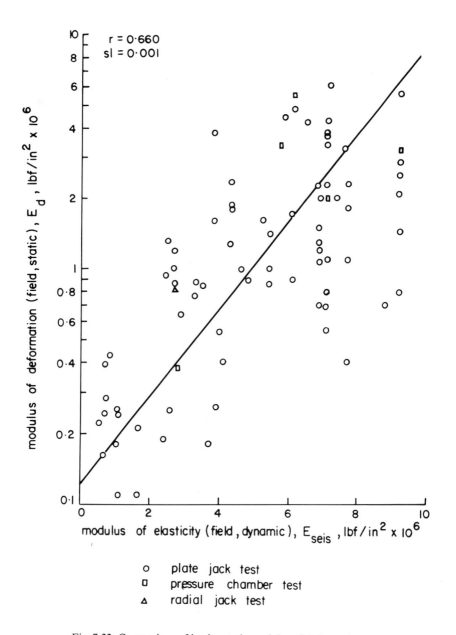

Fig. 7-22: Comparison of in situ static modulus of deformation
and dynamic modulus of elasticity
(after Coon, 1968).

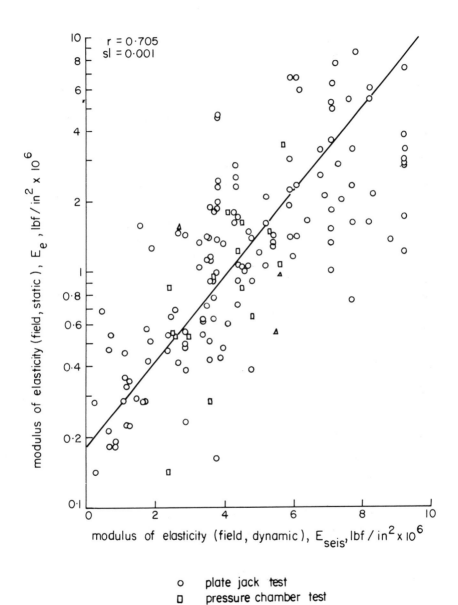

Fig. 7-23: Comparison of in situ static modulus of elasticity
and dynamic modulus of elasticity
(after Coon, 1968).

TABLE 29

Average velocities of propagation of longitudinal waves for some typical rocks

(after RINEHART, FORTIN and BURGIN, 1961)

Compact rocks		Less compact rocks		Unconsolidated rocks	
Dunite	7 km/s	Limestone	4 km/s	Alluvium	1 km/s
Diabase	6.5 km/s	Slate and shale	4 km/s	Loam	1 km/s
Gabbro	6.5 km/s	Sandstone	3 km/s	Sand	1 km/s
Dolomite	5.5 km/s			Loess	0.5 km/s
Granite	5 km/s				

TABLE 30

Dynamic properties of some rocks and minerals

(after RZHEVSKY and NOVIK 1971)

Rock or mineral	Apparent specific gravity, kg/m^3	Velocity of propagation of longitudinal wave V_p, m/s	Specific wave impedance $z \times 10^{-5}$, kg/s m^2
Siltstone	2,600	1,610	42.0
Basalt	2,860	5,400	154.0
Clay	1,500–2,200	1,800–2,400	27.0–52.7
Gabbro	2,900	6,250 (3,380)	181.0
Granite, medium-grained	2,780	4,350 (2,260)	126.0
Gypsum	2,200	1,500	33.0
Halite	2,000–2,200	4,500–5,500	90.0–128.0
Diorite	2,800	4,580	128.0
Limestone	2,300–3,000	3,200–5,500	73.0–165.0
Ferruginous quartzite	3,000	5,600	168.0
Quartz	2,650	5,225 (3,200)	138.0
Fused quartz	2,650	5,570 (3,515)	145.0
Marble	2,880	4,950	142.5
Sand	1,400–2,000	300–1,300	4.2–26.0
Sandstone	2,100–2,900	2,000–3,600 (2,100)	42.0–100.8
Peridotite	2,800	7,000	196.0
Hornfels	2,800	3,640	102.0
Syenite	2,710	4,950	134.0
Skarn	2,800–3,200	6,100	171.0–195.0
Shale	2,510–2,720	2,250	56.5–61.5
Coal	1,200–1,500	1,100–2,800	13.0–36.0

Note: (1) Specific wave impedance is the product of rock density and the velocity of an elastic wave in it.

      (2) Figures in parentheses indicate the velocity of transverse wave, m/s.

TABLE 31
Dynamic wave velocities and moduli of Kolar rocks
(after Ramana and Venkatanarayana, 1973)

Sample No.	Rock type	Density (g/cm³)	Porosity (%)	P-Velocity (km/sec)	S-Velocity (km/sec)	Young's (10¹¹ dyn/cm²)	Rigidity (10¹¹ dyn/cm²)	Bulk (10¹¹ dyn/cm²)	Poisson's ratio
4	Hornblende schist	2.990	9	6.09	3.72	09.95	4.14	5.59	0.20
5	Granulite	3.053	5	6.31	3.39	09.10	3.51	7.49	0.30
6	Hornblende schist	3.052		6.34	3.98	11.36	4.83	5.84	0.18
9	Hornblende schist	2.737	4	6.30	3.92	10.00	4.21	5.27	0.19
13	Hornblende schist	3.011		6.69	3.67	10.42	4.05	8.09	0.29
16	Hornblende schist	2.961	8	6.73	4.00	11.63	4.74	7.11	0.23
17	Dolerite	3.106		5.22	3.34	07.99	3.47	3.86	0.15
20	Dolerite	3.041		6.33	2.78	06.25	2.35	9.06	0.33
26	Uralite diabase	3.162	3	6.13	3.13	08.20	3.10	7.76	0.32
29	Hornblendite	3.247	10	5.63	3.63	09.79	4.28	4.60	0.14
30	Dolerite	3.000		6.37	3.44	09.19	3.55	7.45	0.29
31	Hornblende granulite	3.042	6	6.70	3.59	10.18	3.92	8.44	0.30
32	Hornblende schist	3.198	7	5.84	3.53	09.51	3.99	5.61	0.21
33	Hornblende schist	3.031	6	6.36	3.75	10.51	4.26	6.59	0.23
34	Hornblendite	3.472	3	6.39	3.29	09.92	3.76	9.18	0.32
38	Dolerite	3.136	4	6.48	3.73	10.93	4.36	7.37	0.25
42	Vein quartz	2.796	5	5.21	2.84	05.80	2.26	4.59	0.29
43	Hornblende granulite	3.084	6	6.11	3.71	10.25	4.24	5.87	0.21
51	Uralite diabase	3.031	3	6.01	2.66	05.91	2.14	8.10	0.38
57	Hornblende schist	3.011		5.75	3.70	09.45	4.12	4.47	0.15
61	Uralite basalt	3.062	9	6.58	3.66	10.47	4.10	8.80	0.28
65	Uralite basalt	2.672	7	5.01	3.16	06.24	2.67	3.16	0.17
66	Dolerite	3.111	8	5.59	3.30	08.35	3.39	5.21	0.23
68	Granulite	3.106	7	6.15	3.38	09.11	3.55	7.03	0.28
69	Granulite	3.356		5.42	3.15	08.29	3.33	5.43	0.25
73	Uralite diabase	3.000		6.65	3.71	10.52	4.13	7.78	0.27
75	Dolerite	3.011	7	5.44	3.50	08.46	3.69	4.01	0.15
87	Uralite diabase	3.057		6.10	3.84	10.15	4.51	5.38	0.13
88	Tremolite schist	3.011	6	6.32	3.46	09.27	3.60	7.23	0.29

RAMANA and VENKATANARAYANA (1973) studied the effect of mineral constituents on longitudinal wave velocity for Kolar rocks. Fig. 7-24 shows the plot of longitudinal wave velocity against hornblende content. These results indicate an increase in velocity value with increasing hornblende percentage in these rocks, and likewise in these rocks the plot of velocity against the quartz content shows a decrease in velocity with increasing quartz content (not shown here).

HUGHES and CROSS (1951) found that the more the rock is quartzitic, the lower is its POISSON's ratio.

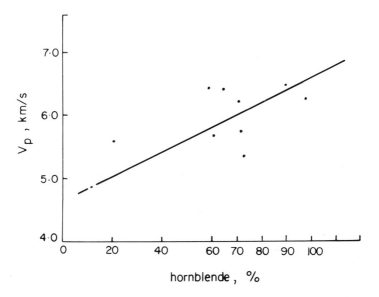

Fig. 7-24: Velocity change due to mineral content
(after RAMANA and VENKATANARAYANA, 1973).

7.6.3. Density

The principal factor which determines the velocity of longitudinal waves is the density and the mean atomic weight of the material tested, velocity being a linear function of density for materials having a common atomic weight (BIRCH, 1960, 1961).

The results of YOUASH (1970) are given in Figs. 7-25 to 7-28. Fig. 7-25 is a plot of rock density versus modulus of elasticity computed by ultrasonic pulse method. As the rock density increases, modulus of elasticity also increases.

Fig. 7-26 is a plot of rock density versus Poisson's ratio computed by ultra-sonic pulse method. While the scatter is high, a rock with high density has high Poisson's ratio.

Fig. 7-27 is a plot of velocity of longitudinal waves computed by pulse method versus rock density. The relationship is reasonably linear.

Fig. 7-28 is a plot of rock density versus velocity of transverse waves computed by pulse method. It is seen, in general, that the velocity of transverse waves increases as the density of rock increases. Although some of the points in Fig. 7-28 are scattered and deviate from the general trend, it seems that the relationship is linear.

The results of Bur and Hjelmstad (1970) for simulated lunar rocks are given in Fig. 7-29. The densities of these rocks ranged from 0.47 to 3.32 g/cm³.

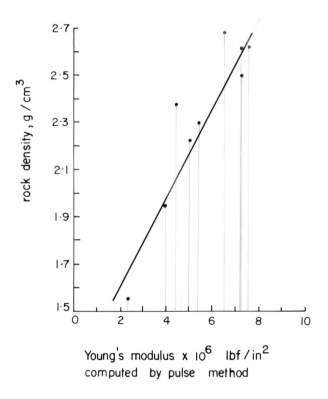

Fig. 7-25: Plot of rock density versus Young's modulus computed by pulse method (after Youash, 1970).

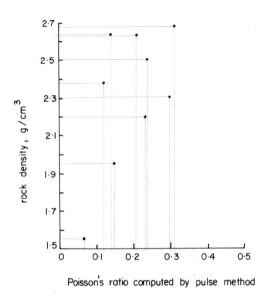

Fig. 7-26: Plot of rock density versus POISSON's ratio computed by pulse method
(after YOUASH, 1970).

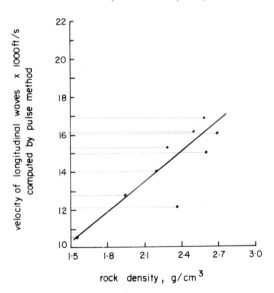

Fig. 7-27: Plot of velocity of longitudinal waves
computed by pulse method versus rock density
(after YOUASH, 1970).

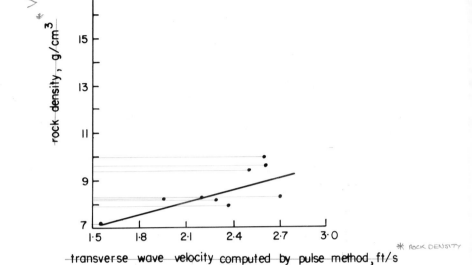

Fig. 7-28: Plot of density versus velocity of transverse waves computed by pulse method
(after YOUASH, 1970).

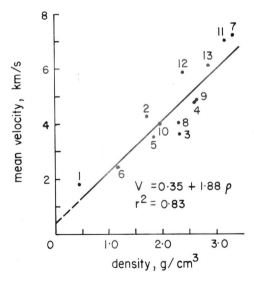

Fig. 7-29: Mean longitudinal pulse velocity versus density.
1. Pumice; 2. Vesicular basalt no. 3; 3. Rhyolite; 4. Serpentinite; 5. Dacite; 6. Semiwelded
tuff; 7. Dunite; 8. Altered rhyolite; 9. Granodiorite; 10. Vesicular basalt no. 1; 11. Duluth
gabbro; 12. Obsidian; 13. Tholeiitic basalt; 14. Vesicular basalt no. 2.
(after BUR and HJELMSTAD, 1970).

The results of RZHEVSKY and NOVIC (1971) for limestones and clays are given in Fig. 7-30(a) and (b). The relationships are curvilinear.

The results of RAMANA and VENKATANARAYANA (1973) on Kolar rocks (Fig. 7-31) also indicate that longitudinal velocity increases with density although there is large scatter in the results.

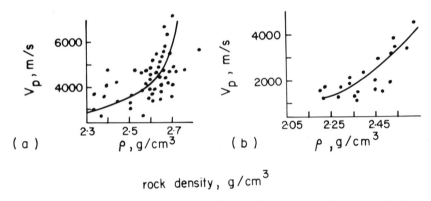

rock density, g/cm³

Fig. 7-30: Plots of density versus velocity of longitudinal waves (a) limestones (b) clays (after RZHEVSKY and NOVIK, 1971).

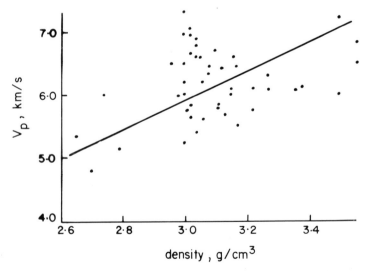

Fig. 7-31: Velocity – density relation in Kolar rocks (after RAMANA and VENKATANARAYANA, 1973).

7.6.4. Porosity

In general, the velocity of the wave decreases as the porosity increases. WYLLIE, GREGORY and GARDNER (1958) determined that two kinds of porosity must be considered: intergranular porosity, and intermediate porosity which is the result of natural fracturing or solution of the primary rock structure by ground waters. Intermediate porosity differs from intergranular porosity in that the vughs formed by solution are much larger and less uniformly distributed than pores in homogeneous rocks such as sandstone.

Porosity, in most cases, refers to intergranular porosity and its influence on propagation velocity is much greater than that of intermediate porosity. That the intermediate porosity of a dolomite does not have much influence upon the propagation velocity of stress waves is shown in Fig. 7-32.

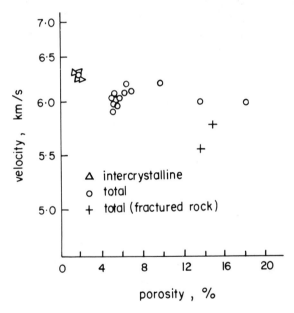

Fig. 7-32: Longitudinal wave velocity (km/s) versus total porosity for specimen of dolomite having the same intergranular porosity. Except the fractured specimen velocity does not vary with vugular porosity
(after WYLLIE, GREGORY and GARDNER, 1958).

Propagation velocity decreases when porosity of the rocks increases. Fig. 7-33 shows experimental values of propagation velocity in dry and saturated sandstone cores for various porosites. Although the kinds of sandstone were very different (the type of material cementing the grains was calcite, silica or clay), there is a relationship between velocity and porosity. Porosity variation considerably outweighs the importance of cementing materials.

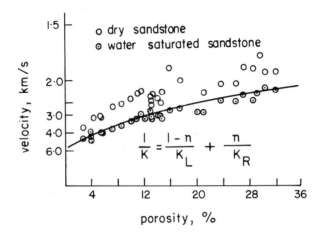

Fig. 7-33: Longitudinal wave velocity (km/s) versus porosity n for sandstone cores. The curve is deduced from the indicated relation. K, K_L, K_R are the respective bulk moduli of sandstone, water and silica. The bulk modulus is related to the velocity by theory of elasticity
(after WYLLIE, GREGORY and GARDNER, 1956).

Velocities were plotted against the porosity of cores obtained from oil wells (WYLLIE, GREGORY and GARDNER, 1956) and it was found that most points tended to fall in a narrow elongated area, associating decreased velocity with increased porosity. Fig. 7-34 shows the average curve deduced from these data and Fig. 7-35 compares the values of the velocities and porosities obtained from measurements in a specific oil well and those deduced from the curve of Fig. 7-34. A marked degree of correlation appears. The experimental data of Fig. 7-34 were established from different kinds of material, sandstone, and limestone, coming from various depths (2,000 to 10,000 ft) (610 to 3,050 m) and from many basins.

Several theories have been proposed to relate propagation velocity to the porosity. A time average relation may give good results (WYLLIE, GREGORY and GARDNER, 1956). On the other hand, GEERTSMA (1960) believes that the agreement between a time average relation and experimental results is purely fortuitous. The various theories are reviewed by PATERSON (1956) who classifies them according to the degree of coupling between the pore filler and the framework of the porous medium. GEERTSMA (1960) discusses BIOT's theory from which a relation between velocity in porous media and

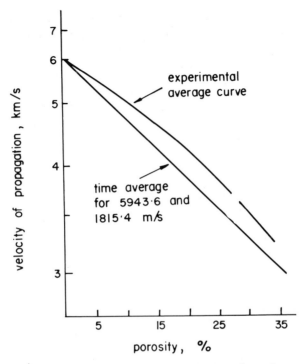

Fig. 7-34: Longitudinal wave velocity (km/s) versus porosity: Experimental average curve
and curve deduced from a time average formula
(after WYLLIE, GREGORY and GARDNER, 1956).

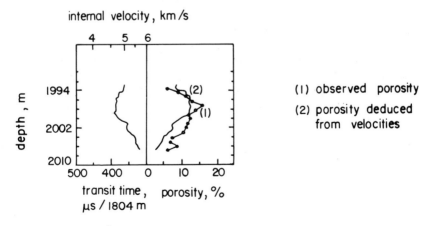

Fig. 7-35: Comparison between measured porosity and porosity deduced from the value of
measured velocity and the experimental average curve of Fig. 7-34
(after WYLLIE, GREGORY and GARDNER, 1956).

its porosoity can be deduced and which is applicable when the deformation properties of the material are known. From data on sandstone which give these deformation properties, GEERTSMA has computed propagation velocity as a function of porosity and has compared it with the values of propagation velocity obtained experimentally (Figs. 7-36 and 7-37). The application of the theory to limestones is more complicated. The problem with lime-

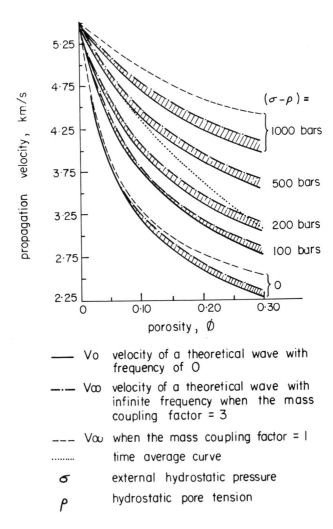

— Vo   velocity of a theoretical wave with frequency of O

—·— V∞   velocity of a theoretical wave with infinite frequency when the mass coupling factor = 3

——— Vαυ   when the mass coupling factor = I

....   time average curve

σ   external hydrostatic pressure

P   hydrostatic pore tension

Fig. 7-36: Longitudinal wave velocity (km/s) as function of porosity in sandstone for different value of σ – ϱ (after GEERTSMA, 1960).

stones is that there are so many different origins of limestone porosity
that a distinction must be made between those with shellshaped pores and
those with spherical pores. The influence of porosity on velocity is much
smaller for limestones with shellshaped pores.

Generally speaking, velocity decreases when intergranular porosity increases. Intermediate porosity has little influence on velocity. A broad
relationship can be obtained between propagation velocity and porosity
but more accurate results are obtained if the nature of the framework and
the shape of the pores of the rock are taken into account. However, the
deformation properties of porous materials must be known in order to
apply these theories relating porosity and velocity. Since the deformation
properties can only be determined experimentally, the usefulness of the
theories is limited.

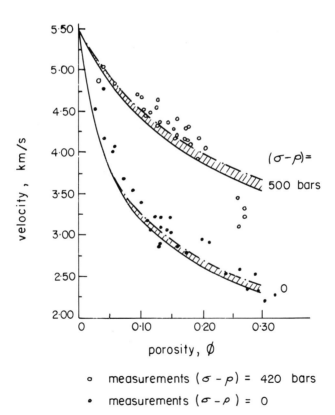

Fig. 7-37: Longitudinal wave velocity (km/s) versus porosity. Comparison between results
given by BIOT's theory and some experimental results for sandstone
(after GEERTSMA, 1960).

The results of YOUASH (1970) are given in Figs. 7-38 and 7-39. Fig. 7-38 is a plot of rock porosity versus YOUNG's modulus computed by the pulse method. While the correlation is poor, the general trend of decreasing YOUNG's modulus with increasing porosity is evident. Fig. 7-39 is a plot of velocity of longitudinal waves computed by resonance method versus rock porosity. Although the data are scattered, the velocity of longitudinal waves decreases with an increase in porosity.

Fig. 7-40 gives the results of RZHVESKY and NOVIK (1971). They found the following relationship between the velocity of longitudinal waves, V_p and porosity, n in percent:

$$V_p = 5430 - 107\ n \text{ m/s} \tag{7.33}$$

The results (Fig. 7-41) of RAMANA and VENKATANARAYANA (1973) also show that the velocity decreases with increasing porosity.

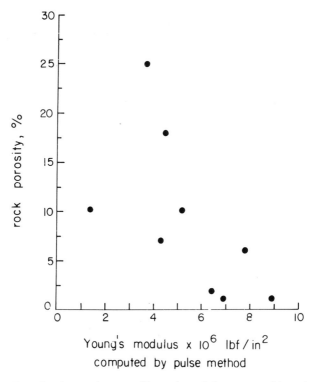

Fig. 7-38: Plot of rock porosity versus YOUNG's modulus computed by pulse method (after YOUASH, 1970).

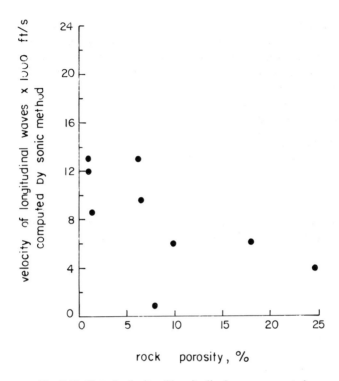

Fig. 7-39: Plot of velocity of longitudinal waves computed
by sonic method versus rock porosity
(after YOUASH, 1970).

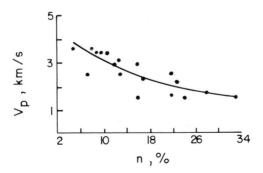

Fig. 7-40: Dependence of the velocity of longitudinal waves versus porosity of rocks
(after RZHEVSKY and NOVIK, 1971).

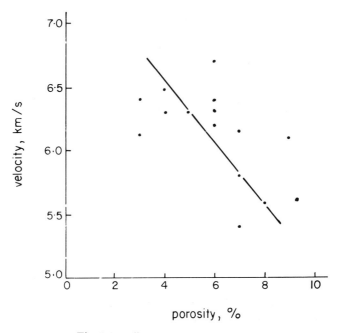

Fig. 7-41: Effect of porosity on wave velocity
(after RAMANA and VENKATANARAYANA, 1973).

7.6.5. Anisotropy

In layered rocks the velocities of elastic waves differ along and across the layers, and the velocity parallel to the layers $V_{\parallel}$ is always greater than the velocity perpendicular to the layers $V_{\perp}$. When the rock consists of layers of two different types, then in the simplest case (in which the two layers have the same specific acoustic impedance*) the velocity of waves can be calculated from the total time taken to travel through the layers:

$$\frac{l}{V_{\perp}} = \frac{l_1}{V_1} \cdot \frac{l_2}{V_2} \qquad (7.34a)$$

where l_1 and l_2 are the thicknesses of the layers with sound velocities V_1 and V_2, respectively ($l = l_1 + l_2$)

---

* Specific acoustic impedance is the product of rock density and the velocity of an elastic wave in it.

The sound velocity along the layers is

$$V_{||} = V_1 s_1 + V_2 s_2 \qquad (7.34b)$$

where s is their cross-sectional area.

When the specific wave impedances of the layers cannot be taken as equal, then the equations for the velocities of elastic waves along and across the layering becomes

$$V_{||} = \sqrt{\frac{\sum V_i^2 \varrho_i v_i}{\sum \varrho_i v_i}} \qquad (7.35)$$

$$\frac{1}{V_\perp} = \sqrt{\sum \frac{v_i}{V_i^2 \varrho} \sum \varrho_i v_i} \qquad (7.36)$$

where ϱ_i is the density of the rock constituting the layer, for which the average volume is v_i and the elastic velocity V_i.

These last formulas are not wholly valid since they are derived without consideration of the effect of shear.

The coefficients of anisotropy* for various rocks are given in Tables 32 and 33. They vary between 1.07 and 1.4. Increase of pressure on rock reduces its anisotropy. When a cleat is present, the velocities are measured in three directions and results for some blocks of coal are given in Table 34. BIRCH (1961) made tests on igneous rocks and prepared oriented cores in three orientations at right angles to each other. For granite, at one atmosphere of pressure, the velocity of longitudinal waves varies as much as 10%. This difference does not exceed 2 to 3% at 10 kilobars pressure. Oriented cores for Yule marble, Lyons sandstone, and Green River shale were prepared by YOUASH (1970). In Yule marble, cores were taken in mutually orthogonal directions X, Y and Z, X being at right angles to the foliation and Y and Z being along foliation. Cores of Lyons sandstone and Green River shale were also prepared in three orthogonal directions, X, Y, and Z, X cores were oriented at right angles to bedding Y, and Z cores were oriented parallel to bedding. The average physical properties for all rocks tested with their indicated orientations are shown in Table 35. In Lyons

---

* The coefficient of anisotropy may be defined as the ratio of the velocity along the layers to the velocity perpendicular to the layers.

TABLE 32
Anisotropy coefficients for various rocks

Rock	Anisotropy coefficient	Reference
Austin chalk	1.17	TOCHER (1957)
Homogeneous anhydrite	1.16	DUNOYER de SEGONZAC and LAHERRERE (1959)
Anhydrite with intercalated limestones	1.12 to 1.14	DUNOYER de SEGONZAC and LAHERRERE (1959)
Limestone	1.08 to 1.10	DUNOYER de SEGONZAC and LAHERRERE (1959)
Arbuke limestone	1.30	UHRIG and VON MELLE (1955)
Salt	no anisotropy	DUNOYER de SEGONZAC and LAHERRERE(1959)
Sandstone	no anisotropy	DUNOYER de SEGONZAC and LAHERRERE (1959)
Eagle Ford shale	1.33	UHRIG and VON MELLE (1955)
Pierre shale (Limon, Colo.)	1.18	UHRIG and VON MELLE (1955)
Pierre shale (Last Chance, Colo.)	1.14	UHRIG and VON MELLE (1955)
Cambridge slate	1.07	TOCHER (1957)
Lorraine shale	1.40	TOCHER (1957)
Gneiss, Hell Gate, N.Y.	1.20	BIRCH (1960)
Micaschist, Woodsville, Vt.	1.36	BIRCH (1960)
Granodiorite gneiss, Bethlehem, N. H.	1.33	BIRCH (1960)
Gneiss, Pelham, Mass.	1.27	BIRCH (1960)

sandstone, the average YOUNG's modulus for cores oriented parallel to bedding is greater than those for cores oriented at right angles to bedding by 14%; average POISSON's ratio is greater by 51%; average velocity of longitudinal waves is greater by 8%; and average velocity of transverse waves is greater by 5%. In Green River shale, for cores oriented parallel to bedding, average YOUNG's modulus is greater than those for cores oriented at right angles to bedding by 7%; average POISSON's ratio is greater by 23%, average velocity of longitudinal waves is greater by 8%; average

TABLE 33
Velocity of longitudinal sound waves in layered rocks
(after RZHEVSKY and NOVIK, 1971)

Rock	Velocity of the longitidinal wave, m/s		
	parallel to stratification V_{II}	perpendicular to stratification $V_\perp$	Coefficient of anisotropy $V_{II}/V_\perp$
Limestone	5300	5100	1.04
Sandstone	3800	3200	1.19
Marl	4300	3900	1.10
Serpentinite	4600	3800	1.18

TABLE 34
Anisotropy coefficients for coal
(after TERRY and SEABORNE, 1957)

	Mean	Standard Error of the Mean
Velocity parallel to bedding plane perpendicular to major cleat	2.03	0.03
Velocity parallel to bedding plane parallel to major cleat	2.01	0.04
Velocity perpendicular to bedding plane	1.76	0.03

velocity of transverse waves is greater by 1%. In Green River shale, for cores oriented parallel to bedding, average YOUNG's modulus is greater than those for cores oriented at right angles to bedding by 97%; average POISSON's ratio is greater by 14%; average velocity of longitudinal waves is greater by 41%; average velocity of transverse waves is greater by 34%. In Green River shale, cores oriented at right angles to bedding in both lean and rich oil content, the average YOUNG's modulus for lean shale is greater than for rich shale by 66%, average POISSON's ratio is less by 21%; average velocity of longitudinal waves is greater by 21%; average velocity of transverse waves is greater by 28%.

From this study it may be concluded that, in general, the dynamic constants are lower at right angles to layering than along layering.

Tables 36(a) and (b) present the data of RAMANA and VENKATANARAYANA (1973) on P-wave anisotropy. Table 36(a) represents the results obtained from a study of rectangular bar-shaped samples. Here the longitudinal and shear velocities obtained from measurements taken on mutually perpen-

TABLE 35
Values of dynamic physical properties for three rocks at different orientation of anisotropy computed by sonic method.
X cores are oriented at right angles to layering Y and Z cores are oriented parallel to layering.
(after YOUASH, 1970)

Rock Type	Kind of cores	YOUNG's Modulus x 10^6 lbf/ in^2	POISSON's Ratio	Velocity of longitudinal waves ft/s	Velocity of transverse waves ft/s
Yule Marble	X	6.645	0.170	14,400	8,800
	Y	7.136	0.315	16,737	8,646
	Z	7.027	0.289	15,930	8,675
Lyons Sandstone	X	5.001	0.076	12,586	8,514
	Y	5.701	0.115	13,557	8,941
Green River Shale					
Lean	X	5.807	0.217	14,472	8,704
Lean	Y	6.174	0.266	15,609	8,812
Rich	X	3.492	0.261	11,923	6,784
Rich	Y	6.882	0.297	16,874	9,064

dicular directions are presented besides the calculated percentage aniso-tropy. The anisotropy has been calculated as the difference of the velocity value between the maximum and minimum over that of the average velo-city in per cent. Table 36(b) presents similar results, but the information recorded is due to a study of two independent cores, taken out of a single rock but from perpendicular directions and wherever possible along and perpendicular to the direction of schistocity.

These results (Tables 36(a) and (b)) very clearly show that Kolar rocks are anisotropic in nature and this is understandable in view of the pre-ferred orientation observed in them due to the metamorphism and schisto-city. The velocity anisotropy extends from 2 to 40 % in them. Over all the results present a picture of S-wave anisotropy being of the same order as P-wave anisotropy, or higher than P-wave anisotropy in some cases. In general, the intrusives are less anisotropic in comparison to the schistose rocks of the region.

The anisotropic behaviour of the schistose rocks (Fig. 7-42) is tending to increase with increasing porosity in them. A similar observation was made by MANGHNANI and WOOLLARD (1968) for basaltic rocks. However, the change noticed in the schistose rocks of Kolar seems to be steeper and higher as compared to the basaltic rocks and this can be interpreted in terms of schistocity and preferential orientation in the fabric.

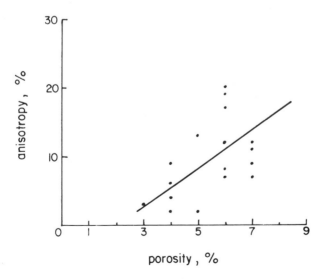

Fig. 7-42: Effect of porosity on anisotropy of schistose rocks, *KGF*
(after RAMANA and VENKATANARAYANA, 1973).

TABLE 36 (a)
Anisotropy studies on rectangular bars
(after RAMANA and VENKATANARAYANA, 1973)

Sample No.	Rock classification	P-velocity in perpendicular directions			Average P-velocity (km/s)	P-anisotropy (%)	S-velocity in perpendicular directions			Average S-velocity (km/s)	S-anisotropy (%)
		A (km/s)	B (km/s)	C (km/s)			A (km/s)	B (km/s)	C (km/s)		
13	Hornblende schist	6.24	6.58	7.08	6.69	12.6	3.57	4.17	3.29	3.67	24
17	Dolerite	5.43	5.01	5.28	5.22	08.0	3.97	3.27	2.78	3.34	36
26	Uralite diabase	5.41	6.23	6.75	6.13	18.5	3.63	2.54	3.23	3.13	35
30	Dolerite	6.40	6.31	6.41	6.38	01.5	3.65	3.60	3.08	3.44	17
36	Granulite (fractured)	4.02	5.16	5.39	4.86	28.2	3.31	3.26	3.79	3.45	15
37	Uralite diabase	6.22	6.18	5.64	6.02	10.1					
38	Dolerite	6.51	6.53	6.39	6.48	02.2	3.57	3.94	3.68	3.73	10
40	Hornblende schist	5.57	5.63	4.31	5.17	25.7	3.13	3.24	4.13	3.50	29
61	Uralite basalt	6.63	6.70	6.40	6.58	04.5	3.46	3.16	4.35	3.66	33
65	Uralite basalt	5.35	4.80	4.88	5.01	11.1	3.00	3.31	—	3.16	10
66	Dolerite	5.49	5.68	5.59	5.59	03.3	3.52	3.03	3.34	3.30	15
76	Hornblende schist	6.95	6.78	7.02	6.91	03.5					
88	Tremolite schist	6.36	5.67	6.93	6.32	19.8	3.49	3.17	3.73	3.46	16

TABLE 36 (b)
Anisotropy of wave velocities in Kolar rocks
(after RAMANA and VENKATANARAYANA, 1973)

Sample No.	Rock particulars	V_{p_1} (km/s)	V_{p_2} (km/s)	P-anisotropy (%)	V_{s_1} (km/s)	V_{s_2} (km/s)	S-anisotropy (%)
4	Hornblende biotite schist	6.213	5.971	04	3.950	3.490	12
5	Granulite	5.895	6.719	13			
6	Hornblende schist	5.770	7.016	20			
7	Amphibolite	6.455	6.350	02			
9	Hornblende schist	6.099	6.479	06			
20	Dolerite (fractured)	6.589	6.076	08			
32	Hornblende schist	6.090	5.581	09	3.800	3.260	15
33	Felsite	5.833	6.888	17	3.480	4.020	14
34	Hornblendite	6.476	6.299	03	3.320	3.270	02
95	Actinolite schist	5.986	7.223	19	4.110	4.910	18
36	Hornblende granulite (ore body)	6.081	6.829	12	3.080	3.510	13
38	Dolerite	6.423	6.194	04	3.500	3.140	11
42	Vein quartz	5.269	5.147	02	2.880	2.810	02
43	Hornblende granulite	6.336	5.885	07	3.850	3.580	07
68	Granulite (lode rock)	5.790	6.509	12	2.920	3.840	28
69	Granulite	5.546	5.291	05			
73	Uralite diabase	7.315	5.980	20	4.110	3.320	21
75	Dolerite	5.643	5.246	07	3.960	3.040	26
80	Hornblende granulite	7.137	6.719	06			
87	Uralite diabase (fractured)	5.611	6.586	16	3.460	4.230	20

THILL, WILLARD and BUR (1969) compared longitudinal wave velocity with structural subfabrics for Yule marble, Newberry Crater pumice and Salisbury granite. Velocities were measured omnidirectionally in spherical specimens by an ultrasonic pulse technique. Microstructural subfabrics were determined from oriented thin sections of the specimens. Two structural subfabrics, namely, crystallographic orientation of constituent anisotropic minerals and shape and orientation of pores or cracks were selected for this study. In the Yule marble, the longitudinal velocity symmetry appears to be axial and corresponds with the preferred orientation of calcite optic axes.

In the Newberry Crater pumice, the velocity symmetry is orthorhombic and corresponds with the preferred orientation of elongate vesicles. The magnitude of velocity anisotropy appears to be dependent upon both the degree of elongation of individual vesicles and their preferred orientation. In Salisbury granite, the velocity symmetry is also orthorhombic and corresponds with the preferred orientation of microcracks in quartz.

The longitudinal velocity anisotropy of these rocks is given in Table 37.

TABLE 37
Longitudinal velocity anisotropy of rock spheres
(after THILL, WILLARD and BUR, 1969)

Rock	V_{max} km/s	V_{min} km/s	V_{aver} km/s	Anisotropy %
Yule marble	6.27	5.24	5.86	17.6
Pumice	3.53	1.90	2.59	62.9
Salisbury granite	5.92	5.20	5.60	12.9

ATTEWELL (1970) reported triaxial anisotropy of dynamic properties of Penrhyn slate from North Wales. Velocities of ultrasonic P- and S-waves were measured on orthogonally cut specimens. The principal axes of wave velocity and dynamic elastic moduli are concordant with the axes of the original principal tectonic stresses as determined from fabric symmetry analysis. The triaxial velocity anisotropy corresponds with the clay mineral orientation in the slate.

BUR and HJELMSTAD (1970) determined the elastic symmetry of 14 igneous and metamorphic rocks. The studies were made by means of quasilongitudinal pulse velocity on spherical specimens. They found that the variations in the elastic properties can be closely approximated by the symmetry of orthorhombic and transversely isotropic systems. The range of velocity anisotropy found is roughly the same as for single crystals. The elastic symmetries agree for most of the rocks, within experimental error. In a few rocks, a significant difference was found and they suggested a separate major contributing cause (subfabric) of the variations in each of these properties. Rocks in which the elastic anisotropy is notably associated with fracture systems have lower than normal mean velocities.

THILL, BUR and STECKLEY (1973) investigated velocity anisotropy in dry- and water-saturated spheres of Barre granite and Tennessee marble by the ultrasonic pulse method. Barre granite was found to be strongly anisotropic in V_p with orthorhombic symmetry at both dry and saturated conditions. Tennessee marble was found to be only weakly anisotropic in V_p and changed in symmetry from axial at dry, to orthorhombic at saturated conditions.

7.6.6. Stress

A specimen may be submitted to a hydrostatic pressure, a triaxial pressure or a uniaxial pressure and the velocity measured in direction parallel or perpendicular to the higher stress. Laboratory tests indicate that the values of propagation velocity in rock under various methods of compression are not significantly different except when the specimen is under uniaxial pressure and the velocity measured in a direction perpendicular to the stress. Fig. 7-43 shows results of TOCHER (1957) for granite and Fig. 7-44 similar results of WYLLIE, GREGORY and GARDNER (1958) for sandstone.

Both the compressional and shear sound velocities of rocks generally in- crease with increasing pressure (Figs. 7-45 and 7-46). The rapid increase at low pressures is due to a decrease in porosity (BIRCH, 1960; HUGHES and MAURETTE, 1956, 1957), closing of cracks and defects (BRACE, 1965; WALSH, 1965), and an increase in the mechanical contact between the grains (SHI- MOZURA, 1960); see Fig. 7-47. The velocity increase at higher pressure results from changes in the intrinsic properties of the rock, such as finite compression of the crystals. In certain specimens, the sound velocity is observed to decrease when the pressure exceeds a certain value (Fig. 7-48). The velocities measured with decreasing pressure are higher than those measured with increasing pressure (Figs. 7-45, 7-46 and 7-49) possibly be- cause some cracks may remain closed as pressure is released.

Some results of BIRCH (1960) are given in Figs. 7-50 and 7-51.

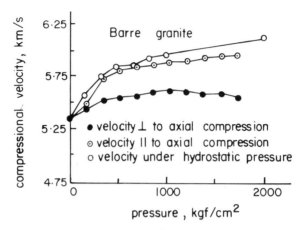

Fig. 7-43: Variation of longitudinal velocity (km/s)
with uniaxial and hydrostatic pressure (kgf/cm²)
(after TOCHER, 1957).

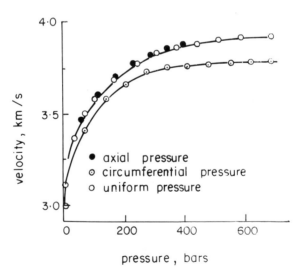

Fig. 7-44: Results of three different methods of compression on the same material.
Longitudinal velocity (km/s) versus pressure (bars)
(after WYLLIE, GREGORY and GARDNER, 1958).

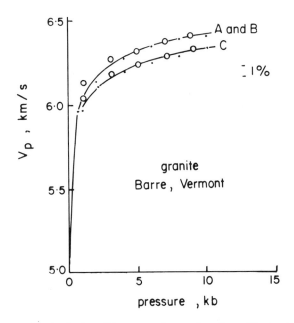

Fig. 7-45: Compressional velocity of Barre granite as a function of hydrostatic pressure. Dots indicate measurements with increasing pressure; circles, with decreasing pressure (after BIRCH, 1960).

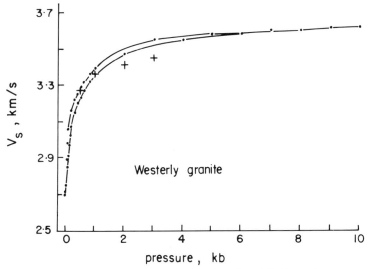

Fig. 7-46: Shear velocity of Westerly granite as a function of hydrostatic pressure. The lower curve was obtained with increasing pressure; the upper curve, with decreasing pressure. Crosses represent data from BIRCH and BANCROFT (1938) (after SIMMONS, 1964).

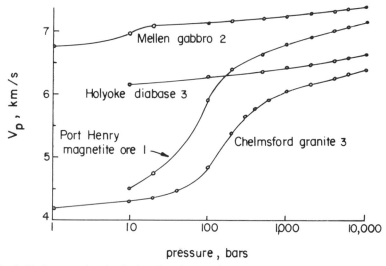

Fig. 7-47: Compressional velocity of various rocks as a function of hydrostatic pressure. V_p is plotted against log P here, emphasising the behaviour below 1 kbar (after BIRCH, 1960).

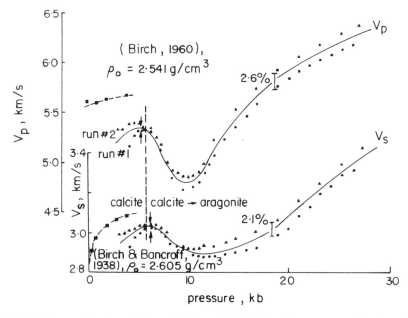

Fig. 7-48: Compressional and shear velocities of Solenhofen limestone as a function of hydrostatic pressure, showing the anomalous decrease of V_p and V_s above 5 to 6 kbar (after AHRENS and KATZ, 1963).

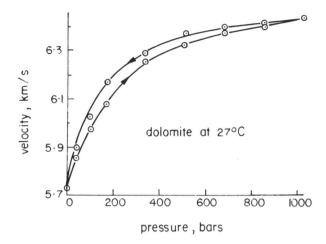

Fig. 7-49: Dilatational velocity (km/s) versus pressure (bars). Velocity measured when decreasing pressure is higher than velocity measured when increasing pressure (after HUGHES and JONES, 1951).

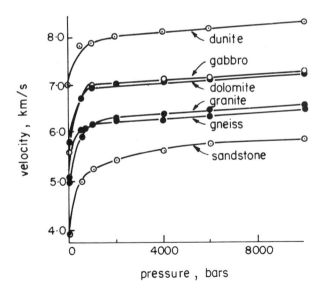

Fig. 7-50: Longitudinal wave velocity (km/s) versus pressure (bars) for rocks showing a large increase at low pressures. Data from BIRCH (1960).
(after RINEHART et al, 1961).

WYLLIE, GREGORY and GARDNER (1958) computed the propagation velocity from the values of E and v in Berea sandstone at different stress levels and compared these values to the measured value at the same stress level. The measured values are very different from the computed values (Figs. 7-52 and 7-53). This result suggests that velocity at different stress levels cannot always be deduced from the stress-strain curves.

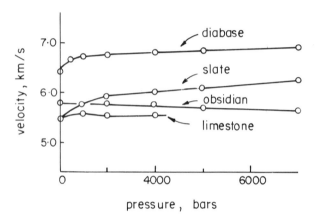

Fig. 7-51: Longitudinal wave velocity (km/s) versus pressure for rocks showing a small
increase at low pressures. Data from BIRCH (1960).
(after RINEHART et al, 1961).

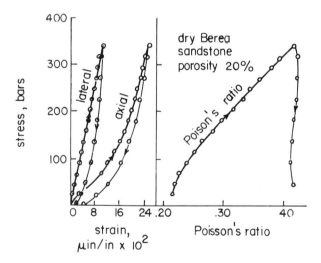

Fig. 7-52: Stress – strain curves and POISSON's ratio versus stress for a sandstone
(after WYLLIE, GREGORY and GARDNER, 1958).

The results of Rzhevsky and Novik (1971) are given in Fig. 7-54. The dependence of velocity of longitudinal waves on loading (hence compaction) is clearer in the case of porous and loose rocks, since pressure has a relatively greater compacting effect on them (Fig. 7-54). In sandy marls, with an initial porosity of 25 %, for example, the velocity of longitudinal waves at pressures up to 1,000 kgf/cm² increases by 50 to 60 %; in less

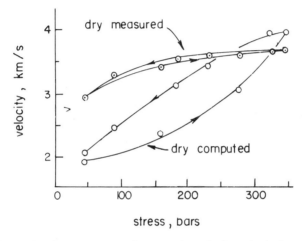

Fig. 7-53: Comparison between measured propagation velocity and velocity computed from the stress – strain curves of Fig. 7-52
(after Wyllie, Gregory and Gardner, 1958).

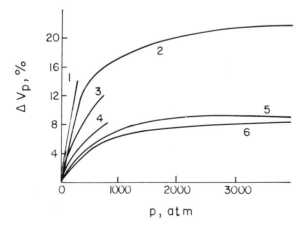

Fig. 7-54: Relative variation of the velocity of longitudinal waves at confining pressures:
1. sandstone 2. syenite 3. dolomite 4. schist 5. basalt 6. gabbro
(after Rzhevsky and Novik, 1971).

porous rocks, it increases only by 10 to 20% (with increase of pressure up to 4,000 atm, V_p rises by 5 to 7% in gabbro, and by 10 to 20% in granite).

A similar pattern is observed in the propagation of transverse waves, but the increase of their velocity with pressure takes place more slowly and up to a definite limit. With further increase of pressure the ratio V_p/V_s remains almost constant.

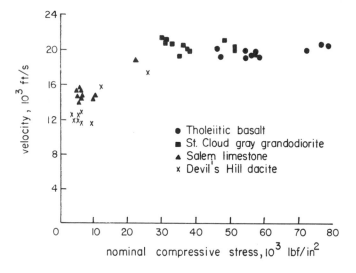

Fig. 7-55: Relationship between longitudinal velocity and compressive stress (after PODNIEKS, CHAMBERLAIN and THILL, 1968).

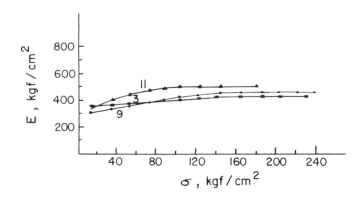

Fig. 7-56: Relation between modulus of elasticity and loading (after POPOVIC and CVETKOVIC, 1972).

Increase of pressure on rock reduces its anisotropy. The results of Podnieks, Chamberlain and Thill (1968) are given in Fig. 7-55. Uniaxial compression is parallel to the direction of propagation.

Popovic and Cvetkovic (1972) studied the effect of stress on dynamic
modulus of elasticity. Their results (Fig. 7-56) show that E increases with
load. By increasing the load up to a 60% compressive strength of the rock
sample the value of E rises, too. Beyond this value the sample breaks and
E becomes constant.

Ramana and Venkatanarayana (1973) tested Kolar rocks in uniaxial compression. They measured strain as well as compressional velocity with axial
stress. The results (Fig. 7-57) indicate an increase in wave velocity with
increasing stress reaching a steady value in certain cases almost near the
breaking point. The stress-velocity changes in these rocks indicate a velocity
change of as much as 6–25%.

In Fig. 7-58 velocity against strain data are plotted. It is difficult to attribute
any particular relation between the parameters velocity and strain.

The effect of reloading in both static and dynamic tests was also investigated
by them in some of the rocks. Fig. 7-59 shows the results of such investiga-

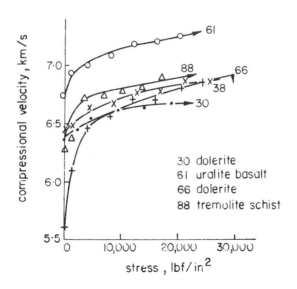

Fig. 7-57: Stress – velocity characteristics of Kolar rocks
(after Ramana and Venkatanarayana, 1973).

tions on a schist sample. The part of the stress-strain curve is indicated by
OABC where *OA*, *AB*, *BC* indicate the first forward loading, first unloading, and second forward loading cycles to which the sample has been subjected in a static test. The corresponding dynamic velocity changes of the
first and second forward cycles of loading are presented in the same figure
for comparison. These figures only indicate the effect of strain hardening

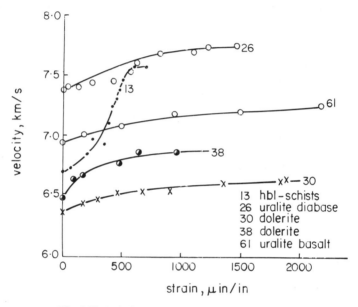

Fig. 7-58: Relation between static and dynamic results
(after RAMANA and VENKATANARAYANA, 1973).

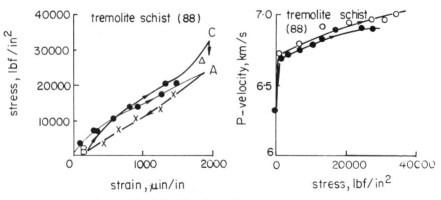

Fig. 7-59: The effect of repeated loading
(after RAMANA and VENKATANARAYANA, 1973).

as a result of which higher velocities are noticed in the second loading and a similar observation follows from other static data also wherein the hysteresis loop is obtained due to a complete cycle of loading.

BARON, HABIB and MORLIER (1964) observed changes in the structure of rocks such as Triel gypsum subjected to triaxial stresses which are high enough to make them ductile. These modifications had unexpected reductions on the velocity of the longitudinal and transverse waves in the material under atmospheric conditions.

On rocks such as potash where the ductile deformation does not create fissures or any other discontinuity, the velocity did not vary. The same applies to iron ore, where aggregates and cements flowed without fissuration in the ductile phase. Apart from these exceptions, all ductile rocks showed the following phenomenon:

1. A reduction by 50% in the velocity of longitudinal sound waves in the case of soft, constrained sandstones (fg. sandstone from the Vosges, sandy limestone from Rouffach).
2. A reduction of the same order of magnitude in a series of compact limestones:

St. Beat limestone V_p in the virgin state: 6,300 m/s
                        V_p for the constrained rock: 3,000 m/s
Buxy limestone V_p in the virgin state: 5,150 m/s
                        V_p for the constrained rock: 2,500 m/s

As the petrographic studies have shown, these variations which are a result from microfissurations only occur in the ductile phase of the deformation. In fact, harder limestones which remain elastic for stresses similar to those exerted on the two previous limestones, have the same velocity for the elastic waves before and after passing through the triaxial cell.

PROSKURYAKOV, LIVENSKII and KUZNETSOV (1975) noted that the longitudinal wave propagation velocity decreases as the stress rises (in uniaxial compression) for salt rocks which have high density and show creep deformation (Fig. 7-60). The relationship between the longitudinal elastic wave propagation velocity and the stress can be approximated by the relation

$$V_p = 4874 \, e^{-0.0018 \, \sigma} \tag{7.37}$$

where

V_p = longitudinal elastic wave propagation velocity, m/s and
σ = stress in the specimen under uniaxial compression at the moment of sound transmission, kgf/cm².

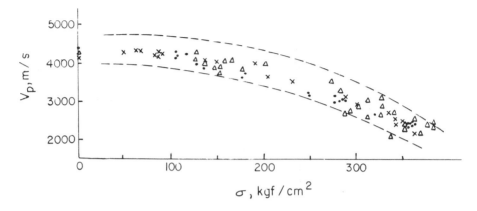

Fig. 7-60: Elastic wave propagation velocity V_p versus stress σ in rock salt.
Size of specimen (cm).
(X) $8 \times 8 \times 8$;　($\triangle$) $10 \times 10 \times 10$;　($\bullet$) $15 \times 15 \times 15$
(after PROSKURYAKOV, LIVENSKII and KUZNETSOV, 1975).

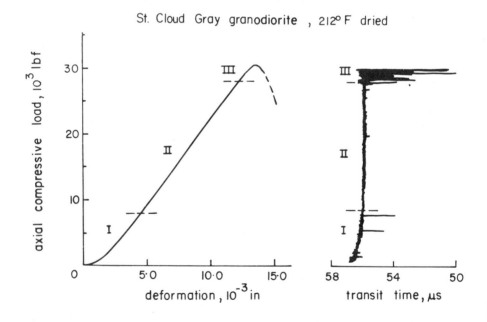

Fig. 7-61: Typical simultaneous recordings of mechanical and acoustic data
(after PODNIEKS, CHAMBERLAIN and THILL, 1968).

The correlation coefficient for this graph, calculated from the results of tests on 76 specimens, was $r = 0.94$, which is highly significant.

PODNIEKS, CHAMBERLAIN and THILL (1968) concurrently recorded load-deformation and load-transit time curves for St. Cloud Gray granodiorite in uniaxial compression. Their results given in Fig. 7-61 show distinct similarity in the relative positions of their linear and curved portions. Curvature in region I of the load-transit time curve is usually attributed to closure of microcracks and sometimes to the partial closure of pores in the rock but also includes slight end effects, if present. Over region II, transit time changes with stress are linear and usually slight. Beginning in region III, transit time commonly increases slightly (velocity decreases) with stress.

THILL (1972) obtained elastic wave velocity data concurrently with stress-strain characteristics of granodiorite and marble specimens during unconfined and conventional triaxial compression tests. The technique used in this investigation monitors longitudinal wave velocity in the directions of greatest σ_1 and least σ_3 principal stresses. The velocity changes in rock under stress can be attributed primarily to changes in crack porosity. When cracks existing in rock are closed by increased hydrostatic or directional stress, the velocities increase in response to the closure. When cracks have initial preferred orientation or are preferentially closed by the stress, velocity anisotropy may occur that corresponds to the directional changes in crack porosity and the associated changes in the effective elastic constants. Rock failure is indicated when the velocity begins to decrease. Characteristically, the decrease in velocity in the σ_3 direction occurs early in the stress-strain cycle and undergoes greater change than that in the σ_1 direction. In brittle rock, the velocity reduction often occurs well in advance of structural collapse or rupture (commonly at 50 to 90 % of the ultimate stress). The velocity anisotropy occurring during failure relates to preferred orientation in the crack distribution and lateral dilatancy in rock, since early forming cracks tend to be localised and extend in the direction of applied stress.

The velocity variation is found to be a more sensitive indicator of brittle fracturing than of ductile behaviour. Whereas an increase in confining pressure to 69 MPa (10,000 lbf/in²) has only slight effect in changing the fracture characteristics and associated acoustic properties of the granodiorite, the marble undergoes transition from brittle to ductile failure with different associated acoustic properties. The reduction in velocities becomes progressively less when the mode of failure in the marble changes from brittle fracturing at unconfined conditions to plastic flow at the confining pressure of 69 MPa (10,000 lbf/in²). The reduction in the velocities also is diminished and velocities are anomalously low during frictional sliding in the post-failure region of the stress-strain curve for the marble at the low confining pressures.

The results also show that at stresses below those required to close cracks and during the process of failure, the velocities cannot be expressed as simple functions of nominal stress or deformation. For example, the velocity is observed to both increase and decrease as stress increases or to change at varying rates as deformation increases, depending primarily upon the mechanism of failure in these rocks. Moreover, the velocity variations are found to be more sensitive indicator of brittle fracturing than of ductile behaviour. This relationship is evident in the results of marble, which undergoes a transition from brittle to ductile behaviour with increasing confining pressure.

RUMMEL (1974) measured velocities of P-waves in specimens of Tennessee marble, Ruhr sandstone and a Greek marble as a function of increasing axial compression throughout the complete stress-strain characteristic. The results show that the P-wave velocity is extremely sensitive to structural changes in the material. Within the elastic stress-strain region, the velocity increase is, except for the porous Ruhr sandstone, only within few percent. The velocity decreases markedly well before the maximum load bearing capacity of the rock is reached. The onset of this velocity decrease characterises the onset of microscopic failure which is also indicated by the onset of dilation. This effect is enhanced in the post-failure region, where the load bearing capacity decreases with increasing axial deformation.

KING (1968) reported measurements of compressional and shear wave velocities in three directions on 3 sandstones, namely, Boise, Bandera and Berea, subjected to hydrostatic confining pressure. Boise sandstone is almost isotropic in behaviour (Fig. 7-62) and the other two, Bandera and Berea, are transversely isotropic, with their axis of symmetry lying perpendicular to the bedding plane (Figs. 7-63 and 7-64). The degree of anisotropy was found to decrease appreciably as the hydrostatic confining pressure was increased. The degree of anisotropy is dependent on the presence of lenticular cracks or clay partings in the bedding planes of the two anisotropic sandstones.

Static and dynamic elastic moduli were calculated for the isotropic sandstone (Boise) from stress-strain data and velocities measured concurrently on the same sample as a function of changes in triaxial loading conditions. While the dynamic moduli are greater in magnitude than the static moduli, the ratio of the two is reduced by an increase in confining pressure (Fig. 7-65). The presence of randomly-oriented lenticular cracks in the rock, which close progressively as the confining pressure is raised can explain these differences.

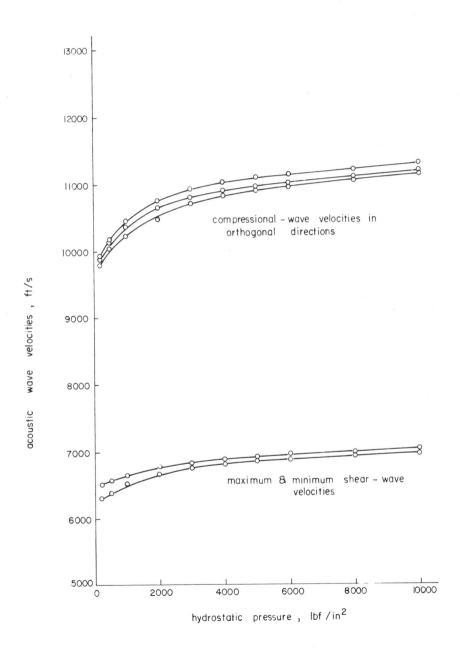

Fig. 7-62: Acoustic wave velocities; Boise sandstone; hydrostatic tests
(after KING, 1968).

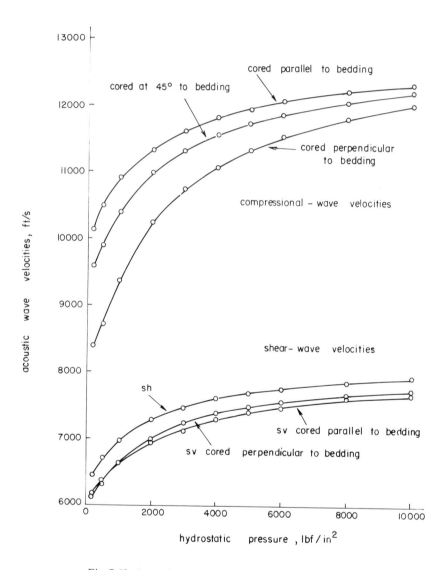

Fig. 7-63: Acoustic wave velocities; Bandera sandstone
(after KING, 1968).

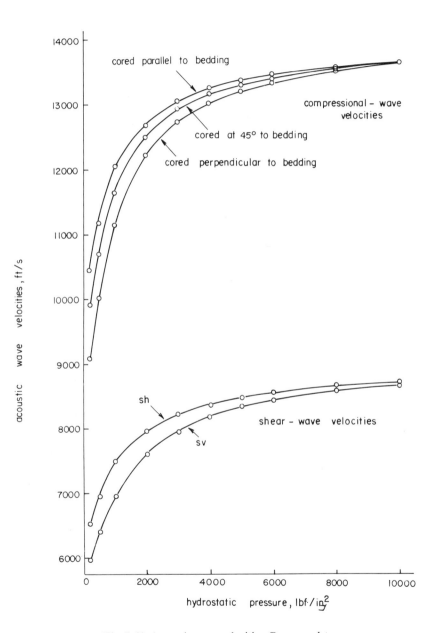

Fig. 7-64: Acoustic wave velocities; Berea sandstone
(after KING, 1968).

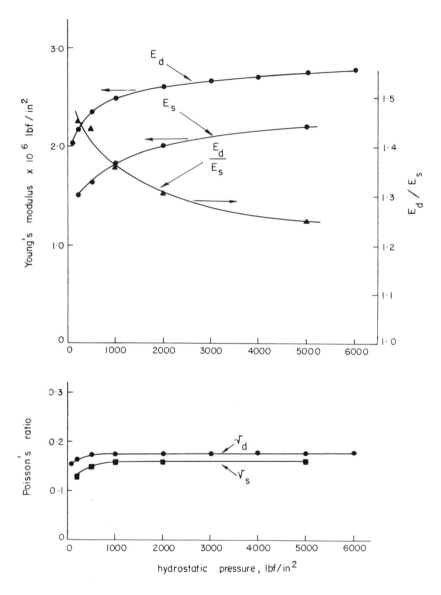

Fig. 7-65: Static and dynamic moduli; Boise sandstone; triaxial tests
(after KING, 1968).

Velocities for both compressional and shear waves in an Alpine gneiss were determined by WANG, LIN and WENK (1975) along its principal fabric directions under high-pressure conditions. Their results are shown in Fig. 7-66. The greatest increase in the velocity of compressional waves with pressure is in the c direction. Changes in velocities in the other directions are much smaller. As a result, the anisotropy in velocity decreases as pressure is raised; it stays fairly constant above 1 kbar. Significant increases also appear in the velocities of shear waves in all directions of propagation and polarisation. The effects of pressure on the ratio between compressional

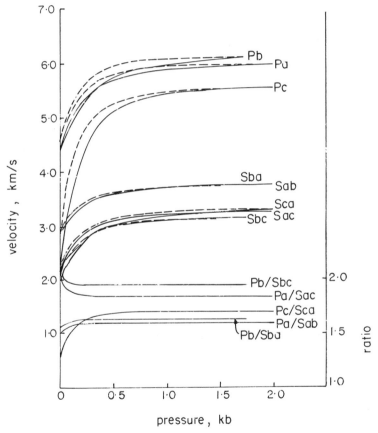

Fig. 7-66: Velocities of elastic waves as a function of pressure in an Alpine gneiss. The subscript c indicates a direction perpendicular to the fabric of the rock and thus perpendicular to the microcracks. The subscripts a and b indicate directions parallel to the fabric of the rock. The quantity P_a is the velocity of compressional waves propagating in the a direction; S_{ab} is the velocity of shear waves propagating in the a direction but polarised in the b direction, etc.; and P_a/S_{ab} is the ratio of the two velocities

(after WANG, LIN and WENK, 1975).

and shear velocities are complicated: along the direction perpendicular to the oriented microcracks the ratio increases significantly; along directions parallel to the oriented microcracks, the ratio changes only slightly, and it either increases or decreases depending on the direction of polarisation of the shear waves.

SPENCER and NUR (1976) made measurements of compressional and shear wave velocities in Westerly granite under conditions of high confining pressure to 5 kbar (500 MPa) (72,500 lbf/in²) and high temperature to 500 °C (932 °F). Measurements in a dry sample show that at a given temperature, confining pressure has a larger accelerating effect on compressional wave velocity (V_p), while at a given confining pressure, temperature has a larger retarding effect on shear wave velocity (V_s) (Fig. 7-67). The combined effects of temperature and pressure act to increase POISSON's ratio in a dry rock.

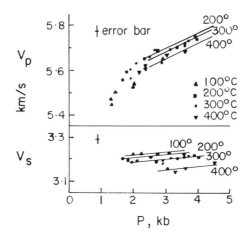

Fig. 7-67: Compressional and shear velocities in dry Westerly granite, as a function of pressure and temperature (after SPENCER and NUR, 1976).

7.6.7. Water Content

The wetting of porous rocks leads to a change of elastic wave velocities in them. If the wetting does not soften the rock, then the velocity of sound in it can be calculated in a simplified way from the time of travel of the elastic wave successively through the mineral skeleton and through the pores (t_m and t_p, respectively):

$$t = t_m + t_p \tag{7.38}$$

i.e.,

$$\frac{1}{V} = \frac{n}{V_p} + \frac{1-n}{V_m} \tag{7.39}$$

or

$$V = \frac{V_p\,V_m}{V_m\,n + V_p\,(1-n)} \tag{7.40}$$

where

n = porosity
V_p = elastic wave velocity in the substance filling the pores
V_m = elastic wave velocity in the mineral skeleton of the rock.

If the difference in the specific acoustic impedances and Poisson's ratios of the mineral skeleton and pore-filling substances is taken into account, then the formula of wave velocity in a heterogeneous medium assumes a more complex form:

$$\frac{1+q_m}{V_m^2\,\varrho_m} = \Sigma\,\frac{v_i\,(1+q_i)}{V_i^2\,\varrho_i}$$

where $q_i = 2\left(\frac{1-2V_i}{1+V_i}\right)$

Thus, the higher the sound velocity in the porefilling material, the greater is its total velocity in the rock sample. Since the velocity of sound in water (V_1) is five times greater than in air (V_a) (Table 38), water saturation leads to a rise in the elastic wave velocity in hard rocks.

TABLE 38
Velocities of longitudinal waves in the constituent phases of rocks
(after Rzhevsky and Novik, 1971)

Phase	Apparent density, kg/m³	Velocity of longitudinal wave, m/s
Water	1,000	1,485
Air	1.29	331
Ice	918	3,200–3,300

The wave velocity in more porous rocks completely saturated with water is lower than in slightly porous rocks, because V_1 is less than the sound velocity in the mineral skeleton.

Transverse waves can only pass through the mineral skeleton; consequently V_s remains almost constant in rocks whatever the degree of wetting.

OBERT, WINDES and DUVALL (1946) observed an abnormal behaviour in some rocks. Their results indicate that for some rocks, for example, marble and granite, the modulus of elasticity increases as the moisture content is increased and for other rocks, for example, limestone, it decreases. The percentage change in going from air-dried to the saturation state is in many cases appreciable (38 % decrease for sandstone; 21 % decrease for limestone, 19 % increase for marble). The modulus of rigidity also changes in a similar way which always agree in direction.

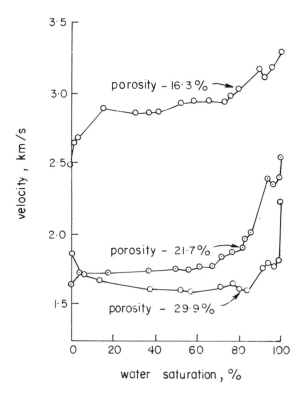

Fig. 7-68: Longitudinal wave velocity (km/s) as function
of water saturation at atmospheric pressure
(after WYLLIE, GREGORY and GARDNER, 1956).

Fig. 7-68 shows variation of velocity in three sandstones as a function of water content at atmospheric pressure. There is a marked decrease in velocity as saturation is reduced from 100 percent to about 70 percent; between 70 percent and 10 percent the velocities are nearly constant; below 10 percent they are erratic.

WYLLIE, GARDNER and GREGORY (1961) studied the effect of fluid saturation on the transverse wave velocity in sandstones. They observed that addition of water or other liquids generally decreases their velocity.

THILL and BUR (1969) studied the effect of wetting on pulse velocity in St. Cloud granodiorite (Fig. 7-69). As illustrated, significant velocity changes can occur even in compact rock having only a minute amount of porosity.

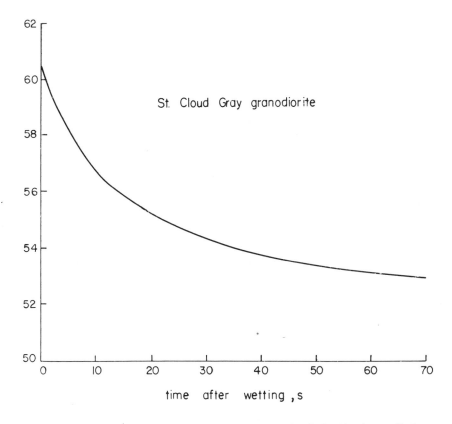

Fig. 7-69: Effects of wetting on longitudinal pulse travel time in St. Cloud granodiorite (after THILL and BUR, 1969).

RAMANA and VENKATANARAYANA (1973) investigated the effect of water saturation on Kolar rocks. The rock samples were suspended in water and observations of weight and longitudinal wave velocity were taken at the end of 4 hr, and the subsequent observations at the end of 24, 48 and 72 hr. The data thus obtained indicates a steep rise in some cases and a

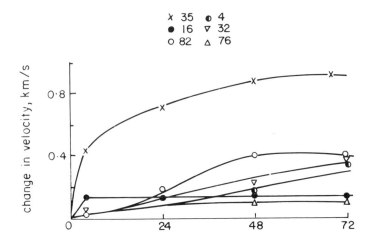

Fig. 7-70: Effect of water saturation on velocity
(after RAMANA and VENKATANARAYANA, 1973).

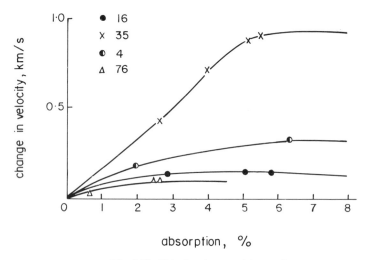

absorption, %

Fig. 7-71: Velocity change with sorption
(after RAMANA and VENKATANARAYANA, 1973).

slight rise in some others of weight as well as velocity. In general both weight and velocity increased with increasing time of saturation. Beyond 48 hr the saturation plots tend to be steady, thereafter showing very little change.

Assuming the wave velocity value obtained in a dry sample at room conditions to be the starting point, the changes in velocity observed with time are plotted in Fig. 7-70. The results indicate an increase in velocity with time in all the samples; the maximum change of 900 m/s (2,952 ft/s) noticed in the case of sample no. 35 which is an intrusive dolerite, whereas the others which are schists showed a change of 100–400 m/s (328–1,312 ft/s). The results of the change in wave velocity against percentage absorption of water are shown in Fig. 7-71, and these results clearly indicate the steady increase in velocity with increasing sorption in these rocks.

THILL, BUR and STECKLEY (1973) investigated the effects of water saturation on compressional wave velocity, V_p as well as elastic property anisotropy omnidirectionally in dry- and water-saturated spheres of Barre granite and Tennessee marble by the ultrasonic method. Barre granite is found to be strongly anisotropic in V_p with orthorhombic symmetry at both dry and saturated conditions. Tennessee marble is only weakly anisotropic in V_p and changes in symmetry from axial at dry, to orthorhombic at saturated conditions. Their results are given in Tables 39 and 40.

Velocities for both compressional and shear waves in an Alpine gneiss were determined by WANG, LIN and WENK (1975) along its principal fabric directions under dry and saturated conditions. Their results are shown in Fig. 7-72. The velocities of compressional waves in all three directions in-

TABLE 39
Properties of Barre granite at different moisture conditions
(after THILL, BUR and STECKLEY, 1973)

Moisture condition	Weight (g)	Apparent density (g/cm³)	Water content per cent	Saturation per cent	L (km/s)	M (km/s)	H (km/s)	Mean[1] (km/s)	Aniso-tropy[2] per cent	Ratio L:M:H
Dry	584.1726	2.647	0	0	3.32	3.92	4.11	3.795	20.6	1:1.18:1.24
Room	584.2226	2.648	0.008	4.1	3.31	3.95	4.13	3.822	21.5	1:1.19:1.25
Saturated	585.3937	2.653	0.209	100.0	4.53	5.10	5.22	4.948	14.0	1:1.13:1.15

[1]) Mean velocity is based on the average from 73 independent measurement directions.
[2]) $(V_{max} - V_{min})/V_{mean}$.

TABLE 40

Properties of the Tennesse marble at different moisture conditions

(after THILL, BUR and STECKLEY, 1973)

Moisture condition	Weight (g)	Apparent density (g/cm³)	Water content per cent	Saturation per cent	Velocity data					
					L (km/s)	M (km/s)	H (km/s)	Mean[1] (km/s)	Aniso-tropy[2] per cent	Ratio $L{:}M{:}H$
Dry	514.3040	2.68834	0	0	5.28	5.44	5.44	5.367	3.0	1:1.03:1.03
Saturated	515.2940	2.69351	0.192	100	5.93	5.98	6.08	6.006	2.5	1:1.01:1.02

[1]) Mean velocity is based on the average from 73 independent measurement directions.

[2]) $(V_{max} - V_{min})/V_{mean}$.

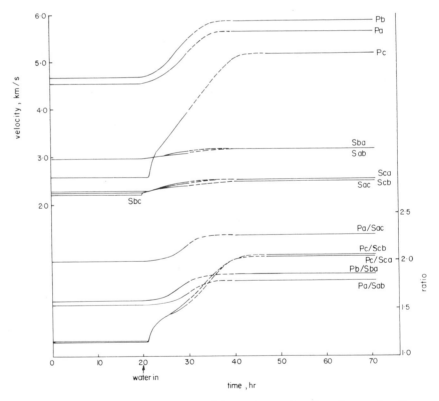

Fig. 7-72: Velocities of elastic waves in an Alpine gneiss as a function of saturation time with water as the fluid medium. The specimen was immersed in water after it was subjected to vacuum for 20 hours. Symbols are those used in Fig. 7-66

(after WANG, LIN and WENK, 1975).

creased significantly as the specimen was immersed in the water, whereas the velocities of shear waves showed little change. After about 27 hours of immersion, velocities did not change, suggesting that the condition of saturation was reached. The greatest change took place in the velocity of compressional waves in the c^* direction, which is nearly three times as large as the corresponding changes in the a and b directions. Also noticeable is that although the velocities of compressional waves along the a and b directions responded to water immersion instantaneously, the velocity along the c direction showed no change until about 1 hour after immersion. This difference can be explained by a model for the paths of diffusion of water mostly parallel to the foliation of the rock. Thus along the c direction the ray paths are perpendicular to the paths of diffusion, so that water must diffuse across part of the specimen before it reaches the ray paths. Along the a and b directions, on the other hand, the ray paths are parallel to the diffusion paths, so that water enters part of the ray paths almost immediately after immersion. The ratio between compression and shear velocities in all directions also increased with saturation. The greatest change is in the c direction; the change is from about 1.2 to 2.0. The changes in the a and b directions are about four times smaller.

Influence of the water content in a rock specimen on the propagation velocity of a stress wave was also studied with the influence of pressure. WYLLIE, GREGORY and GARDNER (1956, 1958) and HICKS and BERRY (1956) observed that it is important to determine the pressure applied both to the rock framework and to the water. The velocity in Berea sandstone as a function of pressure for different water contents is shown in Fig. 7-73. When the external pressure applied to the sample is equal to zero and when the water content increases from zero to saturation, the velocity increases. At high external pressure, the velocity increases only when the specimen is partially saturated. When the specimen is completely saturated, the velocity is much lower. For the curves a and b in Fig. 7-73, the water is under relatively low pressure and the framework under high pressure. For curve c, (Fig. 7-73) the water is under high pressure and the framework under low pressure. In examining the influence of water content, especially if the specimen is submitted to an external pressure, the important factor is the framework pressure. Fig. 7-74 gives the propagation velocity of a stress wave in a sandstone versus external pressure. If the framework or differential pressure Δp remains constant, the velocity does not depend very much on external pressure.

---

* c direction is perpendicular to the foliation of the rock, the b direction is parallel to the lineation of mineral components, and the a direction is perpendicular to both b and c.

Fig. 7-75 shows variation in velocity as a function of water content at a pressure equal to 350 bars. The variation is less than at atmospheric pressure.

Undoubtedly, the presence of cracks and flaws affect the values of velocity. If the pressure on the rock framework remains constant and the water content increases, the cracks are not closed. At atmospheric pressure the pulses are transmitted through the water filled gaps and not through air filled gaps, significantly increasing the velocity. However, at high pressure the water or air content becomes relatively unimportant since, as the cracks are closed, the pulse does not suffer a loss in amplitude and since the velocity in the rock framework is much higher than the velocity in air or water.

PODNIEKS, CHAMBERLAIN and THILL (1968) studied the environmental effects on rock properties. At a constant temperature level, longitudinal velocity

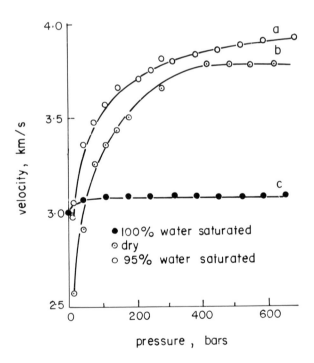

Fig. 7-73: Longitudinal wave velocity (km/s) versus pressure (bars) for different water contents. The great difference in velocity for 95 percent and 100 percent saturated specimen is due to the difference in the conditions of pressure
(after WYLLIE, GREGORY and GARDNER, 1956).

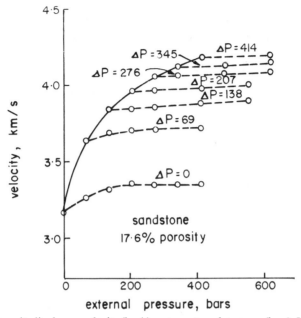

Fig. 7-74: Longitudinal wave velocity (km/s) versus external pressure (bars). It appears that the differential pressure ΔP (difference between external and water pressure) is a parameter more important than the external pressure
(after HICKS and BERRY, 1956).

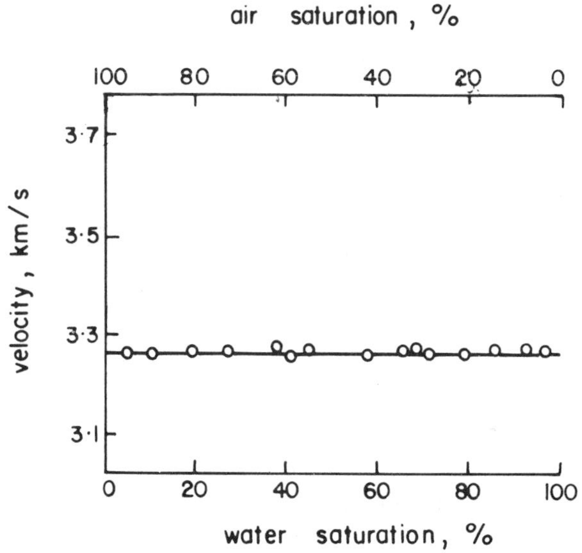

air saturation , %

velocity , km/s

water saturation, %

Fig. 7-75: Longitudinal wave velocity (km/s) as function of water saturation at a pressure $= 350$ bars
(after WYLLIE, GREGORY and GARDNER, 1958).

generally tends to increase with increasing moisture content (Fig. 7-76). The most pronounced velocity changes were recorded for the porous rocks in the cryogenic environment where the mean velocity for saturated specimens exceeds the mean velocity for dry specimens by 33 % for Salem limestone and by 51 % for Devil's Hill dacite. Usually, average velocities are lower at 212 °F than at 72° or –250 °F for constant moisture content.

7.6.8. Temperature

Usually, the velocity of longitudinal waves falls with rise in temperature (Fig. 7-77).

The results of Lewis and Tandanand (1974) on the effect of temperature on dynamic elastic constants are given in Fig. 7-78. The dynamic elastic constants fall with temperature.

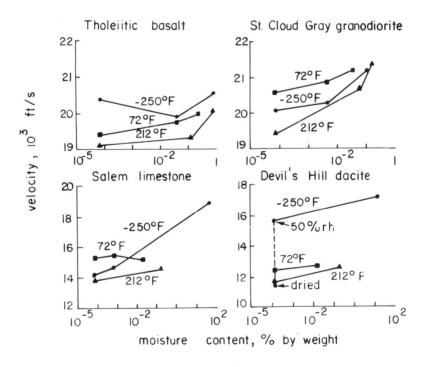

Fig. 7-76: Effect of moisture and temperature on longitudinal velocity (after Podnieks, Chamberlain and Thill, 1968).

Below 400 °C and at atmospheric pressure, the temperature coefficients of variation of velocity have the following values (IDE, 1937):

Rock	º/₀ per 100 °C
Quartzitic sandstone	−1
Solenhofen limestone	−1.2
French Creek norite	−1.6
Danby marble	−3
Sudbury norite	−4
Vinal Haven diabase	−5

Only a part of the decrease is reversible as the material cools. It may be that as the temperature of the sample rises, the unequal expansion of the crystals may be expected to cause some internal cracking and the crystals would be more loosely joined (IDE, 1937). Figs. 7-79 and 7-80 show this phenomena for Sudbury norite and Quincy granite respectively.

BIRCH (1958) pointed out that owing to looseness of grains and to cracks, porosity and other rock defects, only valid measurements of velocity versus temperature are those conducted at elevated pressures. Even when this approach is used, it is important to demonstrate the lack of hysteresis. In view of these considerations only those measurements conducted at pressures greater than 1 kbar are acceptable.

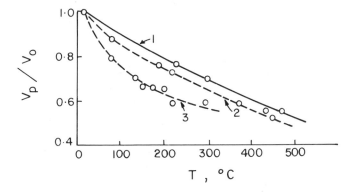

Fig. 7-77: The relative velocity of elastic waves versus temperature
on high-frequency electrical heating of rocks:
1 – epidote-garnet skarn; 2 – pyroxene-garnet skarn; 3 – dolomite
(after RZHEVSKY and NOVIK, 1971).

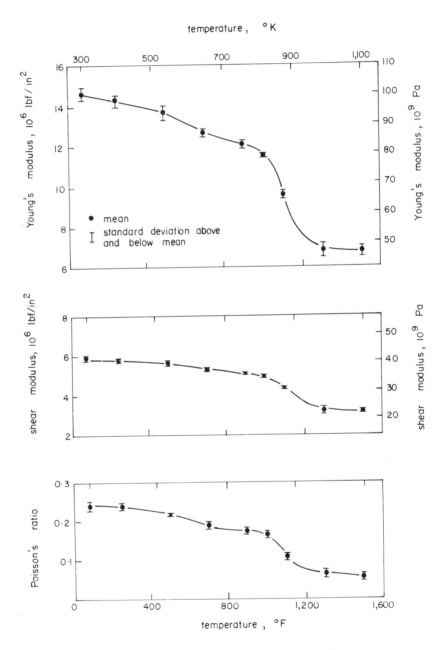

Fig. 7-78: Elastic moduli and POISSON's ratio as a function of temperature
(after LEWIS and TANDANAND, 1974).

The influence of temperature and the added factor of pressure have also been studied (HUGHES and JONES, 1950, 1951; HUGHES and CROSS, 1951; HUGHES and MAURETTE, 1956, 1957). After a run at a constant elevated temperature and increasing pressure, the room temperature velocity is usually higher than before the run (Fig. 7-81). This result can be reasonably ascribed to compaction of the sample due to the pressure. The influence of temperature on the propagation velocity is less at high pressures than at low pressures as shown in Figs. 7-82 and 7-83. As the temperature rises the crystals are more loosely joined, but this tendency is lessened by the action of pressure. Fig. 7-84 shows the compressional and shear velocities as a function of hydrostatic pressure at various temperatures. Fig. 7-85 shows the relative frequencies of the resonant shear-mode vibrations in cylinders of certain rocks as a function of temperature. Measurements were taken at elevated pressures.

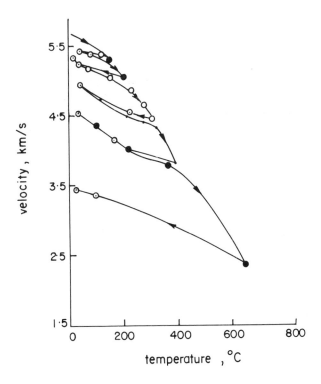

Fig. 7-79: Longitudinal wave velocity (km/s) as function
of temperature (°C) for Sudbury norite
(after IDE, 1937).

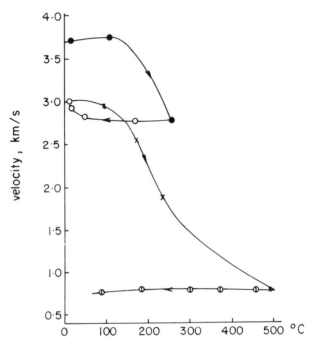

Fig. 7-80: Effect of thermal cycling at room pressure
on the compressional velocity in Quincy granite
(after IDE, 1937).

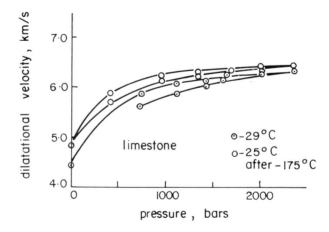

Fig. 7-81: Longitudinal wave velocity (km/s) versus pressure (bars)
at a temperature equal to 25 °C and 29 °C after a run at 175 °C
(after HUGHES and JONES, 1951).

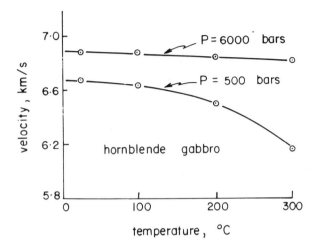

Fig. 7-82: Longitudinal wave velocity (km/s) versus temperature (°C) for two different pressures. Variation in velocity is much greater at low pressure than at high pressure (after HUGHES and MAURETTE, 1957).

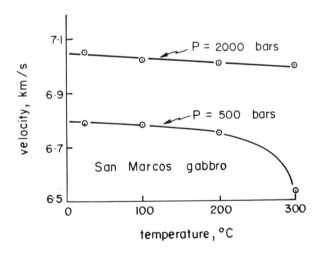

Fig. 7-83: Longitudinal wave velocity (km/s) versus temperature (°C) for two different pressures. Variation in velocity is much greater at low pressure than at high pressure (after HUGHES and MAURETTE, 1957).

Generally, an increase in temperature causes a decrease in velocity. There are some exceptions (HUGHES and JONES, 1951). Several dolomites show an increase in wave velocities with temperatures increasing up to 100 to 125 °C. This suggests that an increase in temperature can change the rock properties in a way not yet explainable. Further increase in temperature causes a decrease in velocity.

In wet rocks, an abrupt rise in sound velocity is observed when their temperature falls to below the freezing point of water (Fig. 7-86). The velocity of sound in ice is about 3,300 m/s. A sharp rise in the velocity of transverse waves is observed in frozen rocks.

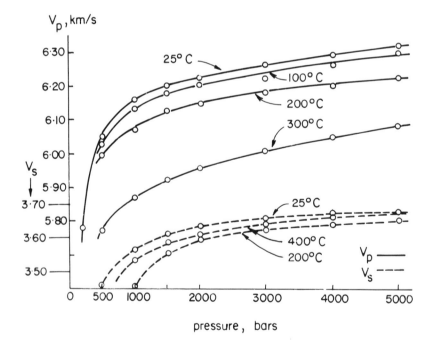

Fig. 7-84: Compressional (V_p) and shear (V_s) velocities as a function of hydrostatic pressure at various temperatures (after HUGHES and MAURETTE, 1957).

7.7. Dynamic Tensile Strength of Rock

Dynamic tensile strength of rock is important in the breakage of rock by explosives.

The principle underlying the determination of this property is to subject the specimen to a known intensity of transient stress waves. As the compressive stress wave moves and reaches the free end of the specimen, it is

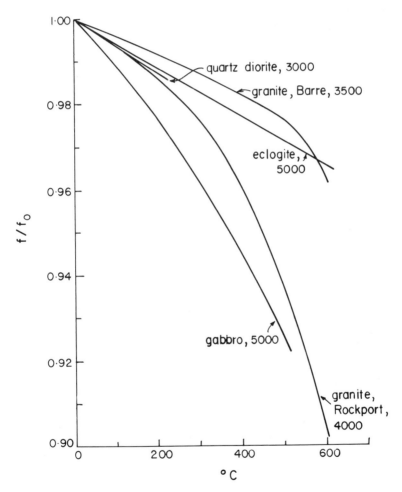

Fig. 7-85: Relative frequencies of the resonant shear-mode vibrations in cylinders of certain rocks as a function of temperature. Measurements were taken at the elevated pressures indicated (in kgf/cm²) (after Birch, 1943).

reflected as a tension wave which on being superimposed upon the tail of the incident compressive wave develops an increasing amount of tension in the material (AB in Fig. 7-87). The tensile stress at failure is the dynamic tensile strength of the material. Experiments are repeated over and over again at slightly increasing or decreasing stress intensities to determine the threshold stress at which slabbing just develops.

BACON (1962) determined the dynamic tensile strength of rocks using a pendulum to impart energy in the form of a sharp pulse to the end of a long rock core suspended from the overhead supports by fine wires. In his later experiments he used projectiles fired from an air-gun. The cores were protected with steel anvils to guard against damage of ends by impact. His results show that the dynamic tensile strength is one to four times the static tensile strength.

RINEHART (1965) used a different arrangement given in Fig. 7-88. It consists

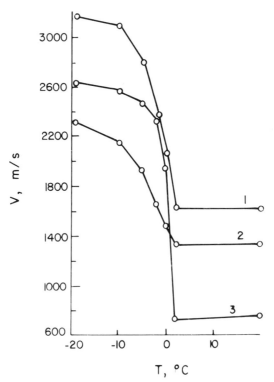

Fig. 7-86: Velocity of ultrasound in frozen rocks: 1 – sandstone; 2 – sand; 3 – clay
(after RZHEVSKY and NOVIK, 1971).

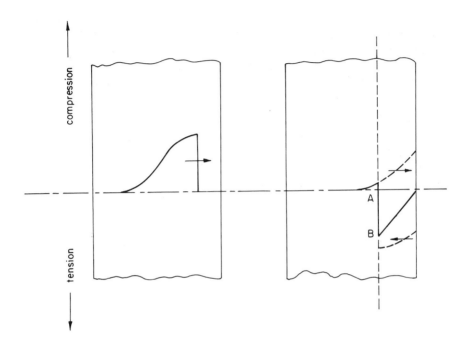

Fig. 7-87: Mechanism of development of tensile stress by the reflection
of the incident (compressive) wave
(after RINHART, 1965).

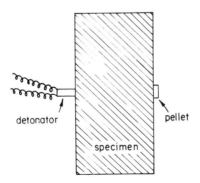

Fig. 7-88: Arrangement for determination of tensile stress developed in the specimen
(after RINEHART, 1965).

basically of a blasting cap to generate a transient stress pulse in the rock specimen about 2.54 cm (1 in) thick and several centimetres (inches) long and a small pellet the throw-off velocity of which when measured enables the determination of stress. He used plas-T-cap blasting caps which have been proved to have high reproducibility of stress pulse from round to round. The pellet material used is identical with the rock specimen to avoid correction required to be made for mis-match in acoustic impedance.

As the stress wave is generated by the blasting cap, it diverges from the point of its origin and its intensity decreases as it passes through the rock specimen and is reflected back as a tensile stress pulse of intensity equal to the intensity of the compressive stress pulse reaching the free end. If the thickness of the specimen is too small, this will spall but if the thickness is too large, no spalling will occur. Thus, for a given shape and intensity of the transient stress pulse, for each rock, there exists a critical thickness at which spalling just starts or just ceases.

Stress distribution, within that transient stress wave which is just capable of generating the spall, can be known by measuring the velocity of the pellet thrown-off the plate of critical thickness. Since the quantity that governs spalling is the maximum stress, this can be determined from the slope of the momentum trapped versus pellet thickness curve. The pellets used were of 1.27 cm (0.5 in) diameter and of thickness varying from 0.04 to 0.64 cm ($^1/_{64}$ to $^1/_4$ in).

The results published by RINEHART on the comparative static and dynamic tensile strengths of rock are given in Table 41. Static tensile strength was obtained by plate and glue method (direct pull test).

These results show that the dynamic tensile strength is 6–10 times the static tensile strength. The probable reason for very high dynamic tensile strength is that with the increase in the rate of loading, the weakest link in the rock may not necessarily have an opportunity to participate in the fracturing process. The situation which causes the spall is highly localised and the strength measured is that of the rock lying in this highly localised region and hence the volume of the rocks subjected to the maximum tensile stress is very small as in the bending tests or ring tests.

TINCELIN, WEBER and DE MONTILLE (1970) published data on the comparative static and dynamic strengths of rock. The principle in their device for measuring the dynamic strengths of rocks is based on the propagation of an impulse along a cylindrical rod. The ratio of dynamic to static values of strength varies from 2.5 to 1 for tension when the rocks are classified by increasing values of static strength.

BIRKIMER (1970) pointed out that the dynamic tensile strength of rock is not constant and varies with straining rate. He used air-fired projectile impact loader. The magnitude and rise time of the pulse was regulated by varying either the projectile configuration or the pressure in the tank. In order to monitor the straining pulse in the vicinity of the fracture location, a system using surface strain gauges, a bridge balance, and a fast rise time oscilloscope was devised.

He tested quartz monzonite specimens with dimensions 5.1 cm (2 in) in diameter and 53.3 cm (21 in) in length. The test results are given in Table 42. The relation between ε_{cr}, the critical fracture strain, and the apparent straining rate has been calculated as:

$$\varepsilon_{cr} = 360 \, \dot{\varepsilon}^{1/3} \tag{7.41}$$

where ε_{cr} = critical fracture strain, μcm/cm (μin/in) and
        $\dot{\varepsilon}$ = apparent straining rate, cm/cm/s (in/in/s).

Fig. 7-89 shows this relationship as well as the limited data points.

TABLE 41
Tensile strength of rocks, MPa (lbf/in^2)
(after RINEHART, 1965)

Rock	Static	Dynamic	Ratio
Bedford limestone	4.1 (600)	26.9 (3900)	6.5
Yule marble (perpendicular to bedding)	2.1 (300)	18.6 (2700)	9.0
Yule marble (parallel to bedding)	6.2 (900)	48.3 (7000)	7.8
Granite	6.9 (1000)	39.3 (5700)	5.7
Taconite	4.8–6.9 (700–1000)	91.0 (13200)	13.0

TABLE 42
Test results of quartz monzonite rock
(after BIRKIMER, 1970)

ε_{cr}	σ_{cr} MPa (lbf/in^2)	$\dot{\varepsilon}$ cm/cm/s or in/in/s
1050×10^{-6}	47.4 (6870)	21.0
740×10^{-6}	36.2 (5250)	9.2
1030×10^{-6}	46.1 (6685)	25.7
$1000 \times 10^{-6}*$	45.9 (6650)	33.3

* No spall formed under this pulse.

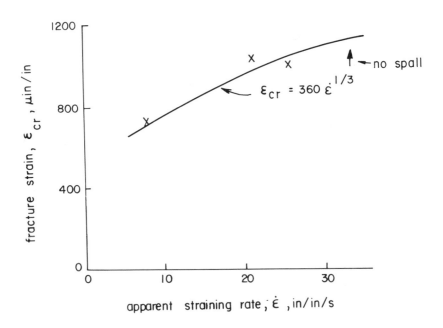

Fig. 7-89: Fracture strain versus apparent straining rate of quartz monzonite rock
(after BIRKIMER, 1970).

7.8. Summary and Conclusions

Dynamic elastic constants are calculated from the elastic wave velocities and density. The resonance and ultrasonic pulse methods are used to determine the elastic wave velocities in laboratory.

The theory, apparatus and limitations of both resonance and ultrasonic pulse methods have been given in the text. Also a comparison between these two methods has been given. The two methods do not give equivalent results even in nearly isotropic rocks. Although the discrepancy between the two methods probably is not of great significance in E and G determinations, it may be very significant regarding v determinations. Consequently, while either method might provide an adequate estimate of E or G, v determination by the resonance method is not recommended because of uncertainty with its determination.

For best results both the resonant and ultrasonic pulse methods should be used to determine, on the same cylindrical specimen, the propagation velocities, V_0 (bar velocity by resonance method), V_p (compression wave velocity by pulse method) and V_s (shear wave velocity by both methods). If both methods are used, two independent measurements of V_s are obtained, one from the resonant torsional frequency and one from the shear pulse travel time. These two independent shear velocity determinations should agree within the experimental error of the measurements. The average value of V_s is used to calculate G by means of Eq. 7.27. The longitudinal bar velocity V_0 is used to calculate E by means of Eq. 7.7. The longitudinal pulse velocity V_p and the longitudinal bar velocity V_0 are used to calculate Poisson's ratio by means of Eq. 7.29.

Seismic wave propagation method is used for field determinations of dynamic elastic constants. The method can also be used to determine the depth of the various rock beds provided they show sufficient variations in wave propagation velocities.

If the environmental conditions in the laboratory and in situ are the same, the results are comparable.

Usually, values obtained by dynamic methods are higher than those obtained by static techniques (both in the laboratory and in situ tests). The greater the degree of compactness, the more nearly dynamic and static elastic constants may agree.

The parameters, rock type, texture, density, porosity, anisotropy, stress, water content, and temperature affecting propagation velocity of waves in rocks have been discussed in detail.

Generally it can be said that velocities are higher for more dense and compact rocks, lower for less dense and compact rocks.

The velocity in a rock may be related with the velocities in rock's various mineral components. Such relationship does not take into account factors such as grain size, or preferred orientation of crystals. Some results indicate an increase in velocity value with increasing hornblende percentage and a decrease in velocity with increasing quartz content.

The velocity of waves is influenced by the size of the grains constituting the rock. The velocity is greater as a rule in fine-grained rocks than in coarse-grained rocks.

In general, the velocity increases as the density of rock increases. The relationship may be linear or curvilinear.

The propagation velocity decreases as the porosity increases. This is true for both dry and saturated rocks.

In layered rocks, the velocities of elastic waves differ along and across the layers, and the velocity parallel to the layers is always greater than the velocity perpendicular to the layers. Increase in porosity increases anisotropy. Anisotropy usually corresponds with the microstructural subfabrics, namely, crystallographic orientation of constituent anisotropic minerals and shape and orientation of pores or cracks. Wetting may change the symmetry.

Velocities generally increase with increasing pressure. There is a rapid increase at low pressures due to a decrease in porosity, the closing of cracks and defects, and an increase in the mechanical contact between the grains. The velocity increase at higher pressure results from changes in the intrinsic properties of the rock, such as finite compression of the crystals. In certain cases, the velocity is observed to decrease when the pressure exceeds a certain value.

The degree of anisotropy decreases appreciably as the hydrostatic confining pressure is increased. Also the ratio E_{dyn}/E_{st} is reduced as the confining pressure is increased.

The wetting of rocks usually leads to a rise in the elastic wave velocities. The wave velocity in more porous rocks completely saturated with water, however, is lower than in slightly porous rocks, because the elastic wave velocity in water is less than the elastic wave velocity in the mineral skeleton.

Transverse waves can only pass through the mineral skeleton; consequently V_s remains almost constant in rocks whatever the degree of wetting.

Usually, the velocities and dynamic elastic constants fall with temperature. As the rock cools, only a part of the decrease is reversible because as the temperature rises, the unequal expansion of the crystals causes some internal cracking and the crystals would be loosely joined. Because of hysteresis, measurements of velocity versus temperature are recommended at elevated pressures, usually greater than 1 kbar (100 MPa) (14,500 lbf/in²).

In wet rocks, an abrupt rise in velocity is observed when their temperature falls to below the freezing point of water.

Determination of the dynamic tensile strength of rock has also been discussed. The dynamic tensile strength is one to thirteen times the static tensile strength. The probable reason for very high dynamic tensile strength is that with the increase in the rate of loading, the weakest link in the rock may not necessarily have an opportunity to participate in the fracturing process. The situation which causes the spall is highly localised and the strength measured is that of the rock lying in this highly localised region and hence the volume of the rocks subjected to the maximum tensile stress is very small. The dynamic tensile strength is not constant and increases with straining rate.

References to Chapter 7

1. AHRENS, T. J. and KATZ, S.: Ultrasonic observation of the calcite-aragonite transition. J. Geophys. Res., Vol. 68, 1963, pp. 529–537.
2. A.S.T.M.: Standard method for laboratory determination of pulse velocities and ultrasonic elastic constants of rock. A.S.T.M. Designation: D 2845-69, 1969, pp. 877–884 (A.S.T.M. Standards, Part 11, 1970, D 2845-69, pp. 877–884).
3. ATTEWELL, P. B.: Triaxial anisotropy of wave velocity and elastic moduli in slate and their axial concordance with fabric and tectonic symmetry. Int. J. Rock Mech. Min. Sci., Vol. 7, No. 2, March, 1970, pp. 193–207.
4. AUBERGER, M. and RINEHART, J. S.: Ultrasonic velocity and attenuation of longitudinal waves in rocks. J. Geophys. Res., Vol. 66, 1961, pp. 191–199.
5. BACON, L.: A method of determining dynamic tensile strength of rock at minimum loading. U.S.B.M.R.I. 6067, 1962, 22 p.
6. BARON, G., HABIB, P. and MORLIER, P.: The effect of stresses and deformations on the velocity of sound in rock. Proc. 4th Int. Conf. Strata Control and Rock Mech., New York, 1964.
7. BERNABINI, M. and BORELLI, G. B.: Methods for determining the average dynamic elastic properties of a fractured rock mass and the variations of these properties near excavations. Proc. 3rd Cong. Int. Soc. Rock Mech., Denver, 1974, Vol. 2, Part A, pp. 393–397.
8. BIRCH, F.: Elasticity of igneous rocks at high temperatures and pressures. Geol. Soc. Am. Bull., Vol. 54, 1943, pp. 263–286.
9. BIRCH, F.: Interpretation of the seismic structure of the crust in the light of experimental studies of wave velocities in rocks. In "Contributions in Geophysics in honour of Beno Gutenberg", Oxford, Pergamon, 1958.
10. BIRCH, F.: The velocity of compressional waves in rocks to 10 kilobars, Part 1. J. Geophys. Res., Vol. 65, 1960, pp. 1083–1102.
11. BIRCH, F.: The velocity of compressional waves in rocks to 10 kilobars, Part 2. J. Geophys. Res., Vol. 66, 1961, pp. 2199–2224.
12. BIRCH, F. and BANCROFT, D.: The effect of pressure on the rigidity of rocks, Part I. J. Geol., Vol. 46, 1938, pp. 59–87.
13. BIRCH, F. and BANCROFT, D.: New measurements of the rigidity of rocks at high pressure. J. Geol., Vol. 48, 1940, pp. 752–766.
14. BIRKIMER, D. L.: A possible fracture criterion for the dynamic tensile strength of rock. Proc. 12th Symp. Rock Mech., Rolla, Missouri, 1970, pp. 573–590.
15. BRACE, W. F.: Some new measurements of linear compressibility of rocks. J. Geophys. Res., Vol. 70, 1965, pp. 391–398.
16. BRUCKSHAW, J. M. and MAHANTA, P. C.: The variation of the elastic constants of rock with frequency. Petroleum, Vol. 17, No. 1, Jan., 1954, pp. 14–18.
17. BUR, T. R. and HJELMSTAD, K. E.: Elastic and attenuation symmetries of simulated lunar rocks. Icarus, Vol. 13, 1970, pp. 414–423.
18. CHENEVERT, M. E.: The deformation-failure characteristics of laminated sedimentary rocks. Ph. D. Thesis, Univ. Texas, 1964, 203 p.
19. CLARK, G. B.: Deformation moduli of rocks. Proc. Symp. Testing Techniques for Rock Mech., Seattle, Washington, 1965.
20. COON, R.: Correlation of engineering behaviour with the classification of in situ rock. Ph. D. Thesis, Univ. Illinois, 1968.
21. DICKSON, E. W. and STRAUCH, H.: Apparatus for the measurement of internal friction and dynamic Young's modulus at kilocycles frequencies. J. Sci. Inst., Vol. 36, 1959.
22. DUNOYER DE SEGONZAC, Ph. and LAHERRERE, J.: Application of the continuous velocity log to anisotropy measurements in northern Sahara; results and consequences. Geophys. Prosp., Vol. 7, 1959, pp. 202–217.

23. DVORAK, A.: Field test of rocks on dam sites. Proc. 4th Int. Conf. Soil Mech. Found. Eng., London, 1957, Vol. 1, pp. 221–224.

24. EVANS, I. and POMEROY, C. D.: The strength, fracture and workability of coal. London, Pergamon, 1966, 277 p.

25. GEERTSMA, J.: Velocity-log interpretation; the effect of rock bulk compressibility. Paper presented to Society of Petroleum Engineers, A.I.M.E., Paper No. 1535 G, Oct. 2–5, 1960, 14 p.

26. HICKS, W. G. and BERRY, J. E.: Application of continuous velocity logs to determination of fluid saturation of reservoir rocks. Geophysics, Vol. 21, 1956, pp. 739–754.

27. HUGHES, D. S and CROSS, J. H.: Elastic wave velocities in rocks at high pressures and temperatures. Geophysics, Vol. 16, 1951, pp. 577–593.

28. HUGHES, D. S. and JONES, H. J.: Variation of elastic moduli of igneous rocks with pressure and temperature. Geol. Soc. Am. Bull., Vol. 61, 1950, pp. 843–856.

29. HUGHES, D. S. and JONES, H. J.: Elastic wave velocities of sedimentary rocks. Trans. Am. Geophys. Union, Vol. 32, 1951.

30. HUGHES, D. S. and MAURETTE, C.: Elastic wave velocities in granite. Geophysics, Vol. 21, 1956, pp. 277–284.

31. HUGHES, D. S. and MAURETTE, C.: Determination des vitesses d'ondes elastiques dans diverses roches en fonction da la pression et de la temperature. Rev. Inst. Franc. Petr. et Ann. Comb. liquides, Vol. 12, 1957, pp. 730–752.

32. IDE, J. M.: Comparison of statically and dynamically determined Young's modulus of rocks. Proc. Nat. Acad. Sci. U.S.A., Vol. 22, 1936, pp. 81–92.

33. IDE, J. M.: The velocity of sound in rocks and glasses as a function of temperature. J. Geol., Vol. 45, 1937, pp. 689–716.

34. JAEGER, C.: Rock mechanics and engineering. Cambridge, Univ. Press, 1972, 417 p.

35. JAMIESON, J. C. and HOSKINS, H.: The measurement of shear wave velocities in solids using axially polarised ceramic transducers. Geophysics, Vol. 28, 1963, pp. 87–90.

36. JOHNSON, C. F.: A pulse technique for the direct measurement of bar velocity. Proc. 12th Symp. Rock Mech., Rolla, Missouri, 1970, pp. 212–226.

37. KING, M. S.: Ultrasonic compressional and shear-wave velocities on confined rock samples. Proc. 5th Can. Rock Mech. Symp., Toronto, Dec., 1968, pp. 127–154.

38. LECOMTE, P.: Methods for measuring the dynamic properties of rocks. Proc. 2nd Can. Rock Mech. Symp., Kingston, Ontario, 1963, pp. 15–26.

39. LEWIS, W. E. and TANDANAND, S.: Bureau of Mines test procedures for rocks. U.S.B.M.I.C. 8628, 1974, 223 p.

40. MANGHNANI, M. H. and WOOLLARD, G. P.: The crust and upper mantle of the Pacific area. In Geophysical Monograph No. 12, Am. Geophys. Union, 1968, pp. 501–516.

41. NICHOLLS, H. R.: In situ determination of the dynamic elastic constants of rock. Proc. Int. Symp. Min. Res., Rolla, Missouri, 1961, Vol. 2, pp. 727–738. (Also U.S.B.M.R.I. 5888, 1961, 13 p.).

42. OBERT, L. and DUVALL, W. I.: Rock mechanics and the design of structures in rock. New York, Wiley, 1967, 650 p.

43. OBERT, L., WINDES, S. L. and DUVALL, W. I.: Standardised tests for determining the physical properties of mine rock. U.S.B.M.R.I. 3891, 1946, 67 p.

44. PATERSON, N. R.: Seismic wave propagation in porous granular media. Geophysics, Vol. 21, 1956, pp. 691–714.

45. PESELNICK, L. and OUTERBRIDGE, W. F.: Internal friction in shear and shear modulus of Solenhofen limestone over a frequency range of 10^7 cycles per second. J. Geophys. Res., Vol. 66, 1961, pp. 581–588.

46. PODNIEKS, E. R., CHAMBERLAIN, P. G. and THILL, R. E.: Environmental effects on rock properties. Proc. 10th Symp. Rock Mech., Univ. Texas, Austin, Texas, 1968, pp. 215–241.

47. POPOVIC, R. and CVETKOVIC, M.: Determination of the dynamic parameters of elasticity for rocksalt using resonant and ultrasonic techniques. In Serbo-Croat. Proc. 3rd Yugoslav. Symp. Rock Mech. Underground Excavations, 1972, Pap. 1–6, 6 p.

48. PROSKURYAKOV, N. M., LIVENSKII, V. S. and KUZNETSOV, Yu. F.: Study of the velocity of propagation of elastic waves in relation to stress in salt rocks under uniaxial compression. Sov. Min. Sci., Vol. 11, No. 1, Jan.–Feb., 1975, pp. 68–69.

49. RAMANA, Y. V. and VENKATANARAYANA, B.: Laboratory studies on Kolar rocks. Int. J. Rock Mech. Min. Sci. & Geomech. Abstr., Vol. 10, 1973, pp. 465–489.

50. RAYLEIGH, Lord: The theory of sound. Vol. 1. New York, Dover, 1945.

51. RINEHART, J. S.: Dynamic fracture strength of rocks. Proc. 7th Symp. Rock Mech., Univ. Park, Penn., 1965, pp. 205–208.

52. RINEHART, J. S., FORTIN, J. P. and BURGIN, L.: Propagation velocity of longitudinal waves in rocks. Effect of state of stress, stress level of the wave, water content, porosity, temperature, stratification and texture. Proc. 4th Symp. Rock Mech., Univ. Park, Penn., 1961, pp. 119–135.

53. RUMMEL, F.: Changes in the P-wave velocity with increasing inelastic deformation in rock specimens under compression. Proc. 3rd Cong. Int. Soc. Rock Mech., Denver, 1974, Vol. 2, Part A, pp. 517–523.

54. RZHEVSKY, V. and NOVIK, G.: The physics of rocks. Moscow, MIR Publishers, 1971, 320 p.

55. SCHREIBER, E., ANDERSON, O. L. and SOGA, N.: Elastic constants and their measurement. New York, McGraw-Hill, 1973, 196 p.

56. SHIMOZURA, D.: Elasticity of rocks and some related geophysical problems. Japan J. Geophys., Vol. 2, 1960, 85 p.

57. SHOOK, W. B.: Critical survey of mechanical property test-methods for brittle materials. Ohio State University, Engineering Experiment Station, Columbus, Ohio, July, 1963, 136 p.

58. SIMMONS, G.: Velocity of shear waves in rocks to 10 kilobars, 1. J. Geophys. Res., Vol. 69, 1964, pp. 1123–1130.

59. SIMMONS, G.: Ultrasonics in geology. Proc. I.E.E.E., Vol. 53, 1965, pp. 1337–1346.

60. SIMMONS, G. and BRACE, W. F.: Comparison of static and dynamic measurements of compressibility of rocks. J. Geophys. Res., Vol. 70, 1965, pp. 5649–5656.

61. SPENCER, J. W. and NUR, A. M.: The effects of pressure, temperature, and pore water on velocities in Westerly granite. J. Geophys. Res., Vol. 81, 1976, pp. 899–904.

62. SUTHERLAND, R. B.: Some dynamic and static properties of rock. Proc. 5th Symp. Rock Mech., Minneapolis, Minn., 1962, pp. 473–490.

63. TERRY, N. B.: Some considerations of the magnetostrictive composite oscillator method for the measurement of elastic moduli. Brit. J. Appl. Phys., Vol. 8, 1957.

64. TERRY, N. B. and SEABORNE, N. F.: The elastic properties of coal. Part 5. An apparatus for measuring elastic pulse velocities in coal specimen. N.C.B., Mining Research Establishment Rept. 2079, 1957.

65. THILL, R. E.: Acoustic methods for monitoring failure in rock. Proc. 14th Symp. Rock Mech., Univ. Park, Penn., June, 1972, pp. 649–687.

66. THILL, R. E. and BUR, T. R.: An automated ultrasonic pulse measurement system. Geophysics, Vol. 34, No. 1, Feb., 1969, 101–105.

67. THILL, R. E., BUR, T. R. and STECKLEY, R. C.: Velocity anisotropy in dry and saturated rock spheres and its relation to rock fabric. Int. J. Rock Mech. Min. Sci. & Geomech. Abstr., Vol. 10, 1973, pp. 535–557.

68. THILL, R. E., McWILLIAMS, J. R. and BUR, T. R.: An acoustical bench for an ultrasonic pulse system. U.S.B.M.R.I. 7164, July, 1968, 22 p.

69. THILL, R. E. and PENG, S. S.: Statistical comparison of the pulse and resonance methods for determining elastic moduli. U.S.B.M.R.I. 7831, 1974, 24 p.

70. THILL, R. E., WILLARD, R. J. and BUR, T. R.: Correlation of longitudinal velocity variation with rock fabric. J. Geophys. Res., Vol. 74, 1969, pp. 4897–4909.
71. TIMOSHENKO, S. and GOODIER, J. N.: Theory of elasticity. 2nd ed. New York, McGraw-Hill, 1951.
72. TINCELIN, E., WEBER, P. and DE MONTILLE, G.: Comparative study of static and dynamic rock strengths. Proc. 2nd Cong. Int. Soc. Rock Mech., Belgrade, 1970, Vol. 3, pp. 71–76.
73. TOCHER, D.: Anisotropy in rocks under simple compression. Trans. Am. Geophys. Union, Vol. 38, 1957, pp. 89–94.
74. UHRIG, L. F. and VON MELLE, F. A.: Velocity anisotropy in stratified media. Geophysics, Vol. 20, 1955, pp. 774–779.
75. U.S. Bureau of Reclamation: Effect of cracks in concrete upon dynamic measurements of elastic modulus. Materials Lab. Rep. No. C – 383, 1948.
76. U.S. Bureau of Reclamation: Physical properties of some typical foundation rocks. Concrete Lab. Rep. No. SP – 39, 1953, 50 p.
77. WALSH, J. B.: The effect of cracks on the compressibility of rock. J. Geophys. Res., Vol. 70, 1965, pp. 381–389.
78. WANG, C., LIN, W. and WENK, H.: The effects of water and pressure on velocities of elastic waves in a foliated rock. J. Geophys. Res., Vol. 80, 1975, pp. 1065–1069.
79. WYLLIE, M. R. J., GARDNER, G. H. F. and GREGORY, A. R.: Some phenomena pertinent to velocity logging. J. Pet. Tech., Vol. 13, 1961, pp. 629–636.
80. WYLLIE, M. R. J., GREGORY, A. R. and GARDNER, L. W.: Elastic wave velocities in heterogeneous and porous media. Geophysics, Vol. 21, 1956, pp. 41–70.
81. WYLLIE, M. R. J., GREGORY, A. R. and GARDNER, G. H. F.: An experimental investigation of factors affecting elastic wave velocities in porous media. Geophysics, Vol. 23, 1958, pp. 459–493.
82. YOUASH, Y. Y.: Dynamic physical properties of rock: Part I, Theory and procedure. Proc. 2nd Cong. Int. Soc. Rock Mech., Belgrade, 1970, Vol. 1, pp. 171–183.
83. YOUASH, Y. Y.: Dynamic physical properties of rocks: Part II, Experimental results. Proc. 2nd Cong. Int. Soc. Rock Mech., Belgrade, 1970, Vol. 1, pp. 185–195.
84. ZISMAN, W. A.: A comparison of the statically and seismologically determined elastic constants of rock. Proc. Nat. Acad. Sci. U.S.A., Vol. 19, 1933, pp. 680–686.

Uncited References to Chapter 7

1. ABE, M.: Measurement of the attenuation of elastic waves in rocks. J. Seismol. Soc. Japan, Vol. 25, No. 3, 1972, pp. 265–266.
2. ATTEWELL, P. B. and BRENTNALL, D.: Internal friction: some considerations of the frequency response of rocks and other metallic and non-metallic materials. Int. J. Rock Mech. Min. Sci., Vol. 1, 1964, pp. 231–254.
3. AVEDISSIAN, Y. M.: Stress relaxation in rocks and their dynamic properties. Proc. 1st Iranian Cong. Civil Eng. and Eng. Mech., Pahlavi Univ., Shiraz, Iran, 1972.
4. BARKER, L. M., HENDRICKSON, R. R., LINGLE, R. and JOHNSON, J. N.: Measurement of low-pressure static elastic moduli of rock and comparison with ultrasonic data. Final report prepared for Defence Nuclear Agency, Oct., 1975, 30 p.
5. BONDARENKO, V. G. and ROGOZHNIKOV, V. I.: A device for emitting longitudinal and transverse elastic waves from a borehole. Sov. Min. Sci., Vol. 8, No. 5, Sept.–Oct., 1972, pp. 577–578.
6. BRADY, B. T.: The effect of mechanical anisotropy on the transmission of low-amplitude stress waves in brittle rock. Int. J. Rock Mech. Min. Sci., Vol. 6, 1969, pp. 439–452.

7. Bur, T. R., McWilliams, J. R. and Hjelmstad, K. E.: Velocity errors in cemented specimens. Int. J. Rock Mech. Min. Sci., Vol. 6, 1969, pp. 203–209.

8. Butcher, B. M. and Stevens, A. L.: Shock wave response of Window Rock coal. Int. J. Rock Mech. Min. Sci. & Geomech. Abstr., Vol. 12, 1975, pp. 147–155.

9. Carroll, R. D.: The determination of the acoustic parameters of volcanic rocks from compressional velocity measurements. Int. J. Rock Mech. Min. Sci., Vol. 6, 1969, pp. 557–579.

10. Crowley, B. K.: Effects of porosity and saturation on shock-wave response in tuffs. Int. J. Rock Mech. Min. Sci. & Geomech. Abstr., Vol. 10, 1973, pp. 437–464.

11. Curran, D. R., Petersen, C. F. and Rosenberg, J. T.: Dynamic tensile failure in rocks. Stanford Research Inst., Menlo Park, Calif., Report AD – 743988, March, 1972, 75 p.

12. Depman, A., Doods, K. and Parrillio, D.: Tocks island project spillway. Rock mechanics studies. Proc. 13th Symp. Rock Mech., Urbana, Illinois, 1971, pp. 443–486.

13. Dvorak, A.: Influence of the testing method on results of loading experiments in rock masses. Proc. 2nd Cong. Int. Soc. Rock Mech., Belgrade, 1970, Vol. 4, pp. 247–250.

14. Fourney, W. L., Dally, J. W. and Holloway, D. C.: Attenuation of strain waves in core samples of three types of rock. Exp. Mech., Vol. 16, No. 4, April, 1976, pp. 121–126.

15. Friedman, M. and Bur, T. R.: Investigations of the relations among residual strain, fabric, fracture and ultrasonic attenuation and velocity in rocks. Int. J. Rock Mech. Min. Sci. & Geomech. Abstr., Vol. 11, 1974, pp. 221–234.

16. Godfrey, C.: Dynamic strength of in situ rock. Proc. 3rd Cong. Int. Soc. Rock Mech., Denver, 1974, Vol. 2, Part A, pp. 398–403.

17. Goldsmith, W. and Sackman, J. L.: Wave traversal and comminution of rock. Final Report to the National Science Foundation, Jan., 1976, 148 p.

18. Goldsmith, W., Sackman, J. L. and Ewert, C.: Static and dynamic fracture strength of Barre granite. Int. J. Rock Mech. Min. Sci. & Geomech. Abstr., Vol. 13, 1976, pp. 303–309.

19. Grady, D. E. and Murri, W. J.: Dynamic unloading in shock compressed feldspar. Geophys. Res. Lett., Vol. 3, No. 8, Aug., 1976, pp. 472–474.

20. Green, R. E.: Ultrasonic investigation of mechanical properties. New York, Academic Press, 1973, 166 p.

21. Gregory, A. R.: Fluid saturation effects on dynamic elastic properties of sedimentary rocks. Geophysics, Vol. 41, No. 5, Oct., 1976, pp. 895–921.

22. Hayashi, M., Kitahara, Y. and Fuziwara, Y.: Dynamic deformability and viscosity of rock masses. Proc. 3rd Cong. Int. Soc. Rock Mech., Denver, 1974, Vol. 2, Part B, pp. 713–718.

23. Ito, I.: Rock Fracture in blasting and dynamic characteristics of rocks under impulsive loads such as blasting. Proc. 2nd Cong. Int. Soc. Rock Mech., Belgrade, 1970, Vol. 4, pp. 448–449.

24. Jeffreys, H.: Damping of P waves. Geophys. J. Roy. Astr. Soc., Vol. 47, No. 2, Nov., 1976, pp. 347–349.

25. Johnson, W.: Impact strength of materials. London, Edward Arnold, 1973, 372 p.

26. Kieffer, S. W., Phakey, P. P. and Christie, J. M.: Shock process in porous quartzite: Transmission electron microscope observations and theory. Contrib. Mineral. Petrol., Vol. 59, No. 1, Dec., 1976, pp. 41–93.

27. Krishnamoorthy, K., Goldsmith, W. and Sackman, J. L.: Measurement of wave processes in isotropic and transversely isotropic elastic rocks. Int. J. Rock Mech. Min. Sci. & Geomech. Abstr., Vol. 11, 1974, pp. 367–378.

28. Lindolm, U. S., Nagy, A. and Yegkley, L. M.: A study of the dynamic strength and fracture properties of rock. Southwest Res. Inst., San Antonio, Texas, Final Tech. Rep., Aug., 1972, 59 p.

29. LINDOLM, U. S., YEAKLEY, L. M. and NAGY, A.: The dynamic strength and fracture properties of Dresser basalt. Int. J. Rock Mech. Min. Sci. & Geomech. Abstr., Vol. 11, 1974, pp. 181–191.

30. MARDISEWOJO, P.: Elastic wave velocity behaviour of rocks at low temperatures. Ph. D. Thesis, Univ. Tulsa, Oklahoma, 1973, 176 p.

31. MCCARTER, M. K.: A correlation of strength and dynamic properties of some elastic sedimentary rocks. Thesis, Univ. Utah, Salt Lake City, Utah, 1972.

32. MCCARTER, M. K., and WILLSON, J. E.: Strength versus energy dissipation in sandstone. Proc. 14th Symp. Rock Mech., Univ. Park, Penn., June, 1972, pp. 223–245.

33. MEDLAND, A. J.: An alternative definition of material properties. Strain, Vol. 12, No. 1, Jan., 1976, pp. 14–19.

34. MEEHAN, R. L.: Dynamic strength of hydraulic fill. J. Geotech. Eng. Div., Am. Soc. Civ. Eng., Vol. 102, No. GT6, June, 1976, pp. 641–646.

35. MEISTER, D.: A new ultrasonic borehole-meter for measuring the geotechnical properties of intact rock. In German. Proc. 3rd Cong. Int. Soc. Rock Mech., Denver, 1974, Vol. 2, Part A, pp. 410–417.

36. MURRI, W. J. and PETERSEN, C. F.: Dynamic properties of rock required for prediction calculations. Final Report Prepared for Defence Nuclear Agency, Feb., 1976, 78 p.

37. MYUNG, J. I. and BALTOSSER, P. E.: Fracture evaluation by the borehole logging method. Proc. 13th Symp. Rock Mech., Urbana, Illinois, 1971, pp. 31–56.

38. NEW, B. M.: Ultrasonic wave propagation in discontinuous rock. Dept. Envir. Transport and Road Res. Lab. Rep. 720, 1976, 19 p.

39. NIWOKOYE, D. N.: Prediction and assessment of concrete properties from pulse-velocity tests. Mag. Conc. Res., Vol. 25, No. 82, March, 1973, pp. 39–46.

40. PERSEN, L. N.: Rock dynamics and geophysical exploration. Amsterdam, Elsevier, 1975, 276 p.

41. PRATT, H. R., SWOLFS, H. S., BRACE, W. F., BLACK, A. D. and HANDIN, J. W.: Elastic and transport properties of an in situ jointed granite. Int. J. Rock Mech. Min. Sci. & Geomech. Abstr., Vol. 14, 1977, pp. 35–45.

42. PRATT, H. R., SWOLFS, H. S., BRECHTEL, C., JOHNSON, J. H. and BLACK, A. D.: Experimental and analytical study of the response of earth materials to static and dynamic loads. Final Report Prepared for Air Force Weapons Laboratory, May, 1976, 82 p.

43. RAMANA, Y. V.: Elastic behaviour of some Indian rocks under confining pressure. Int. J. Rock Mech. Min. Sci., Vol. 6, 1969, pp. 191–201.

44. RAMANA, Y. V.: Ultrasonics in rock mechanics. Proc. Symp. Rock Mech., Dhanbad, India, July, 1972, pp. 278–289.

45. RAO, M. V. M. S. and RAMANA, Y. V.: Dilatant behaviour of ultramafic rocks during fracture. Int. J. Rock Mech. Min. Sci. & Geomech. Abstr., Vol. 11, 1974, pp. 193–203.

46. RICKETTS, T. E.: Generalised Rayleigh wave propagation in anisotropic rock. Int. J. Rock Mech. Min. Sci. & Geomech. Abstr., Vol. 11, 1974, pp. 251–259.

47. RICKETTS, T. E. and GOLDSMITH, W.: Dynamic properties of rocks and composite structural materials. Int. J. Rock Mech. Min. Sci., Vol. 7, 1970, pp. 315–335.

48. SCHULER, K. W., LYSNE, P. C. and STEVENS, A. L.: Dynamic properties of oil shale. Trans. Am. Geophys. Union, Vol. 56, No. 12, 1974, p. 1194.

49. SCHULER, K. W., LYSNE, P. C. and STEVENS, A. L.: Dynamic mechanical properties of two grades of oil shale. Int. J. Rock Mech. Min. Sci. & Geomech. Abstr., Vol. 13, 1976, pp. 91–95.

50. SHOCKEY, D. A., CURRAN, R. D., SEAMAN, L., ROSENBERG, J. T. and PETERSEN, C. F.: Fragmentation of rock under dynamic loads. Int. J. Rock Mech. Min. Sci. & Geomech. Abstr., Vol. 11, 1974, pp. 303–317.

51. SHOCKEY, D. A., PETERSEN, C. F. and CURRAN, D. R.: Failure of rock under high rate tensile loads. Proc. 14th Symp. Rock Mech., Univ. Park, Penn., June, 1972, pp. 709–738.

52. SIKORA, W., LINOWSKI, H. and KIDYBINSKI, A.: Seismic detection of roof rock compaction in longwall face workings. Proc. Symp. Protection against Rock Fall, Katowice, 1973.

53. SILVA, W.: Body waves in a layered inelastic solid. Bull. Seism. Soc. Am., Vol. 66, No. 5, October, 1976, pp. 1539–1554.

54. SIMMONS, G., TODD, T. and BALDRIDGE, N. S.: Toward a quantitative relationship between elastic properties and cracks in low porosity rocks. Am. J. Sci., Vol. 275, No. 3, March, 1975, pp. 318–345.

55. SINGH, V. P.: Investigations of attenuation and internal friction of rocks by ultrasonics. Int. J. Rock Mech. Min. Sci. & Geomech. Abstr., Vol. 13, 1976, pp. 69–74.

56. SKOGLUND, G. R. and MARCUSON, W. F.: Evaluation of resonant column test devices. J. Geotech. Eng. Div., Am. Soc. Civ. Eng., Vol. 102, No. GT11, 1976, pp. 1147–1158.

57. STAS, B., SKRABIS, A. and DLOUHY, L.: Research into variations of stress in the vicinity of mine openings by means of seismic method. In German. Proc. Symp. Protection against Rock Fall, Katowice, 1973, 31 p.

58. SWIFT, R. P. and McKAY, M. W.: Dynamic behaviour of rocks to spherical stress waves. Proc. 13th Symp. Rock Mech., Urbana, Illinois, 1971, pp. 797–823.

59. SWOLFS, H. S. and HANDIN, J.: Dependence of sonic velocity on size and in situ stress in a rock mass. Proc. Symp. Investigation of Stress in Rock – Advances in Stress Measurement, Sydney, 1976, pp. 41–43.

60. TIXIER, M. P., LONELESS, G. W. and ANDERSON, R. A.: Estimation of formation strength from the mechanical properties log. J. Pet. Tech., Vol. 27, March, 1975, pp. 283–293.

61. TOKSOZ, M. N., CHENG, C. H. and TIMUR, A.: Velocities of seismic waves in porous rocks. Geophysics, Vol. 41, No. 4, Aug., 1976, pp. 621–645.

62. WORONKOV, O. K., NIZDRIN, G. I. and MAROW, W. I.: Elastic properties of rocks at sub-zero temperatures. In German. Proc. 3rd Cong. Int. Soc. Rock Mech., Denver, 1974, Vol. 2, Part A, pp. 424–429.

63. YANG, Z. and HATHEWAY, A. W.: Dynamic response to tropical marine limestone. J. Geotech. Eng. Div., Am. Soc. Civ. Eng., Vol. 102, No. GT2, Feb., 1976, pp. 123–138.

64. YOUASH, Y. Y.: Dynamic physical properties of rocks. Proc. 2nd Cong. Int. Soc. Rock Mech., Belgrade, 1970, Vol. 4, pp. 204–205.

Appendix II
Laboratory mechanical properties of rocks

c	cohesion, MPa
cg	coarse grained
E_t	tension modulus, GPa
fg	fine grained
hd	hard
I	irregular specimens
mg	medium grained
P	pressure, bars
p	porosity, %
R	test in bending
T	tangent modulus at 50 % fracture strength
t	direct tensile test
vcg	very coarse grained
vfg	very fine grained
w	weathered
w	*in remarks column:* water, %
# 1	
# 2	direction of loading
# 3	
⊥	right angle to
‖	parallel to
*	dynamic test values
**	saturated or wet
$\varnothing$	angle of internal friction
τ	shear strength, MPa
——	location not known

LABORATORY MECHANICAL PROPERTIES OF ROCKS

Rock	Location & Description	Density ϱ (g/cm³)	Modulus of Elasticity E (GPa)	Modulus of Rigidity G (GPa)	Poisson's Ratio υ	Compressive Strength σc (MPa)	Tensile Strength σt (MPa)	Remarks	Reference
	(1)	(2)	(3)	(4)	(5)	(6)	(7)	(8)	(9)
Actinolite Schist	India, K. G. F.	3.50	148.6* 77.9	59.1*	0.26* 0.29	234.4	–	p=6	RAMANA et al, 1973
Aegirine	USSR, Obmannyi Pass	3.47	146.9*	57.5*	0.28*	–	–	–	BELIKOV, 1967
Agglomerate	Japan, Furen, altered	2.70	–	–	–	39.3	–	p=1.1	YAMANO, 1974
Albite	USSR, Khema Lambina, 8 % anorthite	2.61	75.1*	30.0*	0.27	–	–	–	BELIKOV, 1967
Albitite	USA, Sylmar, Pa., P = 4000	2.61 2.615	69* 80*	28* 31.1*	– 0.29*	–	–	– –	STEWART, 1956 STEWART, 1956
Aleurolite	USSR, Karaganda	–	–	–	–	11.0	0.7	weak	KOIFMAN & CHIRKOV, 1969
	USSR, Kuzbass	–	–	–	–	77.0	5.1	dark gray	KOIFMAN & CHIRKOV, 1969
	USSR, Kuzbass, fg., compact	–	–	–	–	75.0	7.6	dark gray	KOIFMAN & CHIRKOV, 1969
	USSR, Kuzbass, fg., arenaceous	–	–	–	–	37.5	6.8	laminated	KOIFMAN & CHIRKOV, 1969
	USSR, Kuzbass	–	–	–	–	58.0	1,9	fissured	KOIFMAN & CHIRKOV, 1969
	USSR, Kuzbass	–	–	–	–	53.0	3.8	+ plant remains	KOIFMAN & CHIRKOV, 1969
	USSR, Kuzbass	–	–	–	–	65.0	8.7	arenaceous	KOIFMAN & CHIRKOV, 1969
	USSR, Kuzbass, cg	–	–	–	–	27.0	1.2	–	KOIFMAN & CHIRKOV, 1969

319

		ρ							Reference
Aleurolite	USSR, Kuzbass	–	–	–	–	11.0	1.0	–	KOIFMAN & CHIRKOV, 1969
	USSR, Donbass, fg., weak, friable	–	–	–	–	10.0	1.0	–	KOIFMAN & CHIRKOV, 1969
	USSR, Leninsk – Kuznetsk Basin	–	7.0	–	0.28	65.0	13.0	–	YAGODKIN et al, 1969
		–	4.0	–	0.20	37.5	10.7	–	YAGODKIN et al, 1969
		–	5.0	–	0.23	52.5	3.0	–	YAGODKIN et al, 1969
		–	18.0	–	0.20	75.0	12.0	–	YAGODKIN et al, 1969
		–	–	–	–	10.0	1.0	–	YAGODKIN et al, 1969
	USSR, Donbass	–	–	–	–	53.0–82.5	6.0–12.0	–	KOIFMAN, 1969
	USSR, Karaganda	–	–	–	–	22.0	1.4	–	KOIFMAN & CHIRKOV, 1969
		–	–	–	–	21.0	1.3	–	KOIFMAN & CHIRKOV, 1969
Amphibole	mineral (average)	3.15	117.0	–	–	–	–	–	BELIKOV, 1967
	India, siliceous	3.12	–	–	–	520	41.7 (R)	–	WUERKER, 1956
	India, siliceous, banded	3.07	–	–	–	443	–	–	WUERKER, 1956
	India, fg	3.07	105.0	–	0.14	426	51.4(R)	–	WUERKER, 1956
	India, cg	–	–	–	–	428	–	–	WUERKER, 1956
	India, siliceous, cg	–	–	–	–	271	–	–	WUERKER, 1956
	India, siliceous, fg	3.01	–	–	–	358	29.1(R)	–	WUERKER, 1956
	India, siliceous, veined	3.01	–	–	–	211	36.8(R)	–	WUERKER, 1956
	India, biotite rich	–	–	–	–	228	–	–	WUERKER, 1956
	India, siliceous, ‖ to foliations	3.02	46–91*	17–33*	0.39*	276–337	27.7–34.7(R)	–	WUERKER, 1956
	USSR, Karelian Isthmus	–	110.8	–	0.30	204.9	–	p = 0.26	BELIKOV et al, 1967

LABORATORY MECHANICAL PROPERTIES OF ROCKS

Rock	Location & Description	Density ρ (g/cm³)	Modulus of Elasticity E (GPa)	Modulus of Rigidity G (GPa)	Poisson's Ratio ν	Compressive Strength σc (MPa)	Tensile Strength σt (MPa)	Remarks	Reference
	(1)	(2)	(3)	(4)	(5)	(6)	(7)	(8)	(9)
Amphibole	USA, Bridge Canyon Dam, massive, cg	3.06	106.18	–	0.16	153.75	13.75	hornblende crystals 0.2–2 mm	BRANDON, 1974
	USA, Oroville-Feather Dam, Calif.	2.94	92.39	–	0.20	277.86	22.75	–	BRANDON, 1974
	India, Nandydroog	3.07	–	104*	0.46	–	–	–	WINDES, 1949
	USA, Madison, Mont., P=4000	3.07	–	46.7*	–	–	–	–	SIMMONS, 1964
	Canada, Daniel Johnson Dam, altered	–	9	17–35	–	–	–	–	BROWN et al, 1970
	Canada, Daniel Johnson Dam	–	22	4.2–5.3	–	–	–	–	BROWN et al, 1970
	USSR, Karelian Isthmus	–	113.0	–	0.30	209.0	–	p=0.26	BELIKOV et al, 1967
		–	81.0	–	–	288.0	–	p=0.24	BELIKOV et al, 1967
	USSR, hornblendite	3.12	98.6	–	0.26	–	–	p=0.66	BELIKOV et al, 1967
	USA, Auburn Dam	2.98	96.53	–	–	120.66	–	p=0.18	WALLACE et al, 1969
Andesite	USA, Palisades Dam, Idaho, hypersthene, fg, very hard	2.57	44.3	–	0.16	130.6	–	–	BALMER, 1953
	USA, Hoover Dam, Nev., coarse, porous, slightly weathered	2.37	37.02	–	0.23	102.73	7.24	–	BRANDON, 1974

Andesite	Japan, Yokohama, Manazuru machi, augite + magnetite + plagioclase	–	33.2	14.4	0.15	304	10.1	phenocryst = 0.8–1.6 mm	NISHIMATSU, 1970
	Germany, Walkenburg, P = 100–900	–	27	–	–	–	–	–	BREYER, 1930
	USA, Palisades Dam, Idaho, altered	2.57	54	0.18	–	–	–	p=0–70	USBR, 1953
		–	40	0.16	–	–	–	p=0–350	USBR, 1953
	Japan, Funabara	2.56	–	–	–	139.7	10.2	–	MITANI & KAWAI, 1974
	Japan, Soajome, soft	2.21	–	–	–	77.1	4.5	–	MITANI & KAWAI, 1974
	Japan, Funabara	2.43	–	–	–	100.0	7.5	–	MITANI & KAWAI, 1974
	Japan, Seikan Tunnel	2.72	–	–	–	96.4	11.1	–	MOCHIDA, 1974
	Japan	–	26.1	–	–	141.0	–	–	YAMAGUCHI, 1968
	Japan, Komori	–	38.0	–	–	–	–	+chlorite	IIDA et al, 1960
		–	97.0	–	–	–	–		IIDA et al, 1960
	Japan, Shinkomatsu	2.59	21.0	–	–	141.3	6.35	–	YAMAGUCHI & MORITA, 1969
	USSR, Bakuriani, Georgia	–	24.0	–	0.24	90.0	–	p=21.86 basaltic	BELIKOV et al, 1967
	USSR, geosynclines, porphyrites	2.63	63.3	–	0.22	–	–	p=2.7	BELIKOV, 1967
	–porphyritic	2.88	83.98	–	–	199.26	7.1	–	LIVINGSTON, 1961
Anhydrite	Persia	3.02	72–74	28	0.295	–	–	–	RICHARDS, 1933
	Germany	–	75.00	–	–	120.0	4.8	–	HÖFER & MENZEL, 1964
	–	–	75.8*	–	0.27*	–	–	–	GEYER & MYUNG, 1970
	USSR, Ukraine	–	89.24	–	0.33	225.56	–	p=0.4	BELIKOV et al, 1967
Anorthosite	Canada, Que. P=4000	2.708	–	33.2*	0.32*	–	–	–	BIRCH, 1961
	P=70–600	–	82.5	32.8	0.262	–	–	–	ADAMS & COKER, 1906

LABORATORY MECHANICAL PROPERTIES OF ROCKS

Rock	Location & Description	Density ρ (g/cm³)	Modulus of Elasticity E (GPa)	Modulus of Rigidity G (GPa)	Poisson's Ratio ν	Compressive Strength σc (MPa)	Tensile Strength σt (MPa)	Remarks	Reference
	(1)	(2)	(3)	(4)	(5)	(6)	(7)	(8)	(9)
Anortho-site	USA, Stillwater, Mont., P=1	2.770	35.3*	–	–	–	–	–	BIRCH, 1940
	P=4000	2.750	37.1*	0.31	–	–	–	–	BIRCH, 1940
	P=4000		38.9	–	–	–	–	–	SIMMONS, 1964
	USSR, Ukraine	–	91.0	–	0.33	230.0	–	p=0.4	BELIKOV et al, 1967
	USSR, Ukraine	2.75	87.5	–	–	–	–	p=0.28 mono-mineralic	BELIKOV et al, 1967
	USSR, Ukraine, dark	2.90	91.5*	–	–	–	–	p=0.46	BELIKOV et al, 1967
	USSR, Ukranian shield, cg.	2.77	82.3	–	0.34	–	–	p=0.42	BELIKOV, 1967
Argillite	USA, Devil Canyon Dam, Alaska	2.73	67.57	–	0.22	116.52	2.90	–	BRANDON, 1974
	USA, Cliffs shaft, Mich., ferruginous	2.97	–	–	–	95.84	17.24(R)	sericite + chlorite	BLAIR, 1956
	USA, Cliffs shaft, Mich.	2.85	–	–	–	122.74	17.93(R)	sericite + quartz	BLAIR, 1956
	USSR, Leninsk-Kuznetsk Basin	–	–	–	–	28.0	1.9	–	YAGODKIN et al, 1969
	USSR, Leninsk-Kuznetsk Basin	–	–	–	–	69.0	2.9	–	YAGODKIN et al, 1969
	USSR, Leninsk-Kuznetsk Basin	–	–	–	–	80.0	4.5	–	YAGODKIN et al, 1969
	USSR, Leninsk-Kuznetsk Basin	–	–	–	–	44.0	1.8	–	YAGODKIN et al, 1969
	USSR, Leninsk-Kuznetsk Basin	–	–	–	–	52.0	3.2	–	YAGODKIN et al, 1969

	Locality								Reference
Argillite	USSR, Leninsk-Kuznetsk Basin	–	–	–	–	35.0	2.7	–	YAGODKIN et al, 1969
	USSR, Leninsk-Kuznetsk Basin	–	–	–	–	70.0	4.1	–	YAGODKIN et al, 1969
	USSR, Karaganda	–	–	–	–	47.0	3.3	–	KOIFMAN & CHIRKOV, 1969
	USSR, Kuzbass	–	–	–	–	80.0	4.5	fissured	KOIFMAN & CHIRKOV, 1969
	USSR, Kuzbass	–	–	–	–	69.0	2.9	fissured + glide planes	KOIFMAN & CHIRKOV, 1969
	USSR, Kuzbass	–	–	–	–	70.0	4.1	fissured	KOIFMAN & CHIRKOV, 1969
	USSR, Kuzbass	–	–	–	–	52.0	3.2	fissures in 3 directions	KOIFMAN & CHIRKOV, 1969
	USSR, Kuzbass	–	–	–	–	44.0	1.8	strongly fissured + glide planes	KOIFMAN & CHIRKOV, 1969
	USSR, Kuzbass	–	–	–	–	35.0	2.7	fissured	KOIFMAN & CHIRKOV, 1969
	USSR, Kuzbass	–	–	–	–	28.0	1.9	compact	KOIFMAN & CHIRKOV, 1969
Augite	USSR, geosynclines, – porphyrites	2.84	83.4	–	0.27	–	–	$p = 0.94$	BELIKOV, 1967
	USSR, Western Georgia	3.16	143.7*	57.8*	0.24	–	–	–	BELIKOV, 1967
Barite	Germany, Clausthal, Harz	–	57.8	22.8	0.27	–	–	–	DRUDE & VOIGT, 1891
	quasi-isotropic-aggregate	–	61	25	0.21	–	–	–	VOIGT, 1928
Basalt	Germany, Ostritz, P=100–900	–	11.2	–	–	–	–	–	BREYER, 1930
	USA, Champion Mine, Mich.	2.85	61	27	–	–	–	–	WINDES, 1949
		2.97	85*	34*	–	–	–	–	WINDES, 1949

LABORATORY MECHANICAL PROPERTIES OF ROCKS

Rock	Location & Description	Density ϱ (g/cm³)	Modulus of Elasticity E (GPa)	Modulus of Rigidity G (GPa)	Poisson's Ratio υ	Compressive Strength σc (MPa)	Tensile Strength σt (MPa)	Remarks	Reference
	(1)	(2)	(3)	(4)	(5)	(6)	(7)	(8)	(9)
Basalt	USA, Howard Prairie Dam, Ore., altered	2.74	63	–	0.25	–	–	p=0–70	WINDES, 1949
		2.74	61	–	0.22	–	–	p=0–350	WINDES, 1949
	P = 1.0	2.82	76.4	31.5	0.22	–	–	–	FAI, 1961
	P = 1000	2.82	82.5	33.0	0.25	–	–	–	FAI, 1961
	USA, Guadalupe drill site	2.82	48.5*	–	0.38	–	–	–	SOMERTON et al, 1963
	USA, Medford, Ore., P = 0–416 (low strength)	2.72	60	–	0.22	169.8	–	–	WUERKER, 1956
	(high strength)	2.74	–	–	–	221	–	–	WUERKER, 1956
	USA, Mich.	2.85	96*	38*	0.281	232	–	–	WUERKER, 1956
	altered	–	–	–	–	81.9	–	–	WUERKER, 1956
	USA, Mich., amygdaloidal, altered, epidotized	2.70	41*	18*	0.09*	120	14.6	–	WUERKER, 1956
	USA, Mich., amygdaloidal, heavily altered, calcitized	2.8	60.0	–	0.15	344	28.4	–	WUERKER, 1956
	India, Koyna Dam, porous, vesicular	2.54	36.20	–	0.13	68.26	2.21	altered augite as ground mass	BRANDON, 1974
	USA, Black Canyon Dam, Idaho, glossy, dark gray	2.62	32.41	–	0.95	57.9	3.17	augite, 0.04–0.11 mm, fractures filled with opal + chalcedony	BRANDON, 1974

Basalt	USA, South Coulee Dam, Wash., slightly porous, fg.	2.81	50.50	—	0.18	171.68	—	BRANDON, 1974
	USA, South Coulee Dam, Wash., highly vesicular, fg.	2.62	59.29	—	0.21	95.84	—	BRANDON, 1974
	USA, South Coulee Dam, Wash., more highly vesicular, fg.	2.58	38.61	—	0.13	82.05	—	BRANDON, 1974
	USA, South Coulee dam, Wash., vesicular, with montmorillonite, fg.	2.47	42.06	—	0.19	61.36	—	BRANDON, 1974
	USA, AEC, Nev. test site, olivine basalt, hard, fg.	—	57.0	—	0.23*	—	—	YOUASH, 1970
	USA, Dresser Basalt	2.9	100.7	—	0.25	292.3	15.13	LEHNHOFF et al, 1973
	Brazil, Jupia Dam, dense	3.0	73.0	—	0.21	104.8	—	RUIZ, 1966
	Brazil, Barra Bonita Dam, dense	2.97	61.0	—	0.19	137.4	—	RUIZ, 1966
	Brazil, Jurumirim Dam, dense	2.71	42.8	—	0.16	133.0	—	RUIZ, 1966
	Brazil, Mussa Quarry, dense	2.68	44.0	—	0.19	126.8	—	RUIZ, 1966
	USA, AEC, Nev. test site, dense, unweathered, fg.	2.83	33.9	—	0.32	148.0	18.1 (at 50% of failure)	STOWE, 1969
	USA, Howard Prairie Dam, Ore., very dense, fg.	2.73	61.8	—	0.23	194.0	—	BALMER, 1953

LABORATORY MECHANICAL PROPERTIES OF ROCKS

Rock	Location & Description	Density ρ (g/cm³)	Modulus of Elasticity E (GPa)	Modulus of Rigidity G (GPa)	Poisson's Ratio ν	Compressive Strength σc (MPa)	Tensile Strength σt (MPa)	Remarks	Reference
	(1)	(2)	(3)	(4)	(5)	(6)	(7)	(8)	(9)
Basalt	USA, lower granite, massive, compact	–	50.2*	–	0.24*	228.0	12.9	at 50% of failure	MILLER, 1965
	USA, Little Goose, compact, massive	–	77.5*	–	0.27*	296.0	11.1	at 50% of failure	MILLER, 1965
	USA, John Day, massive, compact to vesicular	–	83.8	–	0.27	355.0	14.5	at 50% of failure	MILLER, 1965
	USSR, Bakuriani, Georgia, porous, andesite-basalt	–	23.54	–	0.24	88.26	–	p=21.86	BELIKOV et al, 1967
	USA, Green Peters Dam, amygdaloidal porphyritic	2.58	35.0	–	0.21	92.0	–	–	CORNS & NESBITT, 1967
	USA, AEC, Nev. test site	–	29.5	–	0.37	144.5	13.10	–	STOWE & AINSWORTH, 1968
	Brazil, Bariri Dam	2.97	21.8–46.4*	8.3*–21.1	0.31–0.10*	–	–	–	AVILA, 1966
	Brazil, Cochoeira Dam	2.94	76.5–100.3*	33.7–41.5*	0.14–0.21*	–	–	–	AVILA, 1966
	Brazil, Xavantes Dam	2.44	24.0–51.8*	11.2–23.1*	0.07–0.12*	–	–	–	AVILA, 1966
	Brazil, Jupia Dam	2.77	24.1–66.2*	10.5–30.4*	0.14–0.09*	–	–	–	AVILA, 1966
	Brazil, Ibitinga Dam	2.79	10.7–30.6*	3.8–11.7*	0.31–0.41*	–	–	–	AVILA, 1966

327

Basalt	Brazil, Barra Bonita Dam	2.91	29.0–60.6*	10.7–23.2*	0.31–0.35*	–	–	–	AVILA, 1966
	Brazil, Jupia Dam, compact	2.92	69.9*–74.4	–	0.21	108.7	–	p=2.1	RUIZ, 1966
	Brazil, Barra Bonita Dam, compact	2.82	57.2*–62.2	–	0.19	19.54	–	p=1.4	RUIZ, 1966
	Brazil, Jurimirim Dam	2.52	56.7*–43.6	–	0.16	157.2	–	p=4.2	RUIZ, 1966
	Brazil, Mussa Quarry	2.50	41.6*–44.9	–	0.19	172.1	–	p=5.7	RUIZ, 1966
	Brazil, Melhurb Quarry, pyroclastic	2.44	39.3*	–	–	146.0	–	p=11.6	RUIZ, 1966
	USA, Eniwetok Atoll, hd, fg-cg.	2.86	68.5	–	0.20	194.4	–	black, olivine	BLAIR, 1956
	Australia, Brooklyn, Vic. #1		22.8	–	–	86.2	12.89	–	BAMFORD, 1969
	#2	–	23.5	–	–	41.8	12.89	–	BAMFORD, 1969
	#3	–	15.9	–	–	116.9	12.89	–	BAMFORD, 1969
	Japan, Genbudo	–	55.0	–	–	–	–	+ olivine	IIDA et al. 1960
	USA	2.72	44.8–66.19*	–	–	255.1	9.65	–	RICKETTS & GOLDSMITH, 1970
	USA, Dresser	2.97	100.7	–	0.24	–	–	–	THIRUMALAI, 1970
	—	2.91	28.48	–	–	73.77	–	–	LIVINGSTON, 1961
Basalt + Sandstone	Brazil, Jupia Dam	2.25	28.5*	–	–	66.1	–	p=13.7	RUIZ, 1966
Basic Intrusion	USSR, Baltic shield	2.76	53.4	–	0.26	–	–	–	BELIKOV, 1967
Beschtau-nite	USSR, Caucasus	–	62.0	–	–	177.0	–	p=7.87	BELIKOV et al, 1967
Biotite	USA, West Pt. Dam, gneiss, amphibole	2.75	20.0	–	–	150.0	–	τ=8.8 MPa	CORNS & NESBITT, 1967

LABORATORY MECHANICAL PROPERTIES OF ROCKS

Rock	Location & Description	Density ϱ (g/cm³)	Modulus of Elasticity E (GPa)	Modulus of Rigidity G (GPa)	Poisson's Ratio υ	Compressive Strength σc (MPa)	Tensile Strength σt (MPa)	Remarks	Reference
	(1)	(2)	(3)	(4)	(5)	(6)	(7)	(8)	(9)
Biotite	USSR	3.10*	69.7*	27.7*	0.28*	-	-	-	BELIKOV, 1967
Blastonite	Canada, St. Lawrence, fg.	3.02	45.3	-	-	119.0	-	fluorite + quartz	COATES & PARSONS, 1966
Borax Ore	USA, Jenifer Mine, Calif.	2.14	4.21–7.38	-	-	44.1	2.76(R)	clay & carbonate	BLAIR, 1956
	USA, Jenifer Mine, Calif.	1.74	9.51	-	0.42	7.58	0.69(R)	clear borax	BLAIR, 1956
	USA, Jenifer Mine, Calif.	1.72	10.48	-	0.20	22.75	2.07(R)	clear borax	BLAIR, 1956
Breccia	—	2.71	14.7	-	-	110.3	-	-	JUDD, 1969
	India, Koyna Dam, basaltic breccia, soft, weathered, porous	2.01	13.10	-	0.08	17.93	1.45	-	
	Angola, Cambambe Dam	-	67.5	-	-	135.0	-	-	SARMENTO & VAZ, 1964
	USA, Green	2.35	23.0	0.19	63.0	-	-	p=17.5	CORNS & NESBITT, 1967
Brownstone	Peters Dam USA, Conn.	2.34	24.8	-	0.26	72.4	3.17	-	SINGH & HUCK, 1970
Calcite	mineral average	2.71	85.0	-	-	-	-	-	BELIKOV, 1967
Carbonatite	Canada	-	26.89	-	0.21	66.19	4.62(I)	irregular	EVERELL ET AL, 1970
	Canada, Oka, Quebec, fg	-	34.47	-	0.25	40.61	3.65	-	LADANYI & ROY, 1971
Carnallite	Germany	-	16.0	-	-	10.0	1.2	-	HOFER & MENZEL, 1964
	Germany, Werra	-	-	-	-	6.2–6.8	-	-	SPACKELER ET AL, 1960

Carnallite	Germany, Stassfurt	–	–	–	20.0	–	fragmented	SPACKELER ET AL., 1960
Chalk	Israel, Nareth	1.67	–	–	2.0	1.5	σt = Brazilian	WISEMAN, LEVY & ALSENSTEIN, 1966
	Israel, Haifa	2.12	–	–	4.0	3.8	σt = Brazilian	WISEMAN, LEVY & ALSENSTEIN, 1966
	Israel, Haifa	2.10	–	–	8.0	3.6	σt = Brazilian	WISEMAN, LEVY & ALSENSTEIN, 1966
	USA, Niobrara, Smoky Hill Formation	1.28	1.24	3.03	0.83	–	–	UNDERWOOD, 1964
	USA, Niobrara, Smoky Hill Formation	1.84	2.76	3.51	1.66	–	–	UNDERWOOD, 1964
Chert	USA, Okla., chalcedony	2.56	55.23	–	0.16	359.91	35.85 (R)	secant at 50 %	BLAIR, 1955
	USA, Okla., calcareous	2.39	–	–	–	204.77	11.03 (R)	p = 8.9	BLAIR, 1955
	USA, Pine Creek Mine, Tenn., dolomitic, fg	2.63	35.4* 52.0	16.48*	0.00* 0.19	210.3	23.44 (R)	p = 3.8, 40 % dolomite	BLAIR, 1956
	USA, Sligo Mine, Tenn., dolomitic, fg	2.67	55.2* 62.0	23.72*	0.14* 0.21	202.2	28.3 (R)	p = 4.8, 40 % dolomite	BLAIR, 1956
	Canada, MacLeod Mine	–	95.2	–	0.22	–	–	–	HERGET, 1973
Chlorite	Canada, Wawa, Ont., fg	2.78	60.9	–	–	112.0	–	+ 10–30 % CaCo3	COATES & PARSONS, 1966
Clay Fill	Yugoslavia, compacted	–	0.05	0.01	–	–	–	τ = 0–0.12 MPa	DVORAK, 1967
Claystone	USA, Idaho, Palisades Dam, porous, calcareous, montmorillonite matrix	2.20	2.83	0.04	–	8.27	–	weathered	BRANDON, 1974

LABORATORY MECHANICAL PROPERTIES OF ROCKS

Rock	Location & Description	Density ρ (g/cm³)	Modulus of Elasticity E (GPa)	Modulus of Rigidity G (GPa)	Poisson's Ratio ν	Compressive Strength σc (MPa)	Tensile Strength σt (MPa)	Remarks	Reference
	(1)	(2)	(3)	(4)	(5)	(6)	(7)	(8)	(9)
Claystone	Canada, Sanford Dam (spillwaysite), porous, fg	1.8	0.26	–	–	1.86	–	shows slickensides + shrinkage cracks	Brandon, 1974
Coal	Poland, Upper Silesia, mostly durain	–	5.54	–	0.44	50.5	–	–	Borecki & Chudek, 1972
	Poland, Upper Silesia, bedded	–	5.09	–	0.40	49.5	–	–	Borecki & Chudek, 1972
	Poland, Upper Silesia, saprope-lite	–	1.46	–	0.49	101.0	–	–	Borecki & Chudek, 1972
	Poland, Upper Silesia, saprope-lite	–	2.1	–	–	–	–	–	Kidybinski, 1969
	Poland, Upper Silesia, hard coal	–	3.5	–	–	45.0	2.0	tensile strength, ∥ bedding	Kidybinski, 1969
	Poland, Upper Silesia, bedded coal seam 506, middle section	–	2.04	–	–	20.8	–	–	Lama, 1966
	Poland, Upper Silesia, seam 506, top section	–	5.1	–	–	29.1	–	–	Lama, 1966
	Poland, Upper Silesia, seam 504	–	1.01	–	–	9.7	–	–	Lama, 1966
	UK, Barnsley Hard, ⊥-lamination	–	–	–	–	69.36	–	–	Hobbs, 1964

Coal							
UK, Barnsley Hard	—	—	—	—	51.71	—	HOBBS, 1964
UK, Oakdale, lamination	—	—	—	—	10.34	—	HOBBS, 1964
UK, Deep Duffryn	—	—	—	—	12.85	—	EVANS & POMEROY, 1958
UK, Cwmtillery coal ‖-bedding & cleat	—	—	—	—	5.79	—	MURRELL, 1958
⊥-bedding & cleat	—	—	—	—	10.69	—	MURRELL, 1958
S. Africa, Witbank Colliery, hard coal	—	3.64	—	—	17.4	in situ test on -0.45 m specimen, h/d = 2	BIENIAWSKI & VOGLER, 1970
USA, Old Ben Mine, Herrin 6 seam	# 1	3.83	—	0.42	14.13	—	KO & GERSTLE, 1976
	# 2	3.45	—	0.42	10.0	—	KO & GERSTLE, 1976
	# 3	3.13	—	0.42	10.0	—	KO & GERSTLE, 1976
USA, Bruceton Mine, Pa., Pittsburgh seam	# 1	3.73	—	—	29.65	—	KO & GERSTLE, 1976
	# 2	1.52	—	—	14.40	—	KO & GERSTLE, 1976
	# 3	1.66	—	—	14.48	—	KO & GERSTLE, 1976
	isotropic	2.97	0.55	0.37	—	—	KO & GERSTLE, 1976
India, Jharia, Bhutgoria Colliery, XVI seam ⊥ – bedding	—	—	—	—	15.1	h/d = 1	SINGH, 1965
‖ – bedding	—	—	—	—	8.23	—	SINGH, 1965
India, Jharia, Sendra Bansjora, IX seam ⊥ – bedding	—	—	—	—	22.65	h/d = 1	SINGH, 1965
‖ – bedding	—	—	—	—	24.65	—	SINGH, 1965

LABORATORY MECHANICAL PROPERTIES OF ROCKS

Rock	Location & Description	Density ρ (g/cm³)	Modulus of Elasticity E (GPa)	Modulus of Rigidity G (GPa)	Poisson's Ratio ν	Compressive Strength σc (MPa)	Tensile Strength σt (MPa)	Remarks	Reference
	(1)	(2)	(3)	(4)	(5)	(6)	(7)	(8)	(9)
Coal	India, Raniganj, Jamuria Colliery, Koithi seam, ⊥ – bedding	–	–	–	–	25.75	–	h/d = 1	SINGH, 1965
	‖ – bedding	–	–	–	–	23.3	–	–	SINGH, 1965
	India, Raniganj, Jamuria Colliery, Poniati seam	‖ –, ⊥ –	–	–	–	26.0, 13.8	–	h/d = 1	SINGH, 1965; SINGH, 1965
	India, Raniganj, West Jamuria, Taltore seam	‖ –, ⊥ –	–	–	–	25.1, 13.36	–	h/d = 1	SINGH, 1965; SINGH, 1965
	India, Raniganj, West Jamuria, Jamuria seam	‖ –, ⊥ –	–	–	–	18.13, 15.5	–	h/d = 1	SINGH, 1965; SINGH, 1965
	India, Raniganj, Methani, Dishergarh seam	‖ –, ⊥ –	–	–	–	19.4, 18.1	–	h/d = 1	SINGH, 1965; SINGH, 1965
	India, Kothagudem, King seam, No. 5 Incline	‖ –, ⊥ –	–	–	–	32.6, 15.38	–	h/d = 1	SINGH, 1965; SINGH, 1965
	S. Africa, Cornelia Collieries	–	–	–	–	44.0	–	top seam	DENKHAUS, 1965
	S. Africa, Cornelia Collieries	–	–	–	–	46.0	–	middle seam	DENKHAUS, 1965
	S. Africa, Cornelia Collieries	–	–	–	–	63.0	–	bottom seam	DENKHAUS, 1965
	S. Africa, Coalbrook, Sth. Colliery	–	–	–	–	32.0	–	top seam	DENKHAUS, 1965

Coal								
S. Africa, Coalbrook, Sth. Colliery	—	—	—	—	48.0	—	middle seam	DENKHAUS, 1965
S. Africa, Coal-Sigma Colliery	—	—	—	—	39.0	—	upper middle seam	DENKHAUS, 1965
USSR, Leninsk-Kuznetsk Basin	—	—	—	—	12.0	0.45	—	YAGODKIN et al, 1969
USSR, Leninsk-Kuznetsk Basin	—	—	—	—	13.5	0.60	—	YAGODKIN et al, 1969
USSR, Leninsk-Kuznetsk Basin	—	—	—	—	14.0	0.60	—	YAGODKIN et al, 1969
USSR, Donbass, bituminous	—	—	—	—	—	1.10	weak	KOIFMAN, 1969
USSR, Donbass, bituminous	—	—	—	—	—	3.05	medium	KOIFMAN, 1969
USSR, Donbass, anthracite	—	—	—	—	45.0	5.9	—	KOIFMAN, 1969
USSR, Nesvetai, anthracite	—	—	—	—	—	2.9	—	KOIFMAN & CHIRKOV, 1969
USSR, semilustrous	—	—	—	—	19.5	1.5	+ argillite	KOIFMAN & CHIRKOV, 1969
USSR, semilustrous	—	—	—	—	—	1.1	+ argillite lenses	KOIFMAN & CHIRKOV, 1969
USSR, semilustrous	—	—	—	—	22.0	0.8	+ argillite lenses	KOIFMAN & CHIRKOV, 1969
USSR, semilustrous	—	—	—	—	20.0	0.9	+ argillite lenses	KOIFMAN & CHIRKOV, 1969
USSR, semilustrous	—	—	—	—	21.0	1.1	+ argillite lenses	KOIFMAN & CHIRKOV, 1969
USSR, semilustrous	—	—	—	—	17.0	0.8	+ argillite lenses	KOIFMAN & CHIRKOV, 1969
USSR, semilustrous	—	—	—	—	14.0	0.6	+ argillite lenses	KOIFMAN & CHIRKOV, 1969
USSR, semilustrous, fissured	—	—	—	—	13.5	0.6	+ argillite lenses	KOIFMAN & CHIRKOV, 1969
USSR, semilustrous, strongly fissured	—	—	—	—	12.0	0.45	+ argillite lenses	KOIFMAN & CHIRKOV, 1969

LABORATORY MECHANICAL PROPERTIES OF ROCKS

Rock	Location & Description	Density ϱ(g/cm³)	Modulus of Elasticity E(GPa)	Modulus of Rigidity G(GPa)	Poisson's Ratio ʋ	Compressive Strength σ_c(MPa)	Tensile Strength σ_t(MPa)	Remarks	Reference
(1)		(2)	(3)	(4)	(5)	(6)	(7)	(8)	(9)
Coal	UK, Barnsley Hard	–	18.62	–	0.17	–	–	–	WILSON, 1961
	USA, Beckley Mine	–	–	–	–	8.02	–	76 mm cube	HOLLAND, 1964
	USA, Island Creek	–	–	–	–	32.58	–	76 mm cube	HOLLAND, 1964
	USA, Coalburg	–	–	–	–	47.31	–	77 mm cube	HOLLAND, 1964
	USA, Hernshaw	–	–	–	–	32.59	–	76 mm cube	HOLLAND, 1964
	Poland, sapropelitic	–	29.55	–	0.49	101.0	–	–	ZNAŃSKI, 1960
	Poland, durain	–	–	–	–	32.0	–	–	ZNAŃSKI, 1960
	Poland, banded	–	50.90	–	0.40	49.5	–	–	ZNAŃSKI, 1960
	Netherlands, Wilhelmina Colliery	–	2–4	–	–	10–20	–	–	CREUELS & HERMES, 1956
	UK, Pentremawr	⊥	–	–	–	45.71	1.59 (R)	1.93 ⊥ cleat	POMEROY, 1960
	UK, Pentremawr	∥	–	–	–	43.71	2.41 (R)		POMEROY, 1960
	UK, Deep Duffryn	⊥	–	–	–	14.48	6.89 (R)	6.89 ⊥ cleat	POMEROY, 1960
		∥	–	–	–	14.55	0.97 (R)		POMEROY, 1960
	UK, Oakdale	⊥	–	–	–	9.79	0.48 (R)	0.62 ⊥ cleat	POMEROY, 1960
		∥	–	–	–	14.55	0.75 (R)		POMEROY, 1960
	UK, Cwmtillery	⊥	–	–	–	21.44	0.55 (R)	0.55 ⊥ cleat	POMEROY, 1960
		∥	–	–	–	11.46	0.97 (R)		POMEROY, 1960
	UK, Barnsley Hard	⊥	–	–	–	50.47	1.10 (R)	2.76 ⊥ cleat	POMEROY, 1960
		∥	–	–	–	33.51	4.07 (R)		POMEROY, 1960
	UK, Teversal	⊥	–	–	–	38.4	1.10 (R)	2.29 ⊥ cleat	POMEROY, 1960
		∥	–	–	–	27.23	3.10 (R)		POMEROY, 1960

Coal

					Reference
UK, Markham	⊥ – ‖ –	34.16 28.54	0.62 (R) 1.72 (R)	1.86 ⊥ cleat	POMEROY, 1960 POMEROY, 1960
UK, Rossington	⊥ – ‖ –	37.58 27.37	0.76 (R) 2.48 (R)	1.65 ⊥ cleat	POMEROY, 1960 POMEROY, 1960
UK, Linby	⊥ – ‖ –	34.40 27.72	0.90 (R) 2.90 (R)	2.00 ⊥ cleat	POMEROY, 1960 POMEROY, 1960
Belgium, 68 seam, Beeringen	–	2.50	–	–	STASSEN & HAUSMAN, 1956
Belgium, 19 seam, Hothalen	–	5.35	–	–	STASSEN & HAUSMAN, 1956
Belgium, Fontaine, Brose seam	–	6.0	–	–	STASSEN & HAUSMAN, 1956
Belgium, Limbourge-Meuse, 27 seam	–	6.0	–	–	STASSEN & HAUSMAN, 1956
Belgium, Blumenthal, Karel 1	–	12.5	–	–	STASSEN & HAUSMAN, 1956
Belgium, Fontaine, Folleemprise seam	–	12.8	–	–	STASSEN & HAUSMAN, 1956
Belgium, Houthalen, 10 seam	–	14.3	–	–	STASSEN & HAUSMAN, 1956
Belgium, Limbourg-Meuse, 7 seam	–	15.4	–	–	STASSEN & HAUSMAN, 1956
Belgium, Beeringen, 63/64 N	–	17.1	–	–	STASSEN & HAUSMAN, 1956
Belgium, P. et Beaujone Colliery	–	22.5	–	–	STASSEN & HAUSMAN, 1956
Belgium, Sachsen, 18 seam	–	35.0	–	–	STASSEN & HAUSMAN, 1956
Belgium, André Dumont Colliery	–	33.6	–	–	STASSEN & HAUSMAN, 1956
Belgium, P. et Beaujone, 14 seam	–	35.0	–	–	STASSEN & HAUSMAN, 1956

LABORATORY MECHANICAL PROPERTIES OF ROCKS

Rock	Location & Description	Density ϱ(g/cm³)	Modulus of Elasticity E(GPa)	Modulus of Rigidity G(GPa)	Poisson's Ratio υ	Compressive Strength σc(MPa)	Tensile Strength σt(MPa)	Remarks	Reference
	(1)	(2)	(3)	(4)	(5)	(6)	(7)	(8)	(9)
Coal	Belgium, Ressaix Colliery	–	–	–	–	36.4	–	–	Stassen & Hausman, 1956
	Belgium, E seam, André Dumont	–	–	–	–	37.0	–	–	Stassen & Hausman, 1956
	Belgium, André Dumont, C seam	–	–	–	–	40.0	–	–	Stassen & Hausman, 1956
	Belgium, Fontaine, Veine du Sud	–	–	–	–	40.0	–	–	Stassen & Hausman, 1956
	Belgium, André Dumont, 27 seam	–	–	–	–	44.3	–	–	Stassen & Hausman, 1956
	Belgium, Limbourg-Meuse, 2 seam	–	–	–	–	49.0	–	–	Stassen & Hausman, 1956
	Belgium, Grospierre seam	–	–	–	–	51.7	–	–	Stassen & Hausman, 1956
	Belgium, Meinrich Colliery, Finefrau seam	–	–	–	–	98.0	–	–	Stassen & Hausman, 1956
	Belgium, President seam	–	–	–	–	125.0	–	–	Stassen & Hausman, 1956
	Belgium, Sonnenschein seam	–	–	–	–	153.0	–	–	Stassen & Hausman, 1956
	Bulgaria, Brown coal	–	1.0	–	–	6.65	0.76	–	Paraskevov et al, 1968
Conglomerate	(–)	2.72	13.0	–	0.15	226.0	–	–	Judd, 1969
	Canada, Denison Mine	–	71.06	31.03	0.13	185.4	7.52	–	Morrison, 1970

Conglomerate	Canada, Millikan Mine	–	91.01	43.44	0.10	121.28	7.38	–	MORRISON, 1970
	USA, Flaming Gorge Dam, Utah, mg – cg	2.54	14.13	–	0.03	88.25	2.96	slightly altered, grain 0.13–4.0 mm	BRANDON, 1974
	USA, McDowell Dam, Ariz., porous, cg, iron oxide + clay mineral matrix	2.47	1.26	–	0.12	30.34	–	fragments 1–25 mm	BRANDON, 1974
	India, Bhakra Dam, calcareous, dense	2.70	46.19	–	0.15	105.49	–	fragments 0.08–0.50 mm	BRANDON, 1974
	Iran, Pahlavi Dam	–	–	–	–	16.79 (w) 18.06 (d)	w = wet d = dry		Fox et al, 1964
	Iran, Pahlavi Dam	–	–	–	–	32.37 (d) 31.0 (w)	–		
	India, Bhakra Dam	–	51.7	–	0.13	107.76	–	secant modulus p = 0.8–2.6	BHATNAGAR & SHAH, 1964
	USA, Estancia Valley P = 200	–	21.58 16.9	10.2* 6.0	0.06* 0.43	66.9	–	–	BRATTON & PRATT, 1968
	USA, Estancia Valley P = 200	–	19.8* 16.9	9.0* 6.0	0.10* 0.43	63.1	–	–	BRATTON & PRATT, 1968
	Canada, Elliott Lake, Ont.	2.76	76.0	–	–	222.0	–	bedded	COATES & PARSONS, 1966
Crassin	France, Saizerais Mine, Lorraine	–	22.5* 7.0	–	–	35.0	2.5	–	TINCELIN & SINOU, 1964
Dacite	S. Africa, Pangolapoost Dam	2.67	–	–	–	112.5	–	p = 6.06	PHELINES, 1967
	USSR, geosynclines, porphyrites	2.62	41.0	–	0.22	–	–	p = 3.06	BELIKOV, 1967

LABORATORY MECHANICAL PROPERTIES OF ROCKS

Rock	Location & Description	Density ρ (g/cm³)	Modulus of Elasticity E (GPa)	Modulus of Rigidity G (GPa)	Poisson's Ratio υ	Compressive Strength σc (MPa)	Tensile Strength σt (MPa)	Remarks	Reference
	(1)	(2)	(3)	(4)	(5)	(6)	(7)	(8)	(9)
Diabase	Canada, Sudbury, Ont., olivine P = 70–600	–	94.9	37.0	0.284	–	–	–	ADAMS & COKER, 1906
	Germany, Neuwerk P = 100–900	–	79.3	–	–	–	–	–	BREYER, 1930
	Germany, Nieder-Kunnersdorf P = 100–900	–	87.2	–	–	–	–	–	BREYER, 1930
	USA, Westfield, Mass.	2.95	80–88.5*	–	–	–	–	–	IDE, 1936
	USA, Fredrick	3.017	107.4*	39.1*	–	–	–	–	BIRCH & BANCROFT, 1939
		3.015	107.5*	42.0*	–	–	–	–	BIRCH & BANCROFT, 1939
	P = 4000	3.017	–	44.2*	–	–	–	–	BIRCH & BANCROFT, 1939
	Canada, Noranda Mines, Ont. (750 metres deep)	2.898	82.1*	36.2*	0.103	–	–	–	BIRCH & BANCROFT, 1939
		2.989	8.92*	40.7*	0.228*	–	–	–	BIRCH & BANCROFT, 1939
	Canada, Noranda Mines, Ont.	3.007	91.3*	40.0*	0.18*	–	–	–	BIRCH & BANCROFT, 1939
	USA, Vinalhaven, Maine, olivine, P = 56	2.96	102.0	–	0.27	–	–	P = 11 bars	ZISMAN, 1933
		2.96	101.5	–	0.27	–	–	–	ZISMAN, 1933
	P = 500	2.96	107.0*	42.1*	–	–	–	–	IDE, 1936
		–	–	43.0	–	–	–	–	BIRCH & BANCROFT, 1939

Diabase	Canada, Noranda Mines, Ont., me-ta-diabase	3.04	116.1*	48.1*	0.22	–	–	dynamic	BIRCH & BANCROFT, 1939
		2.99	106.2*	44.2*	0.22*	–	–	dynamic	BIRCH & BANCROFT, 1939
	Canada	2.98	–	36.3*	–	–	–	dynamic	SIMMONS, 1964
	USSR, Ukraine	–	112.78	–	0.26	313.82	–	p = 0.3	–
	USA, Ahmeck mine, Mich.	2.89	72.0	28.0	–	–	–	–	WINDES, 1949
		2.94	76.0	30.0	–	–	–	–	WINDES, 1949
	USSR, Onega	3.04	104.0	43.0	0.23	–	–	–	FAI, 1961
	P = 1000		114.0	46.0	0.25	–	–	–	FAI, 1961
	USA, N.Y., altered + basalt	2.94	95.8	37.3	0.28	321.3	55.1 (R)	–	WUERKER, 1956
	USA, Mich., cg	2.89	71.7*	16.4*	0.28*	227.5	31.02	–	WUERKER, 1956
	cg	2.91	79.3*	31.0*	0.28*	273.7	39.99	–	WUERKER, 1956
	fg	2.94	76.5*	29.99*	0.27*	290.9	43.40	–	WUERKER, 1956
	epidote-rich	–	–	–	–	240.6	–	grain size = 1 mm	WUERKER, 1956
	USA, Ariz., + gabbro + olivine + gabbro, altered	–	–	–	–	211.6	–	grain size = 1.2 mm	WUERKER, 1956
		2.81	–	–	–	115.1	–	grain size = 6 mm	WUERKER, 1956
	USA, Mich., amygdaloidal, al-tered, calcitised	–	68.9	–	0.25	155.8	29.44 (R)	–	WUERKER, 1956
		–	70.3*	28.28*	–	–	–	–	WUERKER, 1956
	Brazil, Chapadao Quarry	2.96	91.8*	–	0.24	161.4	–	p = 1.2	RUIZ, 1966
			92.9						
	Canada, Elliot Lake, Ont.	2.96	94.8	–	–	218.0	–	–	COATES & PARSONS, 1966
	USA, Pa. #1	–	108.6	–	0.19	–	–	–	SELLERS, 1970
	#2	–	105.2	–	0.27	–	–	–	SELLERS, 1970
	USA, Fredrick, fg	3.01	97.9	–	–	–	42.0	0.7	BRACE, 1963
			101.5*						
	USSR	–	–	–	–	290.0	31.5	–	KOIFMAN, 1969
	USSR, Ukraine	–	115.0	–	0.26	320.0	–	p = 0.3	BELIKOV et al, 1967

LABORATORY MECHANICAL PROPERTIES OF ROCKS

Rock	Location & Description	Density ϱ (g/cm³)	Modulus of Elasticity E (GPa)	Modulus of Rigidity G (GPa)	Poisson's Ratio ν	Compressive Strength σ_c (MPa)	Tensile Strength σ_t (MPa)	Remarks	Reference
	(1)	(2)	(3)	(4)	(5)	(6)	(7)	(8)	(9)
Diabase	USSR, Ukrainian shield	3.00	106.4	–	0.28	–	–	p = 0.25	BELIKOV, 1967
	USSR, Baltic shield	3.19	117.0	–	–	–	–	p = 0.4	BELIKOV, 1967
	Czechoslovakia, Primbram	2.69	79.6	–	0.21	243.8	–	–	SIBEK et al, 1964
	Czechoslovakia, Primbram	2.61	85.4	–	0.24	189.9	–	–	SIBEK et al, 1964
Diallage	USSR, South Ural	3.31	145.1*	60.0*	0.21*	–	–	–	BELIKOV, 1967
Diopside	USSR, baikalite	3.28	160.4*	63.6*	0.26*	–	–	–	BELIKOV, 1967
Diorite	USA, Mineville, N.Y.	3.02	87.0	37.0	–	–	–	–	WINDES, 1949
		3.03	55.0	28.0	–	–	–	–	WINDES, 1949
	USA, Dedham, Mass., quartz P = 4000	2.928	–	33.6*	–	–	–	–	SIMMONS, 1964
			–	41.8*	–	–	–	–	SIMMONS, 1964
	USA, Idaho, quartz	2.50	28.96	–	–	87.36	–	–	WUERKER, 1956
	USA, Idaho, quartz	2.50	30.34*	12.76	0.19*	–	–	–	WUERKER, 1956
	USA, N.Y., gneissic-gabbro	2.03	55.3*	27.79*	0.005*	186.16	13.79	–	WUERKER, 1956
	USA, Utah, w. augite + porphyry	2.74	84.12*	33.72*	0.25	333.0	–	–	WUERKER, 1956
	USA, Utah, w. augite + porphyry, altered	2.72	79.78*	31.78*	0.26*	278.0	–	–	WUERKER, 1956

Diorite	USA, Utah, w. augite + porphyry, altered	2.72	66.4*	25.58*	0.30*	215.1	–	–	WUERKER, 1956
	USA, Utah, biotite, porphyry, altered	2.66	60.05*	24.55*	0.22*	179.95	–	–	WUERKER, 1956
	USA, Colo., gneiss, low strength	2.86	71.02	–	–	64.1	–	–	WUERKER, 1956
	USA, Colo., gneiss, high strength	2.87	–	–	–	104.39	–	–	WUERKER, 1956
	USA, Negaunee Mine, Mich., hornblende	3.01	100.8 106.66*	44.26*	0.26 0.29*	274.41	44.88	secant value	BLAIR, 1956
	USA, Mich., intrusive, altered	2.81	87.08*	29.58*	0.32*	180.04	50.33	secant value	BLAIR, 1956
	USA, Bridge Canyon Dam, Ariz., massive, quartzdiorite, cg	2.71	46.88	–	0.07	118.59	8.20	grain, oligoclase = 1-5 mm, quartz = 0.3–2 mm	–
	USA –	2.79	35.16 37.92*	–	–	219.25	5.22	–	RICKETTS & GOLDSMITH, 1970
	USA, Mather A Mine, Mich.	3.01	106.66* 101.09	42.2*	0.29* 0.27	274.4	46.88(R)	–	BOYUM, 1961
	USA, Mather A Mine, Mich.	2.81	87.08*	29.58*	0.32*	180.64	50.33(R)	–	BOYUM, 1961
	USA, slightly fractured, quartz	–	21.37 30.34*	–	0.05 0.19*	–	–	–	RINEHART et al, 1961
Diorite, quartz	Yugoslavia, mylonitised	–	0.470	0.089	–	–	–	τ = 0.0–0.385	DVORAK, 1967
	USA, Cedar City, Utah, cg, massive, altered	–	4.09	–	0.20	31.21	–	p = 9	PRATT et al, 1972

LABORATORY MECHANICAL PROPERTIES OF ROCKS

Rock	Location & Description	Density ϱ(g/cm³)	Modulus of Elasticity E(GPa)	Modulus of Rigidity G(GPa)	Poisson's Ratio υ	Compressive Strength σc(MPa)	Tensile Strength σt(MPa)	Remarks	Reference
	(1)	(2)	(3)	(4)	(5)	(6)	(7)	(8)	(9)
Dolerite	India, KGF	3.11	83.5*	33.9*	0.23*	–	–	p = 8%	RAMANA et al, 1973
	India, KGF	3.14	109.3*	43.6*	0.25*	211.1	–	p = 4%	RAMANA et al, 1973
			93.9		0.20				
	India, KGF	3.01	84.6*	36.9*	0.15*	–	–	p = 7%	RAMANA et al, 1973
	India, KGF	3.00	91.9	35.5*	0.29*	261.3	–	–	RAMANA et al, 1973
			82.0		0.11				
	S. Africa, Western Deep Ltd.	–	–	–	–	410.0	–	–	DENKHAUS, 1965
	USSR, Khantaika, hydropower station, intrusive, vertical	–	37.8	–	0.25	–	–	in situ test	PANOV et al, 1970
	horizontal (cracked)	–	29.6	–	0.34	–	–	–	
	India, KGF	3.11	79.9*	34.7*	0.15*	–	–	–	RAMANA et al, 1973
	India, KGF	3.04	62.5*	23.5*	0.33*	–	–	–	RAMANA et al, 1973
Dolomite	Poland, Beuthen, Silesia P = 100	–	76.0–81.0	–	–	–	–	–	WINDES, 1949
	USA, Bethlehem, Pa.	2.83	71.0	32.3	–	–	–	–	IDE, 1936
	#1		93.0	36.2	–	–	–	–	IDE, 1936
	#2		91.6	39.8	–	–	–	–	IDE, 1936
	#3		78.5*	–	0.30*	–	–	–	GEYER & MYUNG, 1970
	–		46.3	–	–	96.5	–	–	JUDD, 1969
	USA, Cockeyville, Md., marble ⊥	2.87	49.3	26.1	–	–	–	–	OBERT et al, 1946
	=		63.1	28.4	0.16	–	–	–	OBERT et al, 1946
	‖		71.7	28.2	0.17	–	–	–	OBERT et al, 1946

									Reference
Dolomite	USA, Mascot, Tenn.	2.84	85.0	35.0	–	–	–	–	WINDES, 1949
	USA, Tenn.	2.84	–	–	–	321.98	–	–	WUERKER, 1956
	Brazil, Cucupe quarry	2.84	67.2	–	0.14	102.9	–	–	RUIZ, 1966
	USA, Oneota, porous, massive, fg	2.45	43.9*	–	0.34*	86.9	4.41	50% of fracture	MILLER, 1965
	USA, Lockport, porous, massive, vfg	2.58	51.0*	–	0.34*	90.3	3.03	50% of fracture	MILLER, 1965
	USA, Bonne Terre, dense, fg	2.64	66.2*	–	0.35*	151.7	5.03	50% of fracture	MILLER, 1965
	USA, Niagara, Decew dolomite	2.74	45.8 75.8* 21.37 25.86*	–	0.46 0.19*	–	–	–	HOGG, 1964
	USA	–	–	–	–	63.43	–	–	AVEDISSIAN & WOOD, 1968
	USA	–	23.3 34.1*	–	–	83.4	–	–	AVEDISSIAN & WOOD, 1968
	USA	–	32.1 42.1*	–	–	96.5	–	–	AVEDISSIAN & WOOD, 1968
	USA	–	33.4 37.7*	–	–	91.0	–	–	AVEDISSIAN & WOOD, 1968
	Brazil, Cacupe Quarry	2.83	63.8* 88.5	–	0.14	117.0	–	p = 0.1	RUIZ, 1966
	S. Africa, Crown Mines, Western Deep Ltd.	–	–	–	–	410.0	–	–	DENKHAUS, 1965
	USA, Blair, fg	2.84	106.0 109.0	–	–	–	–	p = 0.2	BRACE, 1963
	USA, Webatuk, mg	2.85	51.0 52.0	–	–	–	–	p = 0.5	BRACE, 1963
	USA, Dunham, fg	2.84	72.0	–	–	–	–	p = 0.8	BRACE, 1963
	USA, Jefferson City	2.4	–	–	–	32.82	4.50	p = 9.7	CLARK et al, 1970

LABORATORY MECHANICAL PROPERTIES OF ROCKS

Rock	Location & Description	Density ρ (g/cm³)	Modulus of Elasticity E (GPa)	Modulus of Rigidity G (GPa)	Poisson's Ratio υ	Compressive Strength σc (MPa)	Tensile Strength σt (MPa)	Remarks	Reference
	(1)	(2)	(3)	(4)	(5)	(6)	(7)	(8)	(9)
Dolomite	USA, Minn.	2.50	29.0	–	0.29	46.9	7.02	–	SINGH & HUCK, 1970
	USA, Mankato	–	51.71	–	0.25	105.5	13.1(R)	p = 9.5	HAIMSON & FAIRHURST, 1969
	USA, Gnome nuclear explosion site	–	–	–	–	74.0	–	p = 5.0	KNUTSON & BOHOR, 1964
	Canada ‖	–	60.26*	16.06*	0.81	–	–	–	LECOMTE, 1963
	⊥	–	72.26*	24.89*	0.45*	–	–	–	LECOMTE, 1963
Doreite	Japan, Seikan Tunnel	2.66	–	–	–	57.1	8.4	–	MOCHIDA, 1974
Dunite	New Zealand	–	152.0*	60.0*	0.27*	–	–	–	WUERKER, 1956
	#1	–	162.0*	58.0*	0.40*	–	–	–	WUERKER, 1956
	#2 P = 4000	3.27	64.7*	–	–	–	–	–	SIMMONS, 1964
	USA, Balsam Gap, N.C. #1	3.275	94.6*	56.0*	–	–	–	–	IDE, 1936
	P = 500	–		64.0*	–	–	–	–	BIRCH & BANCROFT, 1938
	P = 4000	–	–	67.0*	–	–	–	–	BIRCH & BANCROFT, 1938
	#2	–	148.4	47.6	–	–	–	–	IDE, 1936
	P = 500	–	65.4	–	–	–	–	–	BIRCH & BANCROFT, 1938
	P = 4000	–	69.4	–	–	–	–	–	BIRCH & BANCROFT, 1938
	#3	–	68.1	–	–	–	–	–	BIRCH & BANCROFT, 1938
	P = 500	–	70.6	–	–	–	–	–	BIRCH & BANCROFT, 1938
	P = 4000	–		–	–	–	–	–	

	Location	Density						Reference
Dunite	USA, Twin Sisters, Mt. Wash. #1	3.312	195.0*	74.0*	—	—	—	Birch, unpublished, 1940
	#2	(mean)	140	66.0*	—	—	—	Birch, unpublished, 1940
	#3		—	72.0*	—	—	—	Birch, unpublished, 1940
	P = 4000	3.326	—	75.7*	—	—	—	Birch, unpublished, 1940
	S. Africa, Moolhock Mine, Tvl.	3.760	50.9*	—	—	—	—	Simmons, 1964
	P = 10 000	—	57.2*	—	—	—	—	Simmons, 1964
Eclogite	USA, Healdburg, Calif. #1	3.44	142.0*	—	—	—	—	Birch, unpublished, 1940
	#2	3.44	123.0*	—	—	—	—	Birch, unpublished, 1940
	P = 4000	3.44	—	62.5*	—	—	—	Birch, unpublished, 1940
	P = 4000	3.44	—	70.7*	—	—	—	
	Norway	3.578	—	49.0*	—	—	—	Simmons, 1964
	P = 4000	3.578	—	75.1*	—	—	—	Simmons, 1964
	USSR	3.51	151.5	—	0.25	—	p = 0.54	Belikov, 1967
	USSR, porphyritised	3.50	151.5*	—	—	—	p = 0.49	Belikov, 1967
	USSR, Polar Urals, fg	3.54	211.8*	—	—	—	p = 0.22	Belikov, 1967
	USSR, Polar Urals, fg	3.49	206.7	—	—	—	p = 0.33	Belikov, 1967
Epidosite	USA, Mount Weather, Va.	3.26	91.0	41.0	—	—	—	Windes, 1949
	USA, Mich.	2.93	—	—	—	159.27	—	Wuerker, 1956
	USA, Mich., epidote-rich	—	—	—	—	240.6	—	Wuerker, 1956
	USA, Mich., baltic lode	3.18	103.4* / 110.3	—	0.03* / 0.16	—	p = 0.4	Blair, 1955

LABORATORY MECHANICAL PROPERTIES OF ROCKS

Rock	Location & Description	Density ρ (g/cm³)	Modulus of Elasticity E (GPa)	Modulus of Rigidity G (GPa)	Poisson's Ratio ν	Compressive Strength σc (MPa)	Tensile Strength σt (MPa)	Remarks	Reference
	(1)	(2)	(3)	(4)	(5)	(6)	(7)	(8)	(9)
Essexite	Canada, Mt. Johnson, Que. P = 70–600	–	67.1	26.7	0.258	–	–	–	ADAMS & COKER, 1906
Felspar	USSR, acidic-mineral (average)	2.57	71.0	–	–	–	–	–	BELIKOV, 1967
Fluorite	Germany, Stolberg, Harz, quasi-isotropic	–	102.5	42.0	0.22	–	–	–	DRUDE & VOIGT, 1891
Gabbro	Canada, St. Lawrence	2.27	51.7	–	–	102.0	–	pure	COATES & PARSONS, 1966
	Canada, New Glasgow, Que., P = 70–600	–	108.0	43.8	0.2	–	–	–	ADAMS & COKER, 1906
	Germany, Neurode P = 100–900	–	91.4	–	–	–	–	–	BREYER, 1930
	Germany, Harzburg P = 100–900	–	96.0	–	–	–	–	–	BREYER, 1930
	USA, French Crack, Pa., hypersthene	3.05	72.7	–	0.162	–	–	–	ZISMAN, 1933
	P = 11 #1		76.7	–	0.169	–	–	–	ZISMAN, 1933
	P = 56 #1		78.1*	34.8*	–	–	–	–	IDE. 1936
	P = 500	3.05	–	44.6*	–	–	–	–	BIRCH & BANCROFT, 1938
	P = 4000	3.05	–	48.0*	–	–	–	–	BIRCH & BANCROFT, 1938

Rock	Locality	Density						Remarks	Reference
Gabbro	P = 11 #2	3.05	58.4	—	0.114	—	—	—	ZISMAN, 1933
	P = 56 #2		61.5	—	0.136	—	—	—	ZISMAN, 1933
			71.1*	30.7*	—	—	—	—	IDE, 1936
	P = 500		44.3	—	—	—	—	—	BIRCH & BANCROFT, 1938
	P = 4000		48.0	—	—	—	—	—	BIRCH & BANCROFT, 1938
	USA, French Crack, Pa., Hypersthene, P = 11 #3	3.05	67.2	14.9	—	—	—	—	ZISMAN, 1933
	P = 56 #3	3.05	70.3	15.4	—	—	—	—	ZISMAN, 1933
	USA, Mellen, Wis.	2.90	33.2*	—	—	—	—	—	BIRCH & BANCROFT, 1938
	#1 P=1 #2 P=1	2.90	32.5*	—	—	—	—	—	BIRCH & BANCROFT, 1938
	USA, Mt. Weather, Va., Greenstone	3.02	105.1	42.1	0.25	—	—	—	OBERT et al, 1946
		2.96	81.0	35.0	—	—	—	—	OBERT et al, 1946
	USA, Mineville, N. Y., altered	2.93	85.0*	34.0*	—	—	—	—	WINDES, 1949
	USA, San Marcos, Calif.	2.87	—	37.0*	—	—	—	—	SIMMONS, 1964
	P = 4000	2.87	—	41.3*	—	126.17	—	—	SIMMONS, 1964
	USA, Ariz., veined		—	—	—	—	—	grain size = 1.0 mm	WUERKER, 1956
	USA, N.Y., gneissic, diorite	3.03	55.3*	27.79*	0.00–0.16*	186.16	13.79	—	WUERKER, 1956
			75.8*	—	0.16*	—	—	—	GEYER & MYUNG, 1970
			67.6*	—	0.17*	—	—	—	GEYER & MYUNG, 1970
			84.1*	—	0.21*	—	—	—	GEYER & MYUNG, 1970
	USSR, Ukraine		122.59	—	0.38	310.88	—	—	BELIKOV et al, 1967
	USSR, Ukraine		125.0	—	0.38	317.0	—	p = 0.15	BELIKOV et al, 1967
	USSR, Ukraine	3.13	119.0*	—	—	—	—	p = 0.23	BELIKOV, 1967

LABORATORY MECHANICAL PROPERTIES OF ROCKS

Rock	Location & Description	Density ρ (g/cm³)	Modulus of Elasticity E (GPa)	Modulus of Rigidity G (GPa)	Poisson's Ratio ν	Compressive Strength σ_c (MPa)	Tensile Strength σ_t (MPa)	Remarks	Reference
	(1)	(2)	(3)	(4)	(5)	(6)	(7)	(8)	(9)
Gabbro	USSR, Ukrainian shield	3.00	106.4	–	0.28	–	–	p=0.25	BELIKOV, 1967
	USSR, Baltic shield	3.19	117.0	–	–	–	–	p=0.4	BELIKOV, 1967
	USA	3.1	–	–	–	223.39	16.41	p=0.3	CLARK et al, 1970
	USSR, Baltic shield, biotitic	2.70	67.2	–	0.25	–	–	p=0.78	BELIKOV, 1967
	USSR, Ukraine	2.98	121.5	–	–	–	–	p=0.12	BELIKOV, 1967
	USA, Auburn Dam, meta–	3.01	96.53	–	–	133.24	–	p=0.42	WALLACE et al, 1969
Garnet	Brazil (spessartite almandine)	4.25	243.5*	96.3*	0.26*	–	–	–	BELIKOV, 1967
	mineral (average)	4.18	243.0	–	–	–	–	–	BELIKOV, 1967
Gaspe Skarn	Canada	–	55.99	–	0.14	219.94	22.75(R)	–	SHIH, 1963
Gneiss	USA, Pelham, Mass., granite– $\perp$	2.64	3.3	–	0.03	–	–	–	ZISMAN, 1933
	P = 1	2.64	28.4*	14.5*	–	–	–	–	IDE, 1936
		2.64	–	7.9*	–	–	–	–	BIRCH & BANCROFT, 1938
	P = 500	2.64	–	26.2*	–	–	–	–	BIRCH & BANCROFT, 1938
	P = 4000	2.64	–	33.4*	–	–	–	–	BIRCH & BANCROFT, 1938
	USA, Pelham, Mass., granite– ‖ P = 11	2.64	14.2	–	–	–	–	–	ZISMAN, 1933

Gneiss									
p = 56	2.64	22.0	–	–	–	–	–	–	ZISMAN, 1933
	2.64	22.0*	–	–	–	–	–	–	IDE, 1936
P = 11	2.64	16.8	–	0.086	–	–	–	–	ZISMAN, 1933
P = 56	2.64	25.5	–	0.146	–	–	–	–	ZISMAN, 1933
USA, Starlake, N.Y. granite–	2.64	50.0*	24.0*	–	–	–	–	–	WINDES, 1949
Switzerland, Maggio, Tessin		30.0*	–	–	–	–	–	–	GASSMANN et al, 1952
USA, Minesville, N.Y., syenite	2.81	55.0*	24.0*	–	–	–	–	–	WINDES, 1949
USA, diorite –, P = 0–70	2.87	68.0	–	0.08	–	–	–	–	USBR, 1960
P = 0–140	2.87	70.0	–	0.11	–	–	–	–	
USA, Montezuma, Quandragle, Colo, biotite –, P = 1	2.91		34.0*	–	–	–	–	–	BIRCH, 1940
USA, Hellgate, N.Y. ‖P = 1	2.64	–	16.4*	–	–	–	–	–	BIRCH, 1940
⊥P = 1	2.64	–	28.7	–	–	–	–	–	BIRCH, 1940
Brazil, Euclides da Cunha Dam	2.79	78.4	–	0.22	32.4	–	–	–	RUIZ, 1966
Brazil, Graminha Dam	2.73	76.3	–	0.27	165.0	–	–	–	RUIZ, 1966
Brazil, CTA Quarry	2.79	52.7	–	0.19	105.6	–	–	–	RUIZ, 1966
Brazil, Jaguari Quarry	2.73	78.3	–	0.24	137.1	–	–	–	RUIZ, 1966
Brazil, Camaruero Quarry	2.63	48.8	–	0.10	93.6	–	–	–	RUIZ, 1966
	–	64.1*	–	0.19*	–	–	–	–	GEYER & MYUNG, 1970
	–	72.4*	–	0.19*	–	–	–	–	GEYER & MYUNG, 1970
USA, Dworshak Dam, mg-fg, foliations at 45°	2.79	53.6*	–	0.34	162.0	6.89	at 50 % fract.		MILLER, 1965

LABORATORY MECHANICAL PROPERTIES OF ROCKS

Rock	Location & Description	Density ρ (g/cm³)	Modulus of Elasticity E (GPa)	Modulus of Rigidity G (GPa)	Poisson's Ratio υ	Compressive Strength σc (MPa)	Tensile Strength σt (MPa)	Remarks	Reference
	(1)	(2)	(3)	(4)	(5)	(6)	(7)	(8)	(9)
Gneiss	Brazil, granite, DER Quarry	2.74	46.6	–	0.23	88.8	–	–	RUIZ, 1966
	USA, Montezuma Tunnel, Colo, diorite, hard, mg–cg	2.86	68.4	–	0.09	84.5	–	–	BALMER, 1953
	USA, N.J. + augite + hornblende	3.36	103.42*	40.68*	0.27	218.56		–	WUERKER, 1956
	USA, N.J.-biotite hornblende	2.91	67.17*	23.37*	0.24	160.65	–	–	WUERKER, 1956
	USA, N.J.-pyroxene + granite	2.71	55.16*	25.79*	–	223.39	15.50(R)	–	WUERKER, 1956
	USA, N.Y. biotite granitic	2.64	49.99*	23.03*	–	250.97	16.55 (R)	–	WUERKER, 1956
	USA, N.Y. hornblende + syenite	2.75	44.88*	22.48*	–	339.59	14.10 (R)	–	WUERKER, 1956
	USA, N.Y. pyroxene + syenite	2.84	64.47*	28.89*	–	204.77	17.24 (R)	–	WUERKER, 1956
	USA, N.Y. pyroxene + biotite	2.76	57.09	25.51	–	157.20	21.37 (R)	–	WUERKER, 1956
	USA, N.Y. granite + syenite	2.75	42.20*	20.68*	–	215.11	15.86 (R)	–	WUERKER, 1956
	USA, N.Y. quartz + magnetite	3.18	36.96	20.48	–	224.7	11.72 (R)	–	WUERKER, 1956
	USA, N.Y. granitic magnetite	3.04	66.67*	28.75*	–	153.06	14.48 (R)	–	WUERKER, 1956

Gneiss	USA, N.Y. granitic pegmatite	2.65	44.61	21.17	—	195.81	—	—	WUERKER, 1956
	USA, Bridge Canyon Dam, Ariz., nonporous, cg, quartz diorite gneiss	2.83	61.71	—	0.11	81.36	13.51	—	BRANDON, 1974
	schistose-gneiss, fg ⊥	—	4.68	—	—	26.33	1.33	poor to moderate schistosity, grain = (0.2–0.7mm)	DEKLOTZ et al, 1966
	=	—	11.99	—	—	26.63	3.93		
	=	—	13.11	—	—	26.14	4.13		
	Australia, T. 1. UG power station, Snowy Mt.	2.72	65.31 / 49.24 (t)	—	0.16–0.22	144.44	8.00	t = tension	MOYE, 1964
	USA, Mass. Northfield Power Station, granitic gneiss ‖	—	57.23	—	0.21	111.7	—	—	WILD & McKITTRICK 1971
	⊥	—	38.61	—	0.15	224.1	—	—	WILD & McKITTRICK, 1971
	USA, Lake Mead, Nev., gneiss	—	67.22	—	0.13	126.2–222.7	—	—	RISING & ERICKSON, 1971
	Brazil, DER Quarry, granitic	2.68	43.5* / 47.6	—	0.23	73.0	—	p = 1.6	RUIZ, 1966
	Brazil, Euclides de Cunha Dam, granitic	2.65	70.9* / 83.3	—	—	162.0	—	p = 0.5	RUIZ, 1966
	Brazil, Anchieta Dam	2.36	8.9– / 35.4*	3.5– / 13.2*	0.28– / 0.33*	—	—	—	AVILA, 1966
	Brazil, Jaguari Dam, schistosic	2.75	31.2– / 39.1*	12.1– / 15.1*	0.29*	—	—	—	AVILA, 1966

LABORATORY MECHANICAL PROPERTIES OF ROCKS

Rock	Location & Description	Density ϱ (g/cm³)	Modulus of Elasticity E (GPa)	Modulus of Rigidity G (GPa)	Poisson's Ratio υ	Compressive Strength σc (MPa)	Tensile Strength σt (MPa)	Remarks	Reference
	(1)	(2)	(3)	(4)	(5)	(6)	(7)	(8)	(9)
Gneiss	Brazil, Capivari-Cachoeira Dam, gneissic	2.88	24.3–95.4*	10.7–40.2*	0.13–0.19*	–	–	–	AVILA, 1966
	Brazil, Graminha Dam, schistosic	2.76	15.3–91.7*	62.6–38.5*	0.19–0.24*	–	–	–	AVILA, 1966
	Brazil, Euclides de Cunha Dam	2.75	70.4* 80.0	–	0.22	136.0	–	p = 0.2	RUIZ, 1966
	Brazil, Graminha Dam	2.63	64.3* 77.8	–	0.27	158.8	–	p = 0.5	RUIZ, 1966
	Brazil, CTA Quarry	2.62	52.8* 53.8	–	0.19	115.5	–	p = 1.4	RUIZ, 1966
	Brazil, Jaguari Quarry	2.70	68.6* 79.9	–	0.24	140.0	–	p = 0.7	RUIZ, 1966
	Brazil, Camaru-eiro Quarry	2.6	40.2* 49.8	–	0.40	98.0	–	p = 1.4	RUIZ, 1966
		3.36	103.4	–	–	218.6	–	–	CLARK & CAUDLE, 1963
	USA, Atlanta, Ga., fg, granitic	2.62	19.31	–	–	193.05	3.10	–	ATCHISON et al, 1961
	USA, Edmonston, Pump Plant, diorite	2.80	15.27	–	–	118.04	–	–	KRUSE et al, 1969
	USA, Edmonston, Pump Plant, diorite	2.86	15.91	–	–	154.72	–	–	KRUSE et al, 1969
	USA, Edmonston, Pump Plant, diorite	2.64	40.24	–	–	78.94	–	–	KRUSE et al, 1969

Gneiss	USA, Edmonston, Pump Plant, diorite	2.68	51.99	–	186.30	–	–	KRUSE et al, 1969
	USA, Edmonston, Pump Plant, diorite weathered	2.60 / 2.62	8.07 / 14.55	– / –	24.51 / 32.46	– / –	– / –	KRUSE et al, 1969 / KRUSE et al, 1969
Gneiss-biotite	Yugoslavia, decomposed	–	0.015	0.002	–	–	τ = 0–0.16	DVORAK, 1967
Gneiss-granitic	Brazil, DER Quarry	2.68	43.5* / 47.5	–	73.0	–	p = 1.6	RUIZ, 1966
	Brazil, Euclides de Cunha Dam	2.65	70.9* / 83.3	–	162.0	–	p = 0.5	RUIZ, 1966
Granite	Scotland, Peterhead, -biotite, P = 70–600	–	57.1	23.4	0.21	–	–	ADAMS & COKER, 1906
	Canada, Lily Lake, N. B., biotite, P = 70–600	–	56.3	23.3	0.20	–	–	ADAMS & COKER, 1906
	USA, Westerly, R. I., quartz monzonite P = 70–600	–	50.9	20.8	0.22	–	–	ADAMS & COKER, 1906
	USA, Quincy, Mass., cg, P = 70–600	–	46.4	19.2	0.21	–	–	ADAMS & COKER, 1906
	USA, Quincy, Mass., cg, P = 70–600	–	56.8	23.7	0.20	–	–	ADAMS & COKER, 1906
	Italy, Baveno, biotite P = 70–600	–	47.1	18.8	0.25	–	–	ADAMS & COKER, 1906
	Canada, Stanstead, Que., biotite-muscovite P = 70–600	–	39.2	15.6	0.26	–	–	ADAMS & COKER, 1906

LABORATORY MECHANICAL PROPERTIES OF ROCKS

Rock	Location & Description	Density ϱ (g/cm³)	Modulus of Elasticity E (GPa)	Modulus of Rigidity G (GPa)	Poisson's Ratio υ	Compressive Strength σc (MPa)	Tensile Strength σt (MPa)	Remarks	Reference
	(1)	(2)	(3)	(4)	(5)	(6)	(7)	(8)	(9)
Granite	Germany, Jannowitz P = 100–900	–	41.5	–	–	–	–	–	BREYER, 1930
	Germany, Dresden P = 100–900	–	43.5	–	–	–	–	–	BREYER, 1930
	USA, Quincy, Mass., surface								
	P = 3 # 1	2.66	27.1	–	–	–	–	–	ZISMAN, 1933
	P = 78 # 1	2.66	50.8	–	–	–	–	–	ZISMAN, 1933
	# 1	2.66	41.6*	19.7*	0.05*	–	–	–	IDE, 1936
	P = 28 # 2	2.67	23.9	–	–	–	–	–	ZISMAN, 1933
	P = 78 # 2	2.67	47.3	–	–	–	–	–	ZISMAN, 1933
	# 2	2.67	37.6*	18.0*	0.04*	–	–	–	IDE, 1936
	USA, Quincy, Mass., 30 m depth P = 11 # 1 # 1	2.67	51.8	–	–	–	–	–	ZISMAN, 1933
		2.67	59.8*	27.8*	0.07*	–	–	–	IDE, 1936
	USA, Quincy, Mass., 70.5 m depth, P = 11, # 4	2.66	34.8	–	–	–	–	–	ZISMAN, 1933
	P = 56, # 4	2.66	47.0	–	–	–	–	–	ZISMAN 1933
	# 4	2.66	46.7*	20.7*	0.13	–	–	–	IDE, 1936
	USA, Quincy, Mass., 22.5 m depth	2.65	47.0	20.9	0.12	–	–	–	BIRCH & BANCROFT, 1938

Rock	Locality	ρ							Reference
Granite	USA, Rockport, Mass., 30 m depth	2.64	34.8 / 51.0*	– / 20.0*	0.08 / 0.14*	– / –	– / –	– / –	ZISMAN, 1933 / IDE, 1936
	USA, Westerly, R.I.	2.64	39.9*	18.0*	0.06*	–	–	–	IDE, 1936
	USA, Westerly, R.I.	2.66	70.5	27.7	0.27	–	–	–	KNOPOFF, 1954
	P = 1	2.61	55.0	25.0	0.13	–	–	–	FAI, 1961
	P = 1000	2.61	66.0	39.0	0.22	–	–	–	FAI, 1961
	P = 4000	2.61	–	32.5*	–	–	–	–	SIMMONS, 1964
	USA, Arbuckle Mts.	2.65	43.7– / 47.2*	–	–	–	–	–	WEATHERBY & FAUST, 1935
	USA, Barre, Vt. ⊥	2.66	30.4*	16.8*	0.04*	–	–	–	OBERT et al, 1946
	=	2.66	27.3*	15.2*	0.10*	–	–	–	OBERT et al, 1946
	=	2.66	44.2*	16.9*	0.31*	–	–	–	OBERT et al, 1946
	Sweden, Soevik	–	54.0*	–	–	–	–	–	GASSMANN et al, 1952
	USA, Stone Mt., Ga.	2.639	15.0*	–	–	–	–	–	SIMMONS, 1964
	USA, Grand Coulee, Wash., P = 0–340	2.63	32.5	–	0.14	147.73	–	–	BALMER, 1953
	USA, Grand Coulee, sligthly altered	2.61	8.96	–	0.11	64.81	–	–	BALMER, 1953
	USA, Pole Hill, Colo., cg. P = 0–170	2.63	26.20	–	0.12	72.12	–	–	BALMER, 1953
	USA, Lithomia, Ga, gneiss	2.64	38.98*	11.79*	0.19*	193.05	2.83	dynamic σ_t = 13.79	WUERKER, 1956
	USA, Thompson, Colo., pegmatite, Group A P = 0–100	2.62	20.69	–	–	42.69	–	–	WUERKER, 1956
	USA, Thompson, Colo., pegmatite, group A, P = 0–100, Group B	2.62 / 2.62	20.69 / –	– / –	– / –	42.69 / 58.19	– / –	– / –	WUERKER, 1956 / WUERKER, 1956

LABORATORY MECHANICAL PROPERTIES OF ROCKS

Rock	Location & Description	Density ϱ (g/cm³)	Modulus of Elasticity E (GPa)	Modulus of Rigidity G (GPa)	Poisson's Ratio υ	Compressive Strength σ_c (MPa)	Tensile Strength σ_t (MPa)	Remarks	Reference
(1)	(1)	(2)	(3)	(4)	(5)	(6)	(7)	(8)	(9)
Granite	USA, granite	2.66	27.37*	15.17*	–	226.15	7.10	R = 18.62–	WUERKER, 1956
		2.66	44.20	16.89	–	244.07	–	25.5	WUERKER, 1956
								–	
	USA, N. Y., quartz + syenite	2.65	33.92*	16.34*	–	293.72	19–99(R)	–	WUERKER, 1956
	USA, Md., quartz + syenite	2.65	52.54	25.44*	–	250.97	20.68(R)	–	WUERKER, 1956
	USA, Nev., quartz + syenite	2.63	51.30*	22.48*	–	272.34	26.89(R)	–	WUERKER, 1956
	USA, NC., quartz + syenite, ⊥	2.60	15.72*	10.20*	–	209.60	11.03(R)	–	WUERKER, 1956
	‖	2.60	30.27*	12.41*	–	–	–	–	WUERKER, 1956
	Spain, Guadarama	2.60	24.4	–	–	62.8	–	–	SALAS, 1968
	USA, Fremont Canyon, homogeneous, cg	2.62	64.1	–	0.14	–	8.62	–	NESBITT, 1960
	USA, Fremont canyon, cg, slightly weathered,	2.61	45.1	–	0.10	–	2.62	–	NESBITT, 1960
	USA, Fremont Canyon, cg	2.63	73.8	–	0.15	–	–	–	NESBITT, 1960
	USA, Salida	2.64	70.7	–	0.18	324.0	–	–	HOSKINS & HORINO, 1969
	Brazil, Mitidieri, Quarry, porphyritic	2.68	69.6	–	0.24	91.7	–	–	RUIZ, 1966

Granite									
Granite	Brazil, Valinhos, Quarry	2.56	–	65.8	0.18	107.2	–	–	RUIZ, 1966
	Brazil, Cantareira, Quarry	2.75	–	64.1	0.25	111.3	–	–	RUIZ, 1966
	Brazil, Piccicaco, Quarry, tourmaline-	2.62	–	75.5	0.16	127.1	–	–	RUIZ, 1966
		–		70.3*	0.30*	–	–	–	GEYER & MYUNG, 1970
		–		58.6*	0.20*	–	–	–	GEYER & MYUNG, 1970
		–		71.0*	0.25*	–	–	–	GEYER & MYUNG, 1970
	USA, AEC Nevada test site, dense, cg unweathered	2.69	–	73.8*	0.22*	141.1	11.7	at 50 % fracture	STOWE, 1969
	USA, Grand Coulee, Wash., mg, slightly altered	2.61	–	7.79	0.13	56.9	–	–	BALMER, 1953
	USA, Pole Hill, pegmatite, vcg.	2.62	–	19.0	0.09	48.8	–	–	BALMER, 1953
	USA, Pikes Peak, cg, weathered	2.67	–	33.4	0.37*	88.9	3.93	at 50 % fracture	MILLER, 1965
	USA, Pikes Peak, dense, mg	2.64	–	70.6	0.18	226.0	11.9	at 50 % fracture	MILLER, 1965
	USA, Barre, Vt., dense, mg	2.64	–	61.4	0.39	194.4	10.7	at 50 % fracture	MILLER, 1965
	USA, Barre, Vt., granite	2.70	–	26.2–29.0	–	112.4–200	–	p = 2.7	MESRI et al, 1976
	USSR, biotite, Ukraine,								
	fg	–	–	66.69	0.24	294.21	–	p = 0.72	BELIKOV et al, 1967
	mg	–	–	62.76	0.22	269.69	–	p = 0.66	BELIKOV et al, 1967
	cg	–	–	62.76	0.24	235.37	–	p = 0.72	BELIKOV et al, 1967
	USSR, North Caucasus, gray granite, fg	–	–	46.09	0.23	209.38	–	p = 0.73	BELIKOV et al, 1967

LABORATORY MECHANICAL PROPERTIES OF ROCKS

Rock	Location & Description	Density ρ (g/cm³)	Modulus of Elasticity E (GPa)	Modulus of Rigidity G (GPa)	Poisson's Ratio υ	Compressive Strength σc (MPa)	Tensile Strength σt (MPa)	Remarks	Reference
	(1)	(2)	(3)	(4)	(5)	(6)	(7)	(8)	(9)
Granite	USA, St. Cloud, Minn., chalcedony gray	2.72	66.88	–	0.32	227.53	13.10	–	REICHMUTH, 1962
	USA, Lithonia belt, Ga., gneissic	2.62 2.64	10.41– 19.17* 34.47	7.10– 11.79	0.19– 0.28 0.19	193.05– 213.05	6.89– 38.61	E at 14.07	BLAIR, 1956
	USA, Lithonia belt, Ga., gneissic	#1 2.65	23.1	10.89	0.06	–	–	–	CLARK, 1966
		#2 2.66 #3 2.68	12.41 18.60	8.55 10.00	0.27 0.07	–	–	–	CLARK, 1966 CLARK, 1966
	USA, Bridge Canyon Dam, Ariz., massive, cg	2.66	53.78	–	0.82	144.79	5.58	–	BRANDON, 1974
	USA, Valencia County, N.M., massive, weathered	2.59	7.07	–	0.65	25.51	–	grain = 0.13–1.45 mm	BRANDON, 1974
	USA, Valencia County, N.M., massive, cg	–	78.26	–	0.19	154.44	–	grain = 0.4–1.4 mm	BRANDON, 1974
	Italy, Mont Blanc, granite, cg, P = 140	–	–	–	–	12.9	6.2	–	FUMAGALLI, 1966
	USA, Colo., Herman, granite, hard, dense	–	–	–	0.18	–	–	–	YOUASH, 1970

Rock	Locality	Density							Reference
Granite	USA, Silverplume, Colo., granite, hard, with high quartz, dense	–	46.1*	–	0.17*	–	–	–	YOUASH, 1970
	Czechoslovakia, Liberec, with partially deteriorated albite	2.60	15.0 / 65.0*	–	–	150–200	–	–	SAMALIKOVA, 1974
	Czechoslovakia, Liberec, with decomposed albite	2.40	15.0 / 60.0*	–	–	50–80	–	–	SAMALIKOVA, 1974
	Australia, T.1., UG power station, Snowy Mt.	–	62.63 / 57.97(Et)	–	0.19 / 0.25	125.9	7.09	–	MOYE, 1964
	USA, Barre, Vt, granite, P = 1.4	2.76	70.3 / 37.2**	–	0.15	195.6 / 167.5**	–	T	MICHALOPOULOS & TRIANDAFILIDIS, 1976
	USA, P = 1.7	2.66	60.7 / 37.2**	–	0.18	253.2 / 217.2**	–	T	MICHALOPOULOS & TRIANDAFILIDIS, 1976
	USA, Rose granite Pedernal Hills, N. M.	2.64	69.64 / 48.26**	–	0.28	308.0 / 242.2**	–	p = 2.5	MICHALOPOULOS & TRIANDAFILIDIS, 1976
	USA, Raymond, Calif., Raymond granite	2.64	58.6 / 33.1**	–	0.26	178.0 / 156.5**	–	p = 2.5	MICHALOPOULOS & TRIANDAFILIDIS, 1976
	Canada, Daniel-Johnson, Dam, altered	–	24.7	–	–	–	–	–	BROWN et al, 1970
	Canada, Daniel-Johnson, Dam	–	29.5	55.0	–	–	–	–	BROWN et al, 1970
	USA, AEC Nevada test site dense, cg	–	81.4	–	0.33	213.7	11.72	–	STOWE & AINSWORTH, 1968
	Japan, Enzan	2.66	–	–	–	180.9	10.9	–	MITANI & KAWAI, 1974
	Japan, Inada	2.68	–	–	–	192.7	6.8	–	MITANI & KAWAI, 1974
	Brazil, Mitidieri, Quarry, porphyritic	2.68	47.0* / 71.0	–	0.24	88.2	–	p = 1.0	RUIZ, 1966

LABORATORY MECHANICAL PROPERTIES OF ROCKS

Rock	Location & Description	Density ρ (g/cm³)	Modulus of Elasticity E (GPa)	Modulus of Rigidity G (GPa)	Poisson's Ratio ν	Compressive Strength σc (MPa)	Tensile Strength σt (MPa)	Remarks	Reference
	(1)	(2)	(3)	(4)	(5)	(6)	(7)	(8)	(9)
Granite	Brazil, Valinhos, Quarry	2.56	58.4* 67.1	–	0.18	100.0	–	p = 1.3	Ruiz, 1966
	Brazil, Cantareira, Quarry	2.75	56.2* 65.4	–	0.25	121.0	–	p = 0.9	Ruiz, 1966
	USA, Unaweep, Colo.	2.67	27.2* 21.3	16.8*	0.19* 0.05	175.8	4.1	p = 0.5	Blair, 1956
	USA, Unaweep, Colo.	2.71	38.2* 18.7	19.1*	0.02* 0.08	158.6	3.6	–	Blair, 1956
	USA, Unaweep, Colo. ‖bedding	2.73	42.2*	19.0*	0.12*	161.3	5.6	p = 0.6	Blair, 1956
	USA, Unaweep, Colo.	2.66	27.1* 29.4	15.5*	0.13* 0.13	174.4	3.4	p = 0.6	Blair, 1956
	Australia, Harcourt, Vic.	–	39.1	–	–	143.9	–	–	Bamford, 1969
	Australia, Stawell, Vic.	–	–	–	–	125.2	8.1	–	Singh, 1969
	Japan	–	55.6	–	–	166.7	–	–	Yamaguchi, 1968
	Canada, Grenville, Que, cg, 80 % orthoclase	2.61	65.8	–	–	172.0	–	–	Coates & Parsons, 1966
	Canada, St. Lawrence, cg, porous, siliceous	2.59	73.8	–	–	276.0	–	–	Coates & Parsons, 1966
	Canada, St. Lawrence, vcg	2.73	71.6	–	–	144.0	–	fluorite matrix	Coates & Parsons, 1966

Granite										
Japan, Taguchi	—			38.0	—	—	—	—	—	IIDA et al, 1960
Japan, Katashirakawa, biotitic	—			54.0	—	—	140.0	—	—	MATSUSHIMA, 1960
USA, Barre, Vt.	—						160.65	3.28(R) 6.62	p = 1	HARTMAN, 1966
Sweden	—			—		—	175.0	16.1	—	LUNDBORG, 1967
Japan, Inada	2.62			56.0	—	—	166.2	10.4	—	YAMAGUCHI & MORITA, 1969
USA, Georgia	#1 —			21.4	—	0.14	—	—	—	SELLERS, 1970
	#2 —			28.1	—	0.15	—	—	—	SELLERS, 1970
USA, Chelmsford, Mass.	#1 —			32.8	—	0.13	150.0	5.85	—	PENG & JOHNSON, 1972
	#2 —			56.0	—	0.10	175.0	6.83	—	PENG & JOHNSON, 1972
	#3 —			56.0	—	0.11	200.0	10.0	—	PENG & JOHNSON, 1972
USA, Barre, Vt., ⊥ bedding	2.64			58.6	—	—	200.0	13.5	p = 0.69	HAWKES et al, 1973
USA, Westerly, mg	2.62			47.5 41.0*	—	—	—	40.0	p = 0.9	BRACE, 1963
—	2.63			35.9	—	—	148.9	—	—	CLARK & CAUDLE, 1963
—	2.63			26.9	—	—	72.1	—	—	CLARK & CAUDLE, 1963
—	2.63			73.1	—	—	160.7	—	—	CLARK & CAUDLE, 1963
USA, mg–cg	#1 2.66		30.4*	15.8*	0.04*	228.91	19.99	p = 0.9	OBERT et al, 1946
	#2 2.66		27.4*	15.2*	0.10	225.15	25.37(R)	p = 0.9	OBERT et al, 1946
	#3 2.66		44.2*	16.9*	0.31*	244.07	18.41(R)	p = 0.9	OBERT et al, 1946
USSR	—			—	—	—	—	16.0–29.0	—	KOIFMAN, 1969
USSR	—			—	—	—	130.0–220.0	10.0–16.0	—	KOIFMAN, 1969
USSR, Ukraine, biotitic, fg	—			68.0	—	0.24	300.0	—	p = 0.72	BELIKOV et al, 1967
	—			64.0	—	0.22	275.0	—	p = 0.66	BELIKOV et al, 1967
	—			64.0	—	0.24	240.0	—	p = 0.72	BELIKOV et al, 1967

LABORATORY MECHANICAL PROPERTIES OF ROCKS

Rock	Location & Description	Density ρ (g/cm³)	Modulus of Elasticity E (GPa)	Modulus of Rigidity G (GPa)	Poisson's Ratio υ	Compressive Strength σc (MPa)	Tensile Strength σt (MPa)	Remarks	Reference
	(1)	(2)	(3)	(4)	(5)	(6)	(7)	(8)	(9)
Granite	USSR, geosynclines	2.63	53.7	–	0.21	–	–	p = 1.96	BELIKOV, 1967
	USSR, Caucasus range, geosynclines, two-mica	2.66	44.9	–	0.22	–	–	p = 0.7	BELIKOV, 1967
	USSR, Ukraine, biotite–	2.66	67.0*	–	–	–	–	p = 0.45	BELIKOV, 1967
	USSR, Ukraine, plagio–	2.73	71.0*	–	–	–	–	p = 0.3	BELIKOV, 1967
	USSR, Ukranian shield, biotitic	2.66	59.4	–	0.20	–	–	p = 0.67	BELIKOV, 1967
	USSR, Ukranian shield, fg, aplitic	2.65	77.0	–	0.22	–	–	p = 0.42	BELIKOV, 1967
	USSR, Ukranian shield, mg, biotitic	2.65	60.6	–	0.24	–	–	p = 0.56	BELIKOV, 1967
	USSR, Ukranian shield, cg, trachytoid texture	2.65	54.5	–	0.19	–	–	p = 0.85	BELIKOV, 1967
	USSR, Ukranian shield, 4th intrusive cycle	2.64 / 2.64	63.7 / 58.7	– / –	0.19 / 0.20	– / –	– / –	p = 0.66 / p = 0.82	BELIKOV, 1967 / BELIKOV, 1967
	USSR, Baltic shield, biotitic	2.70	67.2	–	0.25	–	–	p = 0.78	BELIKOV, 1967
	USA, San Bernardino, Tunnel	–	10.1	–	0.30	44.4	–	depth 9.75 m	WILLIAM & NELSON, 1970

Granite

Location	Density						Notes	Reference
	–	12.1	–	0.38	29.5	–	depth 19.0 m	WILLIAM & NELSON, 1970
	–	7.79	–	0.23	42.82	–	depth 19.0 m	WILLIAM & NELSON, 1970
	–	3.86	–	0.75	14.0	–	depth 10.5 m	WILLIAM & NELSON, 1970
	–	30.34	–	0.22	75.98	–	depth 24.7 m	WILLIAM & NELSON, 1970
	–	4.59	–	0.65	12.82	–	depth 12.1 m	WILLIAM & NELSON, 1970
	–	4.55	–	0.76	15.93	–	depth 19.4 m	WILLIAM & NELSON, 1970
	–	19.96	–	–	38.40	–	depth 8.9 m	WILLIAM & NELSON, 1970
	–	56.57	–	0.10	155.48	–	depth 23.6 m	WILLIAM & NELSON, 1970
	–	50.64	–	0.26	95.49	–	–	WILLIAM & NELSON, 1970
Canada, Lithonia	2.62	–	25.2*	0.26*	–	31.0	–	LAROCQUE & FAVREAU, 1970
USA, Kitledge, pink	2.6	–	–	–	167.54	10.96	p = 0.8	CLARK et al, 1970
USA, Missouri red	2.6	–	–	–	159.27	9.72	p = 0.4	CLARK et al, 1970
USA, Milford pink	2.62	34.8	–	0.22	220.0	7.65	–	SINGH & HUCK, 1970
USA, charcol-	2.72	77.2	–	0.20	–	–	–	THIRUMALAI, 1970
USA, Granitville, Vt., (Barre granite)	2.65	24.1–48.3	–	–	220.6	–	p = 0.23	RAD & MC GARRY, 1970
USA, Coldspring, Minn., charcoal black-cg	–	41.37	11.34	0.20	128.93	9.24	–	HUCK & SINGH, 1972
USA, Quincy, Mass.	–	35.0 43.0*	–	0.11 0.33*	–	–	–	RINEHART et al, 1961

LABORATORY MECHANICAL PROPERTIES OF ROCKS

Rock	Location & Description	Density ϱ (g/cm³)	Modulus of Elasticity E (GPa)	Modulus of Rigidity G (GPa)	Poisson's Ratio ν	Compressive Strength σ_c (MPa)	Tensile Strength σ_t (MPa)	Remarks	Reference
	(1)	(2)	(3)	(4)	(5)	(6)	(7)	(8)	(9)
Granite	USSR, Caucasus, fg	–	47.0	–	0.23	213.5	–	p = 0.73	BELIKOV et al, 1967
	USA, slightly altered	–	5.52 / 15.17*	–	0.04 / 0.10*	–	–	–	RINEHART et al, 1961
	USA, charcoal-, fg	–	66.88	–	0.32	227.53	11.7(R)	p = 2.2	HAIMSON & FAIRHURST, 1969
	USA, Stone Mountain	2.61	–	–	–	57.59	5.82	p = 2.2	SCHWARTZ, 1964
		–	76.53	–	0.28	–	–	–	LEEMAN, 1964
	USA, N.H.,pink-	–	–	–	–	131.69	–	–	BAUER & CALDER, 1967
	Malaya, Jor Power Plant, biotite, cg, altered	2.66	–	–	–	43–90	–	–	KLUTH, 1964
Granite-Gneiss	USA, Lithonia, Ga.	2.62	19.17*	11.79*	0.19	–	–	–	NICHOLLS, 1961
Granite-Tourmaline	Brazil. Piccicacoa Quarry, tourmaline	2.62	60.1* / 77.0	–	0.16	161.0	–	p = 0.8	RUIZ, 1966
Grano-Diorite	USA, Weston, Mass,	2.63	60.0*	25.3*	0.187*	–	–	–	BIRCH & BANCROFT, 1940
	P = 500	2.63	–	31.3*	–	–	–	–	BIRCH & BANCROFT, 1940
	P = 4000	2.63	–	36.6*	–	–	–	–	BIRCH & BANCROFT, 1940
	Australia, Tumut Pont Dam, cg, massive, slightly fractured	2.71	63.09	–	0.16	126.86	1.45	variably altered	BRANDON, 1974

Grano-Diorite	USA, Raymond, Calif. cg	–	39.0	–	0.24	193.3	–	p = 0.27	PRATT et al, 1972
	USSR, Ukranian shield, fg, aplitic	2.74	68.0	–	0.24	–	–	p = 0.52	BELIKOV, 1967
Granulite	UK, Monar Dam, highly metamorphosed, massive, unweathered	–	56.5 / 69.0*	–	–	15.3	–	–	HENKEL et al, 1964
	Yugoslovia, sound	–	2.24	2.61	–	–	–	τ = 0–0.94	DVORAK, 1967
	disturbed	–	0.390	0.137	–	–	–	τ = 0–0.47	DVORAK, 1967
	Brazil, Gaspar Quarry	2.58	22.2* / 42.5	–	0.31	154.5	–	p = 1.7	RUIZ, 1966
		2.63	41.7	–	0.31	110.0	–	–	RUIZ, 1966
	India, KGF	3.01	91.0*	35.1*	0.30*	–	–	p = 5	RAMANA et al, 1973
	India, KGF	3.11	91.1*	33.9*	0.23*	–	–	p = 7	RAMANA et al, 1973
	India, KGF	3.36	82.9*	35.5*	0.28*	–	–	p = 9.2	RAMANA et al, 1973
		2.46	20.1	–	0.08	79.3	–	–	JUDD, 1969
Gray-wacke	USA, subgraywacke	2.66	33.1	–	0.08	80.0	–	p = 3.3	JUDD, 1969
	USA, subgraywacke, cg, P = 0–136	2.49	10.34	–	0.12	30.34	–	–	WUERKER, 1956
	USA, Colo., sub-, cg	2.49	26.20*	12.41*	0.08*	–	–	–	WUERKER, 1956
	USA, Calif., sub-, fg	2.41	12.41	–	–	48.95	–	–	WUERKER, 1956
	USA, Calif., sub-, fg	2.41	25.51*	10.34*	0.23*	–	–	–	WUERKER, 1956
	USA, Calif., sub-, mg	2.44	13.10	–	–	48.81	–	–	WUERKER, 1956
	USA, Calif., sub-, mg	2.44	26.20*	11.03*	0.19*	–	–	–	WUERKER, 1956

LABORATORY MECHANICAL PROPERTIES OF ROCKS

Rock	Location & Description	Density ϱ (g/cm³)	Modulus of Elasticity E (GPa)	Modulus of Rigidity G (GPa)	Poisson's Ratio υ	Compressive Strength σ_c (MPa)	Tensile Strength σ_t (MPa)	Remarks	Reference
(1)		(2)	(3)	(4)	(5)	(6)	(7)	(8)	(9)
Gray-wacke	USA, Monticello Dam, sub-, porous, massive, cg	2.46	11.4	–	0.05	54.5	–	p = 10.3	BALMER, 1953
	USA, Monticello Dam, sub-, porous, massive, mg	2.44	12.3	–	0.06	48.8	–	p = 11.5	BALMER, 1953
	USA, Monticello Dam, sub-, porous, cg, slightly weathered	2.49	9.52	–	0.08	30.5	–	p = 9.7	BALMER, 1953
	USA, Monticello Dam, sub-, porous, massive, mg–cg	2.49	9.93	–	0.05	50.7	–	p = 9.7	BALMER, 1953
	Taiwan, Shihmen Dam, fg, calcareous	2.50	4.14	–	0.06	55.16	2.0	grain = 0.04–0.15 mm	BRANDON, 1974
	USA, Devil Canyon Dam, Alaska, fg, slightly weathered	2.77	68.43	–	0.17	221.32	5.52	grain = 0.02–0.17 mm	BRANDON, 1974
	India, Bhakra Dam, sub-, dense, mg, calcareous	2.67	21.89	–	0.05	91.01	4.83	grain = 0.16–0.33 mm	BRANDON, 1974
	Yugoslavia, Algonkian	–	4.80	2.6	–	–	–	τ = 0–1.5	DVORAK, 1967

Gray-wacke	USA, sub-, mg, hd, massive	–	12.41 / 26.2*	–	–	0.03 / 0.08*	–	porous	RINEHART et al, 1961
	USA, sub-, hd, cg–mg, massive	–	11.03 / 26.20*	–	–	0.02 / 0.06*	–	porous	RINEHART et al, 1961
	USA, sub-, mg–cg, hd, massive	–	9.65 / 24.82*	–	–	0.02 / 0.29*	–	porous	RINEHART et al, 1961
	USA, sub-, cg, mod. weathered	–	8.96 / 26.2*	–	–	0.05 / 0.08*	–	porous	RINEHART et al, 1961
	Czechoslovakia, Primbram	2.63	96.2	–	208.8	0.36	–	–	SIBEK et al, 1964
Greenstone	USA, Va	3.02	104.14* / 101.35	42.06*	281.31	0.25 / 0.15	40.68(R)	secant at 50% p = 0.7	WUERKER, 1956
	USA, PenMar, Pa., amygdaloidal	3.04	48.95*	30.68*	200.64	0.21	17.93(R)	p = 0.5	WUERKER, 1956
	USA, Mich., schistose	2.93	79.29*	30.89*	122.04	–	22.06(R)	p = 0.6	WUERKER, 1956
	USA, Mich. + phyllite	3.10	74.46*	32.47*	114.45	–	26.89(R)	p = 0.4	WUERKER, 1956
	USA, Mich. + phyllite	3.30	84.12*	38.06*	313.71	–	22.75(R)	–	WUERKER, 1956
	USA, Minn.	–	–	–	126.04	–	–	–	WUERKER, 1956
	USA, Minn.	–	–	–	122.04	–	–	–	WUERKER, 1956
Greenstone Phyllite	USA, Cliffs shaft, Mich.	3.10	74.46*	32.47	144.45	0.19	26.89	p = 0.4	BOYUM, 1961
Greenstone-Phillite-Magnetite	USA, Cliffs shaft, Mich.	3.30	84.12	38.06	313.71	0.10	22.75(R)	p = 0.5	BOYUM, 1961
Gypsum	Canada, anhydrous	–	–	–	21.65	–	2.43	–	WEIR - JONES, 1971

368

LABORATORY MECHANICAL PROPERTIES OF ROCKS

Rock	Location & Description	Density ρ(g/cm³)	Modulus of Elasticity E(GPa)	Modulus of Rigidity G(GPa)	Poisson's Ratio υ	Compressive Strength σc(MPa)	Tensile Strength σt(MPa)	Remarks	Reference
	(1)	(2)	(3)	(4)	(5)	(6)	(7)	(8)	(9)
Gypsum	Canada, crystalline in amorphous matrix	–	–	–	–	21.65	2.01	–	WEIR-JONES, 1971
Halite	Canada, Goderich, silurian, vcg.	2.13	18.4	–	–	15.6	–	–	COATES & PARSONS, 1966
Hematite	USA, Ishpeming, Mich.	4.22	142.0*	55.0*	–	–	–	–	WINDES, 1949
	USA, Soudan, Mich.	5.07	200.0*	78.0*	–	607.43	42.06(R)	–	WINDES, 1949
	USA, Bessemer, Ala., ‖ bedding	3.78	69.0*	27.0*	–	138.58	31.72(R)	–	WINDES, 1949
	USA, Ishpeming, Mich.	4.22	142.03*	54.88*	0.20	238.56–329.57	22.06	p=1	WUERKER, 1956
	Germany, calcitic	–	370–670	–	–	67.0	–	σc from in situ test	JAHNS, 1966
	USA, Cliffs shaft, Mich., ore	4.49	–	–	–	176.5	26.9(R)	–	BLAIR, 1956
	USA, Cliffs shaft, Mich., ore	4.51	–	–	–	111.7	22.1(R)	vuggy	BLAIR, 1956
	USA, Cliffs shaft, Mich.	4.6	–	–	–	304.8	–	conglomerate	BLAIR, 1956
	Canada, banded, Shaley	3.46	–	–	–	191.7	–	–	BLAIR, 1956
	Canada, ~ banded	4.26	–	–	–	164.8	–	–	BLAIR, 1956
	Canada, Bell Island	4.19	74.5	–	–	194.5	–	–	COATES & PARSONS, 1966

Hematite	mineral (average)	5.10	212.0	–	–	–	–	–	BELIKOV, 1967
	USA, Cliffs shaft, Mich.	4.22	142.03*	44.88*	0.30*	329.57	22.06	p = 1.0	BOYUM, 1961
	USA, Cliffs shaft, Mich., conglomerate	4.68	106.86	–	0.22	238.56	–	p = 1.1	BOYUM, 1961
	USA, Cliffs shaft, Mich., hd	4.07	97.22	40.6	0.20	293.03	35.16	–	BOYUM, 1961
Hornblende	India, KGF	3.47	99.2*	37.6*	0.32*			p = 3	RAMANA et al, 1973
	USSR, Slyudyanka	3.12	110.2*	43.1*	0.29*			–	BELIKOV, 1967
	USSR	3.18	124.6*	49.8*	0.25*	–	–	–	BELIKOV, 1967
Hornblende Granulite	India, KGF	3.04	101.8*	39.2*	0.30*	–	–	p = 6	RAMANA et al, 1973
	India, KGF	3.08	102.5*	42.5*	0.21*	–	–	p = 6	RAMANA et al, 1973
Hornblende Schist	India, KGF	2.99	99.5*	41.4*	0.2*	–	–	p = 9	RAMANA et al, 1973
	India, KGF	3.05	113.6*	48.2*	0.18*	–	–	–	RAMANA et al, 1973
	India, KGF	2.74	100.0*	42.1*	0.19*	–	–	p = 4	RAMANA et al, 1973
	India, KGF	3.01	104.2* 98.2	40.5	0.29* 0.28	171.2	–	–	RAMANA et al, 1973
	India, KGF	2.96	1:6.3*	47.5*	0.23*	–	–	p = 8	RAMANA et al, 1973
	India, KGF	3.20	95.1*	39.9*	0.21*	–	–	p = 7	RAMANA et al, 1973
	India, KGF	3.03	105.1*	42.6*	0.23*	–	–	p = 6	RAMANA et al, 1973
	India, KGF	3.01	94.5*	41.2*	0.15*	–	–	–	RAMANA et al, 1973
Hornblendite	India, KGF	3.25	97.9*	42.8*	0.14*	–	–	p = 10	RAMANA et al, 1973

LABORATORY MECHANICAL PROPERTIES OF ROCKS

Rock	Location & Description	Density ρ (g/cm³)	Modulus of Elasticity E (GPa)	Modulus of Rigidity G (GPa)	Poisson's Ratio ν	Compressive Strength σc (MPa)	Tensile Strength σt (MPa)	Remarks	Reference
	(1)	(2)	(3)	(4)	(5)	(6)	(7)	(8)	(9)
Hornfels	USA, Bridge Canyon Dam, Ariz. hornblendehornfels, massive, mosaic	3.04	92.36	–	0.16	133.07	14.48	grain = 0.08–0.23 mm	BRANDON, 1974
	Thailand, lime-silicate hornfels, fg-mg, laminae	2.82	58.61	–	0.17	77.91	4.62	–	BRANDON, 1974
	Switzerland, Emosson Dam	–	72.00 60.00*	–	–	123.0	–	–	SCHNITTER & SCHNEIDER, 1970
Hornstone	USA, AEC Nevada test site	3.19	95.84*	40.89*	–	532.96	26.20	–	WUERKER, 1956
	USA, Anaconda Mines, Calif., vfg	2.83	–	–	–	547.4	–	calcitic pool	BLAIR, 1956
Ignimbrite	New Zealand, Maraetai Dam	1.83–2.16	–	–	0.15–0.30	–	–	p = 6.4–15.6	JAMES, 1955
	wet	–	1.7–6.7	–	–	7.74–33.8	0.86–2.65(R)	–	JAMES, 1955
	dry	–	2.34–7.7	–	–	13.8–46.9	0.89–3.5(R)	–	JAMES, 1955
Iron Ore	France, Saizerais Mine, dry	–	9.0* 4.0	–	–	25.0	1.5	–	TINCELIN & SINOU, 1964
	France, Saizerais Mine, damp	–	3.6	–	–	22.0	–	–	TINCELIN & SINOU, 1964
	France, Moutier Mine, red bed roof	–	34.00*	–	–	36.7	2.88(R)	calcareous	TINCELIN, 1951
	France, Angevillers Mine, grey bed roof	–	37.88*	–	–	44.17	4.0(R)	calcareous	TINCELIN, 1951

Iron Ore	France, CF Mine, savage bed roof	–	25.0*	–	33.97	2.37(R)	calcareous	TINCELIN, 1951
	France, St. Pierremont Mine, grey bed roof	–	44.2*	–	92.7	13.2(R)	calcareous	TINCELIN, 1951
	France, SMK Mine, black bed roof	–	31.4*	–	48.47	4.3(R)	siliceous	TINCELIN, 1951
	France, Roncourt Mine, black bed roof	–	21.5*	–	32.57*	2.92(R)	siliceous	TINCELIN, 1951
	France, Bure Mine, savage bed roof	–	33.7*	–	21.2	2.59(R)	calcareous	TINCELIN, 1951
	France, Terres-rouges Mine, mid red bed roof	–	37.4*	–	22.5	2.91(R)	calcareous	TINCELIN, 1951
	France, Terres-rouges Mine, upper red bed roof	–	29.8*	–	23.8	2.65(R)	calcareous	TINCELIN, 1951
Ironstone	UK, Billsthorpe	–	–	–	167.61	65.6	sideritic	HOBBS, 1964
Jasperite	USA, Cliffs shaft, Mich.	3.39	103.4	48.26	342.00	24.82(R)	p = 0.6	BOYUM, 1961
Jasperoid	Canada, Falconbridge Mine, 97% silica	–	88.25	–	441.26	–	–	MORUZI & PASIEKA, 1964
	Canada, Falconbridge Mine, 80% silica	–	84.12	–	311.64	–	–	MORUZI & PASIEKA, 1964
	Canada, Falconbridge Mine, 60% silica	–	73.08	–	203.4	–	–	MORUZI & PASIEKA, 1964
Kiesengite	Italy, Place Moulin Dam	–	78.3	–	106.3	10.1	–	OBERTI & REBAUDI, 1967
	Italy, Place Moulin Dam	–	71.8	–	72.3	10.4	–	OBERTI & REBAUDI, 1967

LABORATORY MECHANICAL PROPERTIES OF ROCKS

Rock	Location & Description	Density ρ (g/cm³)	Modulus of Elasticity E (GPa)	Modulus of Rigidity G (GPa)	Poisson's Ratio ν	Compressive Strength σc (MPa)	Tensile Strength σt (MPa)	Remarks	Reference
(1)	(1)	(2)	(3)	(4)	(5)	(6)	(7)	(8)	(9)
Labradorite	USSR, Golovino	2.70	88.7*	34.6*	0.28*	–	–	–	BELIKOV, 1967
	mineral (average)	2.70	89.0	–	–	–	–	–	BELIKOV, 1967
Lava	S. Africa, Western Deep Ltd.	–	–	–	–	430.0	–	porphyritic	DENKHAUS, 1965
	S. Africa, Western Deep Ltd.	–	–	–	–	250.0	–	amygdaloidal	DENKHAUS, 1965
Leucogranite	S. Africa, Western Deep Ltd.	2.59	30.7 / 28.96*	–	–	230.4	3.1	–	RICKETTS & GOLDSMITH, 1970
Limestone	USA, Knoxville, Tenn., P = 70–600	–	62.1	24.8	0.25	–	–	–	ADAMS & COKER, 1906
	Canada, Montreal, Que., P = 70–600	–	63.5	25.0	0.25	–	–	–	ADAMS & COKER, 1906
	USA, Kansas City, Mo.	2.10	25.7	–	0.20	50.6	–	–	HOSKINS & HORINO, 1969
	Germany, Solenhofen, dry	2.48	54.7	21.5	0.27	–	–	–	PESELNICK, 1962
		2.50	56.7	22.5	0.28	–	–	–	PESELNICK, 1962
		2.55	61.6	23.9	0.29	–	–	–	PESELNICK, 1962
		2.60	66.6	25.6	0.30	–	–	–	PESELNICK, 1962
		2.65	72.0	27.5	0.31	–	–	–	PESELNICK, 1962
		2.70	77.5	29.5	0.32	–	–	–	PESELNICK, 1962
		2.57	55.3**	21.3**	0.30**	–	–	–	PESELNICK, 1962
		2.63	64.8**	24.8**	0.31**	–	–	–	PESELNICK, 1962
		2.67	70.9**	27.0**	0.31**	–	–	–	PESELNICK, 1962
		2.62	63.8	–	0.29	63.8	4.0	at 50% fracture	MILLER, 1965
		2.70	74.6*	29.3*	0.27*	–	–	–	p = 3.5	VAN DER VLIS, 1970

Limestone

USA, Pa., carbonaceous								
P = 11 #1	2.71	33.7	—	0.1	—	—	—	ZISMAN, 1933
P = 11 #2	2.71	76.0	—	0.27	—	—	—	ZISMAN, 1933
P = 11 #3	2.71	60.6	—	0.18	—	—	—	ZISMAN, 1933
Persia	2.64–2.70	53.0–64.0	—	0.23–0.26	—	—	—	RICHARDS, 1933
USA, Bedford, Ind.	2.30	26.96 30.20*	12.0*	0.27	53.09	4.07	—	LADANYI & DON, 1970
USA, Bedford, Ind.	2.37	33.4	14.2	0.18	—	—	—	OBERT et al, 1946
⊥ =	2.37	41.0	15.9	0.29	—	—	—	OBERT et al, 1946
=	2.37	37.2	15.3	0.22	—	—	—	OBERT et al, 1946
USA, Bedford, Pa., porous, oolitic	—	28.5	—	0.29	51.0	1.58	at 50% fracture	MILLER, 1965
USA, Marble Canyon Dam, Ariz.								
fg, P = 0–247	2.71	67.0–68.0	—	0.25	80.39	—	p = 3.4	US BR, 1953
cg, P = 0–175	2.68	34.0–36.0	—	0.19–0.23	127.41	—	p = 4.7	US BR, 1953
cg, P = 70–100	2.44	19.0–21.0	—	0.20–0.22	133.21	—	p = 13.9	USBR, 1953
USA, Marble Canyon Dam, Ariz.	2.71	71.02*	27.58*	0.28*	80.39	—	—	US BR, 1953
USA, Marble Canyon Dam, Ariz., mg-fg	2.44	28.0*	12.0*	0.31	127.41	—	—	USBR, 1953
	2.67	47.3	—	0.18	99.4	—	p = 1.6	BALMER, 1953
USA, Marble Canyon Dam, Ariz., chalcedonic	2.60	60.67	—	0.22	107.42	—	p = 5.4	BALMER, 1953
	2.60	46.88*	18.62*	0.25*	—	—	p = 5.4	BALMER, 1953

LABORATORY MECHANICAL PROPERTIES OF ROCKS

Rock	Location & Description	Density ϱ(g/cm³)	Modulus of Elasticity E (GPa)	Modulus of Rigidity G (GPa)	Poisson's Ratio v	Compressive Strength σc(MPa)	Tensile Strength σt(MPa)	Remarks	Reference
	(1)	(2)	(3)	(4)	(5)	(6)	(7)	(8)	(9)
Limestone	USA, Eniwetok Atoll, breccia, P = 0–136	2.30	37.8	–	0.16	34.20	–	p = 15.3	BALMER, 1953
	USA, Marble Canyon Dam, Ariz., stylolitic, P = 0–238	2.73	44.82	–	0.14	79.50	–	p = 3.9	BALMER, 1953
	P = 0–102		56.54*	22.06*	0.27*	79.50*	–	p = 3.9	BALMER, 1953
	USA, Ohio, stylolitic	2.69	54.95*	–	25.10*	196.5	19.99(R)	p = 0.7	WUERKER, 1956
	USA, Utah, stylolitic	2.78	65.02*	23.37*	0.20*	193.05	15.17(R)	p = 1.3	WUERKER, 1956
	USA, W.Va., stylolitic	2.68	65.91*	27.30*	0.21*	158.58	13.10(R)	p = 6.0	WUERKER, 1956
	USA, Alaska, cg	2.83	52.68*	24.20*	–	165.47	–	p = 0.9	WUERKER, 1956
	USA, Ind., fossiliferous, ⊥ bedding	2.31	33.37*	14.20*	–	75.15	11.03(R)	p = 11	WUERKER, 1956
	‖ bedding	2.31	40.95–44.20*	15.86–16.59*	–	66.88–70.33	11.72–13.10(R)	–	WUERKER, 1956
	Brazil, Taquarussu Quarry	2.79	82.4	–	0.33	90.7	–	p = 2.0	RUIZ, 1966
	Brazil, Perus Quarry	2.75	64.4	–	0.23	62.9	–	p = 1.2	RUIZ, 1966
	Brazil, Piraporinha Quarry	2.73	90.0	–	0.28	67.6	–	p = 0.4	RUIZ, 1966

Rock / Location									Reference
Limestone	USA, AEC Nevada test site, dense, fg, stylolite	2.72	77.4	–	0.26	77.1	–	p = 0.5 at 50% fracture	STOWE, 1969
	USA, Ozark Tavervalley Dam, fg	–	55.8	–	0.30	97.9	3.92	at 50% fracture	MILLER, 1965
	USA, Colo., kerogeneous, magnesium (marl stone) ⊥ bedding	2.25	12.41*	6.89*	–	114.45	2.76(R)	–	WUERKER, 1956
	∥ bedding	2.18	21.37*	6.67*	–	–	35.85(R)	–	WUERKER, 1956
	USA, Ala., limonite ∥ bedding	2.92	66.05*	28.20*	–	171.68	26.20(R)	p = 0.6	WUERKER, 1956
	USA, Nev., + marble	2.79	78.6*	31.30*	–	153.75	17.93(R)	p = 0.4	WUERKER, 1956
	USA, Utah, mineralised with shale	2.92	65.71*	28.06*	0.17	239.94	16.55(R)	p = 0.62	WUERKER, 1956
	USSR, Caucasus & Carpathians, marmorised	–	72.57	–	0.30	165.74	–	p = 1.22	BELIKOV, 1967
	USSR, Platform limestone + dolomite, Moscow syneclise	2.68	33.34	–	0.27	73.55	–	p = 16.2	BELIKOV, 1967
	USSR, chemogenous limestone + dolomite, Moscow syneclise	2.57	53.94	–	0.32	142.20	–	p = 9.22	BELIKOV, 1967
	USSR, Moscow syneclise, organogenic clastic limestone + dolomite	2.30	36.29	–	0.25	88.26	–	p = 16.0	BELIKOV, 1967

LABORATORY MECHANICAL PROPERTIES OF ROCKS

Rock	Location & Description	Density ρ (g/cm³)	Modulus of Elasticity E (GPa)	Modulus of Rigidity G (GPa)	Poisson's Ratio ν	Compressive Strength σc (MPa)	Tensile Strength σt (MPa)	Remarks	Reference
	(1)	(2)	(3)	(4)	(5)	(6)	(7)	(8)	(9)
Limestone	USSR, Moscow syne clise, fragmented limestone + dolomite	2.16	26.48	-	0.25	51.98	-	p = 21.8	BELIKOV, 1967
	USSR, Turkmenia, limestone + dolomite	2.70	76.49	-	-	209.87	-	p = 1.81	BELIKOV, 1967
	USSR, Turkmenia, shell limestone (algal)	1.73	8.83	-	-	71.77	-	p = 34.5	BELIKOV, 1967
	Canada, Quebec	-	49.64	-	0.25	80.67	4.41	-	LADANYI & DON, 1970
	USA, Prairie du Rocher Mine, Ill., calcareous	2.68	81.36 / 68.05*	26.48	-	153.75	17.93	p = 0.8	BLAIR, 1955
	USA, Smoky Hill, S.D., chalky, dry,								
	⊥ bedding	1.41	2.90* / 4.48	1.59*	0.30*	8.27	1.38(R)	-	BLAIR, 1955
	∥ bedding	1.31		-	-	8.96	-	-	BLAIR, 9155
	chalky dry	1.89	11.1* / 8.69	4.14	0.02* / 0.12	28.96	4.14(R)	-	BLAIR, 1955
	partly wet	2.0	9.17* / 1.72	2.90*	0.11* / 0.14	12.41	2.07(R)	-	BLAIR, 1955
	wet	2.15	8.27* / 2.48	4.62*	0.10* / 0.04	10.34	2.76(R)	-	BLAIR, 1955
	USA, Bonne Terre, Mo., dolomitic	2.66–2.69	69.64*– 97.22* / 33.16– 60.26	28.54*– 37.58*	0.22*– 0.29* / 0.14– 0.31	186.85	12.41– 22.06(R)	p = 2.6– 3.3	BLAIR, 1955

Limestone								
USA, Bonne Terre, Mo., dolomitic-glauconite	2.67	38.68* 66.19	21.03	0.07* 0.20	146.17	10.34(R)	p = 3.6	BLAIR, 1955
USA, Bonne Terre, Mo., dolomitic, galena bearing	3.30	—	—	—	92.39	—	—	BLAIR, 1955
non-mineralised	2.73 2.80	— —	— —	— —	260.62 164.78	— —	— p = 1.8	BLAIR, 1955 BLAIR, 1955
USA, Ore Grande, Calif., metamorphic lineated	2.72	78.60* 82.74	30.13*	0.30* 0.27	165.47	16.55(R)	secant at 50%	BLAIR, 1955
calcite + brucite, interlocked	2.68	65.22* 55.02	27.99	0.16 0.20	59.98	15.86(R)	secant at 50%	BLAIR, 1955
magnesian marble	2.72	88.40* 76.53	25.23*	0.17* 0.29	57.23	15.86(R)	secant at 50%	BLAIR, 1955
calcite marble	2.70	16.62* 49.44	—	0.48 0.30	46.19	—	secant at 50%	BLAIR, 1955
UK, Portland limestone	2.16	40.0	—	—	85.0	13.98	—	HOBBS, 1970
UK, Portland limestone, lamination	—	—	—	41.44	13.98	13.98	—	HOBBS, 1970
UK, Breedon limestone	—	47.9	—	—	12.6	—	—	HOBBS, 1964
lithographic	—	74.34	0.29	—	203.0	25.8(R)	—	SELDENRATH & GRAMBERG, 1958
Canada, Quebec	—	77.22	—	0.33	293.72	—	—	MORRISON, 1970
USA, Hungry Horse Dam, Mont., dense, fg, argillaceous	2.75	42.06	—	0.18	46.19	—	grain – 0.01mm	BRANDON, 1974
USA, Leuder limestone, heterogeneous, fossiliferous,	—	30.95*	—	0.22*	—	—	—	YOUASH, 1970

LABORATORY MECHANICAL PROPERTIES OF ROCKS

Rock	Location & Description	Density ϱ (g/cm³)	Modulus of Elasticity E (GPa)	Modulus of Rigidity G (GPa)	Poisson's Ratio ν	Compressive Strength σ_c (MPa)	Tensile Strength σ_t (MPa)	Remarks	Reference
	(1)	(2)	(3)	(4)	(5)	(6)	(7)	(8)	(9)
Limestone	⊥ bedding	–	24.18*	–	0.21*	–	–	–	CHENEVERT. 1964
	‖ bedding	–	28.8*	–	0.21*	–	–	–	CHENEVERT, 1964
	Euville limestone	2.65	35.1*	12.6*	0.39*	–	–	p = 12.3	VAN DER VLIS, 1970
	Vaurian limestone	2.70	59.1*	23.0*	0.29*	–	–	p = 10.9	VAN DER VLIS, 1970
	Belgian limestone	2.70	87.8	33.8	0.30*	–	–	p = 0.5	VAN DER VLIS, 1970
	Poland, Maestrichtion, clay limestone ⊥ bedding	–	2.1		–	15.6	–		LOZIŃSKA–STEPIEŃ, 1966
	‖ bedding	–	1.3		–	14.6	–		LOZIŃSKA–STEPIEŃ, 1966
	Poland, Turonian, clay limestone ‖ bedding	–	1.9		–	228.0	–	–	LOZIŃSKA–STEPIEŃ, 1966
	⊥ bedding	–	1.6		–	13.4	–	–	LOZIŃSKA–STEPIEŃ, 1966
	⊥ bedding	–	3.2		–	20.2	–	–	LOZIŃSKA–STEPIEŃ, 1966
	Poland, Turonian , Opoka limestone ⊥ bedding	–	1.3		–	18.1	–	–	LOZIŃSKA–STEPIEŃ, 1966
	‖ bedding	–	1.2		–	14.1	–	–	

	Location/description								Reference
Limestone	USA, Niagara, N.Y., Gasport limestone	2.64	56.72 / 47.54*	—	0.21 / 0.08*	118.6	—	—	HOGG, 1964
	USA, Niagara, N.Y., Irondequoit limestone	2.63	51.7 / 79.3*	—	0.20 / 0.75*	117.9	—	—	HOGG, 1964
	USA, Niagara, N.Y., Reynales limestone	2.68	41.82 / 86.46*	—	0.16 / 0.19*	—	—	—	HOGG, 1964
	USA, Bloomington, Ind., Indiana limestone, saturated	2.56	31.72 / 19.31*	—	0.27	55.9 / 46.6*	—	p = 18.8	MICHALOPOULOS & TRIANDAFILIDIS, 1976
	USA, limestone, saturated	2.69	47.6 / 33.1*	—	0.27	76.8 / 74.8*	—	p = 1.5	MICHALOPOULOS & TRIANDAFILIDIS, 1976
	USA, Salem limestone, Spregen, Ind., breeciated, bioclastic	2.66	11.72 / 15.86	—		50.3	—	p = 12.6	MESRI et al, 1976
	USA, Millers Ferry Dam chalky, hard	1.78	49.0	—		—	—	—	CORNS & NESBITT, 1967
	USA	—	19.3 / 29.5*	—		55.26	—	—	AVEDISSIAN & WOOD, 1968
	USA	—	13.2 / 19.3*	—		54.8	—	—	AVEDISSIAN & WOOD, 1968
	USA	—	21.5 / 35.16*	—		81.0	—	—	AVEDISSIAN & WOOD, 1968
	USA	—	13.1 / 18.6*	—		50.3	—	—	AVEDISSIAN & WOOD, 1968
	USA	—	45.2 / 72.4*	—		131.9	—	—	AVEDISSIAN & WOOD, 1968
	USA	—	34.5 / 54.1*	—		120.0	—	—	AVEDISSIAN & WOOD, 1968

LABORATORY MECHANICAL PROPERTIES OF ROCKS

Rock	Location & Description	Density ϱ (g/cm³)	Modulus of Elasticity E (GPa)	Modulus of Rigidity G (GPa)	Poisson's Ratio υ	Compressive Strength σ_c (MPa)	Tensile Strength σ_t (MPa)	Remarks	Reference
	(1)	(2)	(3)	(4)	(5)	(6)	(7)	(8)	(9)
Limestone	USA	–	8.4 13.8*	–	–	49.1	–	–	AVEDISSIAN & WOOD, 1968
	USA	–	15.0 25.5*	–	–	75.8	–	–	AVEDISSIAN & WOOD, 1968
	USA	–	42.33* 22.9	–	–	113.8	–	–	AVEDISSIAN & WOOD, 1968
	USA	–	24.0 43.1*	–	–	108.9	–	–	AVEDISSIAN & WOOD, 1968
	USA	–	18.6 30.2*	–	–	75.7	–	–	AVEDISSIAN & WOOD, 1968
	USA	–	20.7 40.0*	–	–	98.6	–	–	AVEDISSIAN & WOOD, 1968
	USA	–	34.5 64.1*	–	–	130.3	–	–	AVEDISSIAN & WOOD, 1968
	USA	–	23.8 44.8*	–	–	91.0	–	–	AVEDISSIAN & WOOD, 1968
	USA	–	30.1 39.0*	–	–	91.0	–	–	AVEDISSIAN & WOOD, 1968
	USA	–	37.2 51.0*	–	–	108.9	–	–	AVEDISSIAN & WOOD, 1968
	USA, Estancia Valley	–	71.9*	29.7*	0.21*	–	–	–	BRATTON & PRATT, 1968
	USA, Estancia Valley	–	49.6*	21.7*	0.14*	–	–	–	BRATTON & PRATT, 1968
	USA, Estancia Valley, P = 200	–	77.6* 86.2	29.8* 33.9	0.30* 0.27	–	–	–	BRATTON & PRATT, 1968
	USA, Estancia Valley	–	86.2	–	0.48	129.1	–	secant at 21 MPa	BRATTON & PRATT, 1968
	USA, Estancia Valley	–	206.8	–	0.11	139.6	–	secant at 21 MPa	BRATTON & PRATT, 1968

Limestone							secant at 21 MPa	BRATTON & PRATT, 1968
USA, Estancia Valley	-	115.1	-	0.32	147.9	-		-
Brazil, Taquarussu Quarry	2.74	50.0* 84.0	-	0.33	85.0	-	p = 0.4	RUIZ, 1966
Brazil, Perus Quarry	2.72	22.8 65.7*	-	0.23	87.0	-	p = 0.6	RUIZ, 1966
Brazil, Piraporinha Quarry	2.71	75.8* 91.7	-	0.28	75.0	-	p = 0.1	RUIZ, 1966
USA, St. Genevieve Mine, Mo.	2.67	66.7*	26.8*	0.24	164.1	22.0(R)	fossiliferous	BLAIR, 1956
USA, St. Genevieve Mine, Mo.	2.56	56.7* 55.7	27.4	0.20* 0.26	115.8	40.0	oolitic	BLAIR, 1956
USA, St. Genevieve Mine, Mo.	2.46	-	-	-	125.4	18.5(R)	oolitic	BLAIR, 1956
USA, St. Genevieve Mine, Mo.	2.65	70.8*	27.3*	0.29	142.7	18.6(R)	fossiliferous	BLAIR, 1956
USA, Jonathan Mine, Ohio, fg	2.73	65.5* 68.2	21.9*	0.24* 0.25	146.9	24.1	p = 0.9 detrital	BLAIR, 1956
USA, Jonathan Mine, Ohio, fg #1	2.81	59.9	-	0.24	180.0	33.1(R)	fossiliferous	BLAIR, 1956
#2	2.81	61.7	-	0.22	140.55	31.0(R)	fossiliferous	BLAIR, 1956
USA, Jonathan Mine, Ohio, fg	2.69	70.2	-	0.25	148.9	17.24(R)	fossiliferous	BLAIR, 1956
USA, Jonathan Mine, Ohio, sandy, fg	2.59	45.3* 29.5	20.3*	0.09* 0.15	155.8	14.5(R)	detrital	BLAIR, 1956
USA, Jonathan Mine, Ohio, fg	2.41	37.7*	16.6*	0.14*	108.9	17.2(R)	-	BLAIR, 1956
USA, Pinecreek Mine, Tenn.	2.74	53.02* 61.23	23.30*	0.14* 0.24	173.1	6.89(R)	p = 2.4	BLAIR, 1956
USA, Saligo Mine, Tenn., fg	2.73	27.17* 30.7	11.7*	0.22* 0.28	173.0	5.52(R)	p = 3.4	BLAIR, 1956

LABORATORY MECHANICAL PROPERTIES OF ROCKS

Rock	Location & Description	Density r (g/cm³)	Modulus of Elasticity E (GPa)	Modulus of Rigidity G (GPa)	Poisson's Ratio γ	Compressive Strength s_c (MPa)	Tensile Strength s_t (MPa)	Remarks	Reference
	(1)	(2)	(3)	(4)	(5)	(6)	(7)	(8)	(9)
Limestone	USA, Bedford, Ind., porous, massive, hard, calcitic	2.19	–	–	–	36.6	–	p = 15.4	BLAIR, 1956
	USA, Marble Cliff, Columbus, Ohio, massive, hard, fg	2.60	–	–	–	123.4.	–	p = 5.4	BLAIR, 1956
	USA, Eniwetok Atoll, fg, hd	2.25	–	–	–	54.9	–	–	BLAIR, 1956
	USA, Eniwetok Atoll, porous, mg	1.83	–	–	–	17.9	5.52(R)	–	BLAIR, 1956
	USA, Eniwetok Atoll, fg, cavernous	2.51	63.29*	28.3*	0.12*	135.1	28.3(R)	recrystallised	BLAIR, 1956
	USA, Eniwetok Atoll, semiporous, fossiliferous	2.39	51.8* 50.4	22.9*	0.13* 0.23	97.2	16.6(R)	p = 4	BLAIR, 1956
	USA, Eniwetok Atoll, porous, mg	1.84	67.6	–	0.26	–	–	friable	BLAIR, 1956
	USA, Eniwetok Atoll, hd, dense, well cemented	2.53	73.91*	36.5*	0.01*	122.0	22.1	–	BLAIR, 1956
	USA, Eniwetok Atoll, porous, fossiliferous + dolomite	1.89	14.4	–	–	0.21	35.9	–	–	BLAIR, 1956
	USA, Eniwetok Atoll, sandy, friable	1.21	–	–	–	–	1.38	–	weakly cemented	BLAIR, 1956

Limestone							
USA, Eniwetok Atoll, hd, cavernous	2.31	–	–	94.5	–	–	BLAIR, 1956
Australia, Geelong, Vic.	–	–	–	95.9	14.7	–	SINGH, 1969
Australia, Lilydale, Vic. #1	–	–	–	102.1	12.8	–	SINGH, 1969
#3	–	–	–	113.8	13.8	–	SINGH, 1969
Australia, Lilydale, Vic. #4	–	–	–	166.0	12.2	–	SINGH, 1969
#2	–	–	–	204.2	19.3	–	SINGH, 1969
#3	–	–	–	177.1	14.9	–	SINGH, 1969
UK, Portland	–	–	–	41.4	18.0	–	HOBBS, 1964
UK, Portland	–	–	–	77.6	19.2	–	HOBBS, 1964
UK, Bulwell	–	–	–	142.9	38.8	impure	HOBBS, 1964
UK, Breedon	–	–	–	190.6	44.3	crinoidal	HOBBS, 1964
Canada, Ottawa, Ont., fg	2.81	71.0	–	134.0	–	–	COATES & PARSONS, 1966
Canada, Gagnon, Que., + dolomite + serpentine	2.69	58.9	–	278.0	–	–	COATES & PARSONS, 1966
USA, Ind.	2.23	–	–	28.27	10.41(R) 4.2(t)	p = 13	HARTMAN, 1966
USA, Bedford, Ind.	–	17.9 28.9*	–	28.3	3.59	–	RICKETTS & GOLDSMITH, 1970
USA, Ind.	–	36.13	0.26	–	–	–	SELLERS, 1970
USA, Ind.	–	35.85	0.25	–	–	–	SELLERS, 1970
USA, Pa. #1	–	92.25	0.29	–	–	–	SELLERS, 1970
USA, Pa. #2	–	80.00	0.29	–	–	–	SELLERS, 1970
USA, Ohio #1	–	50.4	0.29	–	–	–	SELLERS, 1970
USA, Ohio #2	–	44.06	0.20	–	–	–	SELLERS, 1970

LABORATORY MECHANICAL PROPERTIES OF ROCKS

Rock	Location & Description	Density ρ (g/cm³)	Modulus of Elasticity E (GPa)	Modulus of Rigidity G (GPa)	Poisson's Ratio ν	Compressive Strength σc (MPa)	Tensile Strength σt (MPa)	Remarks	Reference
	(1)	(2)	(3)	(4)	(5)	(6)	(7)	(8)	(9)
Limestone	Bulgaria, fg	2.54	–	–	–	210.0	6.6	p = 3.7	AVRAMOVA-TACHEVA & CHESHANKOVA, 1971
	Bulgaria, fg	2.59	–	–	–	200.0	4.7	p = 3.3	AVRAMOVA-TACHEVA & CHESHANKOVA, 1971
	Bulgaria, fg	2.61	–	–	–	182.0	7.1	p = 2.9 dolomitic	AVRAMOVA-TACHEVA & CHESHANKOVA, 1971
	Bulgaria, fg	2.62	–	–	–	150.0	5.9	p = 1.8 dolomitic	AVRAMOVA-TACHEVA & CHESHANKOVA, 1971
	Bulgaria, fg	2.50	–	–	–	65.0	4.4	p = 6.9 dolomitic	AVRAMOVA-TACHEVA & CHESHANKOVA, 1971
	Bulgaria, cg, sandy	2.55	–	–	–	92.0	–	p = 4.4	AVRAMOVA-TACHEVA & CHESHANKOVA, 1971
	Bulgaria, fg, argillaceous, calcareous	2.25	–	–	–	34.0	–	p = 16.4	AVRAMOVA-TACHEVA & CHESHANKOVA, 1971
	Bulgaria, fg, argillaceous, calcareous	2.38	–	–	–	55.0	–	p = 11.3	AVRAMOVA-TACHEVA & CHESAHANKOVA, 1971
	Hungary, Budapest, nummulitie	2.58	72.0	–	0.31	120.0	–	–	BODONYI, 1971
	Hungary, Tardos	2.69	77.0	–	0.32	80.0	–	–	BODONYI, 1971
	Hungary, Sütto, porous	2.60	68.5	–	0.33	70.0	–	–	BODONYI, 1971
	Hungary, Mecsek Mt., fg, + silica	2.54	27.5	–	0.30	110.0	–	–	BODONYI, 1971
	Egypt, Wade-el-Rayon Tunnel	–	–	–	–	3.31	–	+ marl beds	FAYED & KHATTAB, 1973
	Egypt, Wade-el-Rayon Tunnel	–	–	–	–	0.69	–	+ marl + clay	FAYED & KHATTAB, 1973

Limestone								
Egypt, Wade-el-Rayon Tunnel	–	–	–	–	4.24	–	marl + clay	FAYED & KHATTAB, 1973
Egypt, Wade-el-Rayon Tunnel	–	–	–	–	0.93	–	+ shale	FAYED & KHATTAB, 1973
USA, Ind., ⊥	2.29	30.7	–	–	42.8	5.86	p = 14.1	HAWKES et al, 1973
Germany, Solenhofen, vfg	2.54	55.5 / 58.8*	–	–	–	–	p = 4.7	BRACE, 1963
USA, fg–cg	2.82	52.7	–	–	101.4	–	–	CLARK & CAUDLE, 1963
#1	2.37	33.4*	14.2*	0.18*	75.20	10.69(R)	p = 11	OBERT et al, 1946
#2	2.37	40.9*	15.9*	0.29*	66.88	13.31(R)	p = 11	OBERT et al, 1946
#3	2.37	37.2*	15.3*	0.22*	70.33	11.93(R)	p = 11	OBERT et al, 1946
Canada, Normond, cg	–	53.02	–	0.21	–	6.89(I)	–	EVERELL et al, 1970
Canada, Normond, fg	–	53.78	–	0.22	97.9	7.24(I)	–	EVERELL et al, 1970
Canada, Normond	–	54.06	–	0.28	90.67	6.00(I)	–	EVERELL et al, 1970
Canada, St. Marc	–	49.64	–	0.25	80.67	5.38(I)	–	EVERELL et al, 1970
Canada, Miron	–	58.40	–	0.24	139.27	9.45(I)		EVERELL et al, 1970
Canada, Miron	–	62.12	–	0.26	157.2	8.55(I)		EVERELL et al, 1970
Canada, Montreal East	–	63.22	–	0.25	193.05	9.1(I)		EVERELL et al, 1970
Canada, Trenton	–				293.72	–	–	EVERELL et al, 1970
USSR	–				24.0–160.0	3.0–15.0		KOIFMAN, 1969
USSR, Estonoslanets	–				44.5	3.1	fissured	KOIFMAN & CHIRKOV, 1969
USSR, Estonoslanets	–				46.0	2.7	fissured	KOIFMAN & CHIRKOV, 1969
USSR, Estonoslanets	–				53.0	3.1	fissured	KOIFMAN & CHIRKOV, 1969
USSR, Estonoslanets	–				55.0	3.5	strongly fissured	KOIFMAN & CHIRKOV, 1969
USSR, Estonoslanets	–				45.0	2.9	compact	KOIFMAN & CHIRKOV, 1969

LABORATORY MECHANICAL PROPERTIES OF ROCKS

Rock	Location & Description	Density ϱ (g/cm³)	Modulus of Elasticity E (GPa)	Modulus of Rigidity G (GPa)	Poisson's Ratio υ	Compressive Strength σc (MPa)	Tensile Strength σt (MPa)	Remarks	Reference
	(1)	(2)	(3)	(4)	(5)	(6)	(7)	(8)	(9)
Limestone	USSR, Caucasus & Carpathia, marmorised	–	74.0	–	0.30	151.0	–	p = 1.22	BELIKOV et al, 1967
	USSR, Moscow syneclise, dolomitic	–	55.0	–	0.32	145.0	–	p = 9.22	BELIKOV et al, 1967
	USSR, Moscow syneclise, dolomitic	–	34.0	–	0.27	75.0	–	p = 16.2	BELIKOV et al, 1967
	USSR, Moscow syneclise, organogenic-dolomitic	–	37.0	–	0.25	90.0	–	p = 16.0	BELIKOV et al, 1967
	USSR, Moscow syneclise, fragmented, dolomitic	–	20.0	–	–	39.0	–	p = 21.7	BELIKOV et al, 1967
	USSR, Moscow syneclise, detrital, dolomitic	–	27.0	–	0.25	53.0	–	p = 21.8	BELIKOV et al, 1967
	USSR, Turkmenia, dolomitic	–	78.0	–	–	214.0	–	p = 1.81	BELIKOV et al, 1967
	USSR, Turkmenia, shell (algal)	–	9.0	–	–	120.0	–	p = 34.5	BELIKOV et al, 1967
	USSR, marmorised	2.70	83.6	–	0.31	–	–	p = 0.46	BELIKOV, 1967
	USSR, Carpathian, marmorised	2.68	75.2	–	–	–	–	p = 1.05	BELIKOV, 1967
	USSR, Turkmenia, dolomitic	2.70	73.9	–	–	–	–	p = 1.81	BELIKOV, 1967

Rock	Location / description								Reference
Limestone	USSR, Turkmenia, algal	1.73	11.0	–	–	–	–	p = 34.0	BELIKOV, 1967
	USSR, Estonia, Low-Paleozoic	2.49	31.7	0.21	–	–	–	p = 13.37	BELIKOV, 1967
	USSR, Moscow syneclise, detrital	2.16	23.9	–	–	–	–	p = 21.32	BELIKOV, 1967
	USSR, Moscow syneclise, organo-clastic	2.30	37.1	0.25	–	–	–	p = 15.98	BELIKOV, 1967
	USSR, Moscow syneclise, chemogenic, dolomitic	2.57	50.5	0.32	–	–	–	p = 9.22	BELIKOV, 1967
	USA, oolitic, fg-mg, hd	–	45.5 / 53.78*	0.18 / 0.21*	–	–	–	–	RINEHART et al, 1961
	USA, hd, healed fractures	–	16.55 / 28.27*	0.18 / 0.20*	–	–	–	med. porous	RINEHART et al, 1961
	USA, mg, hd	–	33.78 / 52.4	0.17 / 0.31*	–	–	–	–	RINEHART et al, 1961
	USA, Bucyrus, Ohio, hd	2.56	53.78	–	113.76	4.83	–	–	ATCHISON et al, 1961
	UK, Portland	–	37.92	0.28	–	–	–	oolitic, hd	WILSON, 1961
	USA, Ind.	2.19	–	–	41.37	2.49	–	p = 24.0	SCHWARTZ, 1964
	USA, Gnome nuclear explosion site	–	–	–	81.3	–	–	p = 3.6	KNUTSON & BOHOR, 1964
	USA, Gnome nuclear explosion site	–	–	–	55.0	–	–	p = 0.1	KNUTSON & BOHOR, 1964
	Canada, Queenstown	–	–	–	75.15	–	–	–	BAUER & CALDER, 1967
Loamy Stony Debris	Yugoslavia	–	0.036	0.006	–	–	–	τ = 0–0.16	DVORAK, 1967
Lujaurite	USSR, Khibiny	–	53.0	–	0.26	–	–	alkali intrusion	BELIKOV et al, 1967

LABORATORY MECHANICAL PROPERTIES OF ROCKS

Rock	Location & Description	Density ρ (g/cm³)	Modulus of Elasticity E (GPa)	Modulus of Rigidity G (GPa)	Poisson's Ratio υ	Compressive Strength σc (MPa)	Tensile Strength σt (MPa)	Remarks	Reference
	(1)	(2)	(3)	(4)	(5)	(6)	(7)	(8)	(9)
Magnesite	Chechoslavakia, mines	–	110.0	–	–	137.3	–	–	HEREL, 1966
Magnetite	Australia, raw	–	–	–	–	234.0	17.1	–	SINGH, 1969
	USA, Mineville, N.Y. quasi-isotropic	4.23	31.44*	18.62*	–	141.34	10.34(R)	p = 1.4	WINDES, 1949
	USA, Mich. ore, siliceous	3.64	86.18*	36.27*	–	232.35	35.16(R)	–	WINDES, 1949
	Canada, Noranda Mines, Ont., massive	4.74	103.8*	44.7*	0.160*	–	–	–	BIRCH & BANCROFT, 1940
	USA, Ishpeming, Mich.	4.01	109.0*	44.0*	–	–	–	–	WINDES, 1949
	Canada, Wabush, Que., specularite	3.84	87.2	–	–	236.0	–	+ chamosite	COATES & PARSONS, 1966
	mineral (average)	5.17	219.0	–	–	–	–	–	BELIKOV, 1967
	Canada	3.54	–	41.5*	0.30*	–	12.0	–	LAROCQUE & FAVREAU, 1970
	USA, Cliffs shaft, Mich., siliceous	3.64	86.18	36.27	0.19	232.35	35.16(R)	p = 0.8	BOYUM, 1961
	Canada, ore, specularite	–	–	–	–	162.03	–	–	BAUER & CALDER, 1967
	Sweden, Kiruna	–	–	–	–	96.1	6.6	Lab.	HANSAGI, 1965
	Sweden, Kiruna	–	–	–	–	66.3	4.55	In situ	HANSAGI, 1965
	Sweden, Kiruna	–	–	–	–	83.9	5.60	Lab.	HANSAGI, 1965
	Sweden, Kiruna	–	–	–	–	39.4	2.63	In situ	HANSAGI, 1965

Marble								
USA, Proctor, Vt.	2.71	34.3	–	0.14	–	–	–	Zisman, 1933
P = 11, #1								
P = 56, #1	2.71	46.0	–	0.19	–	–	–	Zisman, 1933
#1	2.71	49.5*	21.7*	–	–	–	–	Ide, 1936
P = 11, #2	2.71	23.2	–	0.10	–	–	–	Zisman, 1933
P = 56, #2	2.71	39.9	–	0.19	–	–	–	Zisman, 1933
#2	2.71	28.0*	–	–	–	–	–	Ide, 1936
P = 11, #3	2.71	38.3	–	0.50	–	–	–	Zisman, 1933
P = 56, #3	2.71	49.5	–	0.22	–	–	–	Zisman, 1933
#3	2.71	50.9*	22.3*	–	–	–	–	Ide, 1936
USA, Proctor, Vt.	2.71	–	31.8	–	–	–	–	Birch & Bancroft, 1938
P = 500, #3								
P = 4000, #3	2.71	87.0	33.3	0.305	–	–	–	Birch & Bancroft, 1938
Belgium, Dinant, fine, black P = 70–600	–	72.4	29.8	0.278	–	–	–	Adams & Coker, 1906
Italy, Carrara P = 70–600	–	55.4	21.7	0.274	–	–	–	Adams & Coker, 1906
USA, Rutland, Vt., gray	2.74	52.4	20.7	0.263	–	–	–	Regula, 1940
USA, Rutland, Vt., white	2.76	37.4	17.0	0.11	–	–	–	Regula, 1940
USA, Star Lake, N.Y., white	2.72	54.0	23.0	–	126.86	11.72(R)	–	Windes, 1949
USA, Marble, Md. ⊥ bedding	2.87	49.30*	26.06*	–	212.36	5.95 19.31(R)	p = 0.6	Wuerker, 1956
USA, Marble, Md. ∥ bedding	2.87	63.03*	28.41*	–	215.12	22.75(R)	–	Wuerker, 1956
USA, N.Y., pyroxene mixed rock	3.02	103.42*	33.72*	–	129.62	–	p = 0.2	Wuerker, 1956
USA, Carthage	2.64	47.9	–	0.17	106.0	–	–	Hoskins & Horino, 1969
USA, taconic-, white massive, fg	2.71	47.9	–	0.40	62.0	1.17	at 50% fracture	Miller, 1965

LABORATORY MECHANICAL PROPERTIES OF ROCKS

Rock	Location & Description	Density ϱ (g/cm³)	Modulus of Elasticity E (GPa)	Modulus of Rigidity G (GPa)	Poisson's Ratio ν	Compressive Strength σc (MPa)	Tensile Strength σt (MPa)	Remarks	Reference
	(1)	(2)	(3)	(4)	(5)	(6)	(7)	(8)	(9)
Marble	USA, Cherokee-, mg-cg	2.71	55.8	–	0.25	66.9	1.79	at 50% fracture	MILLER, 1965
	USA, Imperial, Danby massive, mg	2.71	60.4	–	0.34	64.8	2.21	at 50% fracture	MILLER, 1965
	USSR, Karelia, dolomitic, dense	–	78.46	–	0.28	274.6	–	p = 0.83	BELIKOV et al, 1967
	USSR, Gazgan, Uzbek, fg	–	95.13	–	0.40	170.64	–	p = 0.16	BELIKOV et al, 1967
	USSR, Sultanuizdat, Uzbek, cg	–	73.55	–	–	148.09	–	p = 0.50	BELIKOV et al, 1967
	USA, Yule, Colo., thermally altered from limestone								
	⊥ bedding	–	45.8*	–	0.17*	–	–	–	YOUASH, 1970
	‖ bedding	–	49.2*	–	0.32*	–	–	–	YOUASH, 1970
	‖ bedding	–	48.5*	–	0.29	–	–	–	YOUASH, 1970
	Japan, Yamaguchi Mine, (dynamic σc = 120–200 dynamic σt = 20–40)	2.7	–	–	–	90–110	5.0– 9.0	–	ITO & TERADA, 1970
	USA, West Rutland, Vt., Vermont marble, cg = 2 mm	2.75	17.24 22.1	–	–	48.3 96.5	–	p = 2.1	MESRI et al, 1976
	Japan, Akiyoshi	2.68	–	–	–	63.8	4.4	–	MITANI & KAWAI, 1974

	Locality								Reference
Marble	USA, Wombeyan	–	–	–	–	51.4	6.07	–	SINGH, 1969
	USA, Wombeyan	–	–	–	–	57.1	6.72	–	SINGH, 1969
	Italy, Carrara	–	–	–	–	79.21	8.52	–	SINGH, 1969
	Australia, Buchan #1	–	–	–	–	117.5	13.4	–	SINGH, 1969
	Australia, Buchan #2	–	–	–	–	143.8	10.1	–	SINGH, 1969
	Hungary, Ruskica, cg	2.66	77.0	–	0.32	80.0	–	–	BODONYI, 1971
	mg	2.69	24.0 / 25.0*	–	–	–	7.2	p = 0.7	BRACE, 1963
	USA, cg, #1	2.87	49.3*	26.1*	0.06	212.40	18.1(R)	p = 0.6	OBERT et al, 1946
	#2	2.87	63.1*	28.41*	0.16	215.10	22.68(R)	p = 0.6	OBERT et al, 1946
	#3	2.87	71.7	28.4	0.27	227.50	–	p = 0.6	OBERT et al, 1946
	USSR	–	–	–	–	80.0–160.0	8.0–29.0	–	KOIFMAN, 1969
	USSR, Karelia, dolomitic	–	80.0	–	0.28	280.0	–	p = 0.83	BELIKOV et al, 1967
	USSR, Uzbek, fg	–	97.0	–	0.40	174.0	–	p = 0.16	BELIKOV et al, 1967
	USSR, Uzbek, cg	–	75.0	–	–	151.0	–	p = 0.50	BELIKOV et al, 1967
	USSR, Karelia, fg, dolomitic	2.82	80.7	–	0.26	–	–	p = 0.83	BELIKOV, 1967
	USSR, Ural	2.71	83.8	–	0.31	–	–	p = 0.46	BELIKOV, 1967
	USSR, Uzbek	2.72	88.3	–	–	–	–	p = 0.24	BELIKOV, 1967
	USA, Tenn. mg–cg	2.69	42.0	–	–	120.0	9.5	–	D'ANDREA & CONDON, 1970
	USA, Tenn.	2.70	48.26 / 57.2*	–	–	105.8	6.53	–	HAIMSON & FAIRHURST, 1970
	USA, Baltimore, Md., dolomitic	2.80	82.74	–	–	151.68	6.55	–	ATCHISON et al, 1961
	USA, Tenn., fg	–	71.71	–	0.28	124.11	12.41(R)	p = 2.3	HAIMSON & FAIRHURST, 1969

LABORATORY MECHANICAL PROPERTIES OF ROCKS

Rock	Location & Description	Density ρ (g/cm³)	Modulus of Elasticity E (GPa)	Modulus of Rigidity G (GPa)	Poisson's Ratio υ	Compressive Strength σc (MPa)	Tensile Strength σt (MPa)	Remarks	Reference
	(1)	(2)	(3)	(4)	(5)	(6)	(7)	(8)	(9)
Marble	USA, Ga.	2.69	–	–	–	82.74	4.48	p = 2.8	SCHWARTZ, 1964
		–	–	–	–	38.06	–	–	BAUER & CALDER, 1967
Marl	France, Saizerais Mine, dry	–	10.0* 6.5	–	–	37.0	2.3	–	TINCELIN & SINOU, 1964
	France, Saizerais Mine, damp	–	1.7	–	–	14.0	–	–	TINCELIN & SINOU, 1964
Marlstone	USA, Colo., calcareous dolomite ‖ bedding	2.36	41.02* 33.23	15.31*	0.33* 0.26	172.37*	–	grain = 0.005–0.15mm	WUERKER, 1956
	USA, Colo., calcareous dolomite	2.31	24.89*	11.10*	0.11*	151.0	12.41(R)	grain = 0.005–0.15 mm	WUERKER, 1956
	USA, Colo., kerogenaceous	2.22	12.48 24.06*	10.20*	0.17* 0.29	81.36	2.76(R)	p = 1.1 secant at 50%	BLAIR, 1955
	Holland, Limbargian marl	2.69	46.0	34.0	0.28	–	–	–	VAN DER VLIS, 1970
	USA, Colo., shaly, calcareous	2.39	30.27	–	0.11	–	33.09	–	BLAIR, 1955
	Poland, Opoka marl Maestrichtian ⊥ bedding	–	1.8–2.3	–	–	17.2–25.8	–	–	LOZIŃSKA - STEPIEŃ, 1966
	‖ bedding	–	1.8–3.7	–	–	17.1–20.4	–	–	LOZIŃSKA - STEPIEŃ, 1966
	Poland, Opoka marl, Maestrichtian ⊥ bedding	–	2.0	–	–	19.4	–	CaCo₃ = 50–75%	LOZIŃSKA - STEPIEŃ, 1966

Marlstone

Material	Density						Composition	Reference
‖ bedding	–	2.0	–	–	18.1	–		LOZIŃSKA-STĘPIEŃ, 1966
USA, Jonathan Mine, Ohio, fg, very porous	2.19	25.0*	11.0*	0.13*	55.8	6.2(R)	calcite + kaolinite	BLAIR, 1956
Hungary, clayey ⊥	2.01	–	–	–	7.10	–	p = 39.5 w = 24.0	MARTOS, 1965
Hungary, clayey‖	2.01	–	–	–	9.40	–	p = 39.5 w = 24.0	MARTOS, 1965
Hungary, clayey ⊥	2.01	–	–	–	5.70	–	p = 46.0 w = 30.5	MARTOS, 1965
Hungary, clayey‖	2.01	–	–	–	5.0	–	p = 46.0 w = 30.5	MARTOS, 1965
Hungary, clayey ⊥	2.12	–	–	–	2.30	–	p = 39.5 w = 26.3	MARTOS, 1965
Hungary, clayey‖	2.12	–	–	–	5.30	–	p = 39.5 w = 26.3	MARTOS, 1965
Hungary, clayey ⊥	2.00	–	–	–	4.80	–	p = 39.5 w = 23.0	MARTOS, 1965
Hungary, clayey‖	2.00	–	–	–	5.10	–	p = 39.5 w = 23.0	MARTOS, 1965
Hungary, clayey ⊥	2.16	–	–	–	7.40	–	p = 31.5 w = 19.5	MARTOS, 1965
Hungary, clayey‖	2.16	–	–	–	10.25	–	p = 31.5 w = 19.5	MARTOS, 1965
Hungary, clayey ⊥	2.15	–	–	–	4.0	–	p = 29.13 w = 14.0	MARTOS, 1965

LABORATORY MECHANICAL PROPERTIES OF ROCKS

Rock	Location & Description	Density ϱ (g/cm³)	Modulus of Elasticity E (GPa)	Modulus of Rigidity G (GPa)	Poisson's Ratio ν	Compressive Strength σ_c (MPa)	Tensile Strength σ_t (MPa)	Remarks	Reference
	(1)	(2)	(3)	(4)	(5)	(6)	(7)	(8)	(9)
Marlstone	Hungary, clayey ∥	2.15	–	–	–	7.30	–	p = 29.13 w = 14.0	MARTOS, 1965
	Hungary ⊥	1.86	–	–	–	8.50	–	p = 39.0 w = 17.2	MARTOS, 1965
	Hungary ∥	1.86	–	–	–	8.45	–	p = 39.0 w = 17.2	MARTOS, 1965
	Bulgaria, Bistrica	–	0.75	–	–	2.78	0.33	–	PARASKEVOV et al, 1968
	Bulgaria, Bistrica	–	0.13	–	0.28	5.50	0.71	–	PARASKEVOV et al, 1968
	Bulgaria, Bistrica	–	0.51	–	0.3	2.14	0.23	–	PARASKEVOV et al, 1968
	Czechoslovakia	2.25	9.0	–	0.17	55.0	2.75(R)	–	HOUSKA, 1968
Martite	USSR	–	–	–	–	60.0– 480.0	7.0– 50.0	–	KOIFMAN, 1969
Metadiorite	Canada, Macleod Mine	–	77.2	–	0.31			–	HERGET, 1973
Mica	mineral (average)	2.89	71.0	–	–	–	–	–	BELIKOV, 1967
Microcline	USSR, Chernaya Salma-perthite	2.55	74.8*	29.3*	0.28*	–	–	–	BELIKOV, 1967
Migmatite	USSR, Ukraine	2.65	65.71	–	0.22	307.94	–	p = 0.57	BELIKOV et al, 1967
	USSR, Baltic shield, Ladoga Formation	2.69	72.57	–	0.17	–	–	p = 0.66	BELIKOV et al, 1967
	Migmatite	–	80.0	–	0.17– 0.22*	–	–	–	GEYER & MYUNG, 1970
	USSR, Ukraine	–	67.0	–	0.22	314.0	–	p = 0.57	BELIKOV et al, 1967
	USSR, Ukranian shield	2.65	59.0	–	0.20	–	–	p = 0.52	BELIKOV, 1967

Migmatite	USSR, Baltic shield	2.69	71.4	–	0.17	–	–	p = 0.66	BELIKOV, 1967
Monazite	USA, Grand Coulee, Wash., porphyry	2.57	42.75*	23.44*	0.18	124.73	–	p = 2.32	WUERKER, 1956
			56.54		0.21*	170.51			
		2.58	44.13	–	0.17		–	p = 2.40	WUERKER, 1956
Monazite	USA, San #1 Manuel Mine, Ariz., quartz monazite, cg	2.67	47.6	–	0.26	75.7	10.2	–	KENDORSKI & MEHTAB, 1976
	#2	2.67	48.3	–	0.13	101.4	8.7	–	KENDORSKI & MEHTAB, 1976
	#3 porphyritic	2.67	57.9	–	0.20	105.5	9.7	–	KENDORSKI & MEHTAB, 1976
	USA, Anaconda Mines, Calif., fg – cg	2.75	–	–	–	112.4	–	–	BLAIR, 1956
Monzonite	Bulgaria, Vitosha Mt.	2.67	23.5	–	–	61.5	–	p = 9.0	ILIEV, 1966
	Bulgaria, Vitosha Mt.	2.67	38.5	–	–	135.0	–	p = 4.5	ILIEV, 1966
	Bulgaria, Vitosha Mt.	2.67	50.0	–	–	174.0	–	p = 2.3	ILIEV, 1966
	Bulgaria, Vitosha Mt.	2.66	6.4	–	–	46.0	–	p = 7.0	ILIEV, 1966
	Bulgaria, Vitosha Mt.	2.65	7.0	–	–	50.0	–	p = 12.0	ILIEV, 1966
	Bulgaria, Vitosha Mt.	2.68	31.3	–	–	100.0	–	p = 6.5	ILIEV, 1966
	Bulgaria, Vitosha Mt.	2.66	24.0	–	–	100.0	–	p = 6.5	ILIEV, 1966
	USA, porphyry, altered	–	41.37	–	0.18	–	–	–	RINEHART et al, 1961
			56.54*		0.21*				
Mudstone	UK, Bilsthorpe, silty	–	47.9	–	–	126.0	–	–	HOBBS, 1970
	UK, Bates mudstone, finely laminated ⊥ bedding	2.52	–	–	–	73.15	–	–	HOBBS, 1964

LABORATORY MECHANICAL PROPERTIES OF ROCKS

Rock	Location & Description	Density ϱ (g/cm^3)	Modulus of Elasticity E (GPa)	Modulus of Rigidity G (GPa)	Poisson's Ratio υ	Compressive Strength σ_c (MPa)	Tensile Strength σ_t (MPa)	Remarks	Reference
	(1)	(2)	(3)	(4)	(5)	(6)	(7)	(8)	(9)
Mudstone	UK, Macguyan Cap, massive, shaly ⊥ bedding	2.72	–	–	–	76.74	35.6	σ_t parallel to laminations (Brazilian)	HOBBS, 1964
	UK, Ellington mudstone, shaly ⊥ bedding	2.55	–	–	–	23.86	11.79	σ_t ‖ laminations (Brazilian)	HOBBS, 1964
	UK, Scammonden Dam	–	5.86	–	–	52.4	–	at 31.4 m depth	PENMAN & MITCHELL, 1970
	Japan, Nokanam silty	2.35	–	–	–	10.8	–	p = 6.0	YAMANO, 1974
	Australia, Vic., Silurian, fully weathered	1.95	1.72	–	–	–	–	p = 21	LEARMONTH & GARRETT, 1969
	Australia, Vic., Silurian, highly weathered	2.02	71.03	–	–	–	–	p = 17	LEARMONTH & GARRETT, 1969
	Australia, Vic., Silurian mod. weathered	2.06	27.6	–	–	–	–	p = 13.5	LEARMONTH & GARRETT, 1969
	Australia, Vic., Silurian, slightly weathered	2.14	79.2	–	–	–	–	p = 10	LEARMONTH & GARRETT, 1969
	Australia, Vic., Silurian fresh	2.26	96.5	–	–	–	–	p = 5	LEARMONTH & GARRETT, 1969

Mudstone	Locality							Reference
	UK, Cannock Chase Colliery, dry	—	—	—	200.6	—	massive, calcareous p = 2.5	PRICE, 1960
	UK, Kirby Colliery, dry	—	—	—	44.8	—	p = 3.5 carbonaceous	PRICE, 1960
	UK, Cannock Chase Colliery, dry, laminated	—	—	—	100.6	—	p = 3.5 carbonaceous	PRICE, 1960
	UK, Babbington, massive	—	—	—	142.5	36.3	—	HOBBS, 1964
	Hungary, 4 seam, Lyuko pit ⊥	2.15	—	—	12.8	—	p = 27.5 w = 12.0	MARTOS, 1965
	Hungary, 4 seam, Lyuko pit ∥	2.15	—	—	12.4	—	p = 27.5 w = 12.0	MARTOS, 1965
	Hungary, 4 seam, sandy ⊥	2.13	—	—	14.9	—	p = 26.5 w = 13.0	MARTOS, 1965
	Hungary, 4 seam, sandy ∥	2.13	—	—	10.4	—	p = 26.5 w = 13.0	MARTOS, 1965
	Hungary, 4 seam, cg ∥	1.96	—	—	9.0	—	p = 34.5 w = 11.0	MARTOS, 1965
	Hungary, 4 seam, cg ∥	1.96	—	—	8.6	—	p = 34.5 w = 11.0	MARTOS, 1965
	Hungary, calcareous ⊥	2.16	—	—	13.2	—	p = 30.0 w = 11.5	MARTOS, 1965
	Hungary, calcareous ∥	2.16	—	—	11.7	—	p = 30.0 w = 11.5	MARTOS, 1965
	Hungary, calcareous ⊥	2.23	—	—	10.6	—	p = 28.0 w = 13.2	MARTOS, 1965
	⊥	2.23	—	—	10.6	—	p = 28.0 w = 13.2	MARTOS, 1965
	∥	2.23	—	—	9.5	—	p = 28.0 w = 13.2	MARTOS, 1965
	∥	2.23	—	—	9.5	—	p = 28.0 w = 13.2	MARTOS, 1965
	⊥	2.66	—	—	12.0	—	p = 32.5 w = 14.0	MARTOS, 1965

LABORATORY MECHANICAL PROPERTIES OF ROCKS

Rock	Location & Description		Density ϱ (g/cm³)	Modulus of Elasticity E (GPa)	Modulus of Rigidity G (GPa)	Poisson's Ratio ν	Compressive Strength σ_c (MPa)	Tensile Strength σ_t (MPa)	Remarks	Reference
			(2)	(3)	(4)	(5)	(6)	(7)	(8)	(9)
Mudstone		$=$	2.66	–	–	–	9.4	–	p = 32.5 w = 14.0	MARTOS, 1965
		$\perp$	2.3	–	–	–	16.2	–	p = 23.5 w = 12.0	MARTOS, 1965
		$=$	2.3	–	–	–	16.4	–	p = 23.5 w = 12.0	MARTOS, 1965
		$\perp$	2.20	–	–	–	12.05	–	p = 31.5 w = 13.5	MARTOS, 1965
		$=$	2.20	–	–	–	10.40	–	p = 31.5 w = 13.5	MARTOS, 1965
	Hungary, 4 seam, Lyuko pit	$=$	1.95	–	–	–	6.4	–	p = 34.5 w = 14.5	MARTOS, 1965
		$\perp$	1.95	–	–	–	7.0	–	p = 34.5 w = 14.5	MARTOS, 1965
		$\perp$	2.20	–	–	–	12.8	–	p = 21.0 w = 10.5	MARTOS, 1965
		$=$	2.20	–	–	–	12.4	–	p = 21.0 w = 10.5	MARTOS, 1965
	Hungary, 4 seam, floor		2.0	–	–	–	10.2	–	p = 31.0 w = 00.8	MARTOS, 1965
	Hungary, banded		1.93	–	–	–	13.3	–	p = 34.5 w = 19.0	MARTOS, 1965
			1.93	–	–	–	8.3	–	p = 34.5 w = 19.0	MARTOS, 1965
	Hungary, shale, 4 seam		2.00	–	–	–	7.6	–	p = 32.5 w = 16.5	MARTOS, 1965
	Hungary, calcareous		1.82	–	–	–	6.2	–	w = 28.0	MARTOS, 1965
		$=$	1.82	–	–	–	6.7	–	w = 28.0	MARTOS, 1965
		$\perp$	1.98	–	–	–	3.5	–	p = 28.0	MARTOS, 1965

Mineral	Location		Density						Reference	Author
Mudstone	Hungary, fossiliferous	=	1.98	–	–	–	2.1	–	p = 28.0	MARTOS, 1965
		⊥	2.10	–	–	–	9.50	–	p = 30.0, w = 18.6	MARTOS, 1965
		=	2.10	–	–	–	7.85	–	p = 30.0, w = 18.6	MARTOS, 1965
	Hungary, calcareous + sandy bands	⊥	1.81	–	–	–	7.2	–	p = 28.0, w = 19.0	MARTOS, 1965
		=	1.81	–	–	–	8.80	–	p = 28.0, w = 19.0	MARTOS, 1965
	Hungary	⊥	2.40	–	–	–	7.65	–	p = 43.0, w = 23.0	MARTOS, 1965
		=	2.40	–	–	–	2.5	–	p = 43.0, w = 23.0	MARTOS, 1965
	Hungary	⊥	2.10	–	–	–	10.0	–	p = 28.5, w = 15.5	MARTOS, 1965
		=	2.10	–	–	–	12.40	–	p = 28.5, w = 15.5	MARTOS, 1965
	Hungary, carbon	⊥	2.18	–	–	–	17.0	–	p = 28.5, w = 15.5	MARTOS, 1965
		=	2.18	–	–	–	12.10	–	p = 28.5, w = 15.5	MARTOS, 1965
	Hungary, Tuffy	⊥	2.00	–	–	–	7.10	–	p = 31.5, w = 16.5	MARTOS, 1965
		=	2.00	–	–	–	4.40	–	p = 31.5, w = 16.5	MARTOS, 1965
	Bulgaria, Bistrica	–	–	0.54	–	–	1.00	0.14	–	PARASKEVOV et al, 1968
	Bulgaria, Bistrica	–	–	0.41	–	0.85	2.30	0.19	–	PARASKEVOV et al, 1968
	Bulgaria, Bistrica		–	0.55	–	0.28	1.41	0.15	–	PARASKEVOV et al, 1968
Muscovite	USSR, Slyudyaska		2.79	80.4*	32.3*	0.25*	–	–	–	BELIKOV, 1967
Nepheline	USSR, Vishnevya Mts.		2.62	77.4*	31.2*	0.24*	–	–	–	BELIKOV, 1967
	USSR, Southeast Tuva		2.62	70.1*	28.2*	0.24*	–	–	–	BELIKOV, 1967

LABORATORY MECHANICAL PROPERTIES OF ROCKS

Rock	Location & Description	Density ϱ (g/cm³)	Modulus of Elasticity E (GPa)	Modulus of Rigidity G (GPa)	Poisson's Ratio υ	Compressive Strength σ_c (MPa)	Tensile Strength σ_t (MPa)	Remarks	Reference
	(1)	(2)	(3)	(4)	(5)	(6)	(7)	(8)	(9)
Nepheline	mineral (average)	2.62	73.0	–	–	–	–	–	BELIKOV, 1967
Nephrite	USSR, Siberia	–	109.84	–	0.24	539.39	–	p = 0.42	BELIKOV et al, 1967
	USSR, Siberia	–	112.0	–	0.24	550.0	–	p = 0.42	BELIKOV et al, 1967
	USSR, Siberia	3.00	114.3*	–	–	–	–	p = 0.42	BELIKOV, 1967
Nickel Sulphide Ore	Canada, Falconbridge Mine	–	57.57	–	0.18	221.32	–	–	MORUZI & PASIEKA, 1964
Norite	Canada, Sudbury, Ont. P = 36, #1	2.87	80.7	–	0.224	–	–	–	ZISMAN, 1933
	#2	2.86	90.6*	28.7*	–	–	–	–	IDE, 1936
	S. Africa, Bushveld complex, Tvl.	2.984	–	37.8*	–	–	–	–	SIMMONS, 1964
		–	97.0	–	–	310.0	–	–	BIENIAWSKI, 1969
	Canada, Sudbury, Ont.	–	62.05	–	0.17	190.16	–	–	MORRISON, 1970
	S. Africa, Black Hill #1	–	119.4	–	–	212.3	12.96	–	BAMFORD, 1969
	S. Africa, Black Hill #1	–	73.9	–	–	247.1	12.96	–	BAMFORD, 1969'
	S. Africa, Black Hill #1	–	100.3	–	–	213.0	12.96	–	BAMFORD, 1969
	S. Africa, Black Hill #3	–	54.8	–	–	223.4	12.96	–	BAMFORD, 1969
	S. Africa, Black Hill #2	–	72.7	–	–	262.1	12.96	–	BAMFORD, 1969
	Canada, Sudbury, Ont.	–	83.6 / 88.2*	–	0.22 / 0.27*	–	–	–	RINEHART et al, 1961

Rock	Location								Reference
Norite	Canada, Falconbridge Mine	—	72.39	—	0.18	170.99	—	—	MORUZI & PASIEKA, 1964
Obsidian	USA, Modoc, Calif.	2.446	65.6	30.3	0.08	—	—	—	BIRCH & BANCROFT, 1939
Oligoclase	USSR, Balshoi Medredok	2.61	77.3*	29.8*	0.29*	—	—	29% anorthite	BELIKOV, 1967
	mineral (average)	2.61	77.0	—	—	—	—	—	BELIKOV, 1967
Olivine	Burma (forsterite)	3.32	216.7*	86.7*	0.24*	—	—	—	BELIKOV, 1967
	mineral (average)	3.32	217.0	—	—	—	—	—	BELIKOV, 1967
Orthoclase	USSR, Murzinka	2.54	63.0*	24.4*	0.29*	—	—	—	BELIKOV, 1967
Opal	—	—	38.0	18.0	0.06	—	—	—	VOIGT, 1928
Pegmatite	USA, N.Y.	2.59	61.64*	22.7*	—	213.74	22.06(R)	—	WUERKER, 1956
Peridotite	USSR, Moncha-Tundra Khibiny	—	164.76	—	0.26	—	—	—	BELIKOV et al, 1967
	Canada, Thetford, Que.	2.62	55.2	—	—	197.0	—	olivine + pyroxene + serpentine	COATES & PARSONS, 1966
	USSR, Moncha-Tundra	—	168.0	—	0.26	—	—	—	BELIKOV et al, 1967
	USSR, garnet-peridotite	3.36	134.7	—	0.24	—	—	p = 0.43	BELIKOV, 1967
	USSR, Moncha-Tundra	3.25	157.0*	—	—	—	—	p = 0.29	BELIKOV, 1967
Phlogopite	USSR, Aldan	2.82	68.0*	26.7*	0.29*	—	—	—	BELIKOV, 1967
	USSR, Slyudyaska	2.85	66.8*	26.6*	0.30*	—	—	—	BELIKOV, 1967
Phyllite	USA, Sly Park Dam, Colo., graphitic, porous, fg, P = 35	2.35	9.45 / 26.85	6.89*	—	6.69	—	p = 15.3	BALMER, 1953

LABORATORY MECHANICAL PROPERTIES OF ROCKS

Rock	Location & Description	Density ϱ (g/cm³)	Modulus of Elasticity E (GPa)	Modulus of Rigidity G (GPa)	Poisson's Ratio ν	Compressive Strength σc (MPa)	Tensile Strength σt (MPa)	Remarks	Reference
	(1)	(2)	(3)	(4)	(5)	(6)	(7)	(8)	(9)
Phyllite	USA, Sly Park Dam, Colo., quartose, slightly weathered, porous, fg P = 0–35	2.18	0.62 18.32*	4.83*	0.02	9.38	–	p = 22.0	BALMER, 1953
	USA, Sly Park Dam, Colo., seri-citic, at 30°, fg	2.34	17.3	–	–	9.80	–	p = 17.4	BALMER, 1953
	USA, Mich., green	3.24	76.53*	32.75*	–	126.17	22.75(R)	–	BALMER, 1953
	USA, Mich., + greenstone, grain size = 0.1–0.7 mm	3.10 3.30	74.46* 84.12*	32.47* 38.06*	– –	114.45 313.71	26.89 22.75	p = 0.4% p = 0.5%	BALMER, 1953 BALMER, 1953
	India, River Mahi Dam, Kadana, fresh	2.8	7.0	–	0.25	16.0	–	wet	PARASARTHY et al, 1974
	altered	2.77	5.0	–	0.25	5.0	–	wet	PARASARTHY et al, 1974
	USA, Fontana Dam	–	–	–	–	153.0	–	–	BLEE & MEYER, 1955
	USA, Cliffs shaft, Mich.	3.24	76.53	22.75	0.16	126.17	22.75(R)	p = 1.1 green	BOYUM, 1961
	USA, fg, hd-soft, slightly weathered	–	7.58 18.62*	–	0.03	–	–	porous, laminated	RINEHART et al, 1961
	USA, fg, hd, med. weathered, granite	–	9.65 26.89*	–	–	–	–	–	RINEHART et al, 1961

Plagiogranite	USSR, Ukranian shield	2.73	68.1	–	0.22	–	–	p = 0.49	BELIKOV, 1967
Porphyry	USA, Ariz., altered	–	–	–	–	222.01	–	grain = 0.05–9.5 mm	WUERKER, 1956
	Canada, Ont. (syenite)	2.7	71.02*	30.34*	0.2*	434.37	–	–	WUERKER, 1956
	Canada, Lakeshore	–	64.81	26.89	0.21	250.14	13.10	–	MORRISON, 1970
	Canada, Sigma Mine	–	75.84	–	0.22	173.33	12.41	–	MORRISON, 1970
Potash	Canada, Estherhazy, Sask., vcg	2.10	7.3	–	–	12.7	–	porous, sylvite + carnallite	COATES & PARSONS, 1966
	USA, N.M. #1	–	24.1	–	0.26	–	–	–	SELLERS, 1970
	USA, N.M. #2	–	23.5	–	0.26	–	–	–	SELLERS, 1970
Potash Salt	Germany	–	25.0	–	–	50.0	1.8	–	HÖFER & MENZEL, 1964
Pumice	USA, Ore., Newberry Caldera, mg–fg	0.43	1.59 / 2.76*	–	–	4.21	0.59	–	RICKETTS & GOLDSMITH, 1970
Pyrite	Canada, Noranda Mines, Ont., massive	4.85	164.9*	70.2*	0.17	–	–	–	BIRCH & BANCROFT, 1939
Pyroxene	mineral (average)	3.32	149.0	–	–	–	–	–	BELIKOV, 1967
Pyroxenite	USA, Stilwater, Mont., bronzitite #1	3.27	151.3*	65.4*	0.16	–	–	–	BIRCH & BANCROFT, 1939
	#2	3.27	152.6*	64.8*	0.18	–	–	–	BIRCH & BANCROFT, 1939
	S. Africa, Bushveld, Tvl., bronzitite	3.28	155.7*	62.8*	0.239	–	–	–	BIRCH & BANCROFT, 1939

LABORATORY MECHANICAL PROPERTIES OF ROCKS

Rock	Location & Description	Density ρ (g/cm³)	Modulus of Elasticity E (GPa)	Modulus of Rigidity G (GPa)	Poisson's Ratio ν	Compressive Strength σc (MPa)	Tensile Strength σt (MPa)	Remarks	Reference
	(1)	(2)	(3)	(4)	(5)	(6)	(7)	(8)	(9)
Pyroxenite	USA, Star Lake, N.Y., bronzitite, fresh	3.43	124.0*	56.0*	–	–	–	–	WINDES, 1949
	slightly altered	3.31	113.0	41.0	–	–	–	–	WINDES, 1949
	USA, N.Y., fresh	3.48	124.1*	50.26*	–	182.02	22.75	p = 0.6	WUERKER, 1956
	USA, N.Y., moderately altered	3.31	113.07*	40.61*	–	122.04	18.62	p = 0.5	WUERKER, 1956
	USA, N.Y., heavily altered	2.53	21.99*	7.58*	–	58.61	13.79	p = 5.3	WUERKER, 1956
	USA, N.Y., garnet	3.40	122.73*	44.68*	–	84.81	15.17	p = 1.0	WUERKER, 1956
	USSR	3.29	143.4	–	0.20	–	–	p = 0.39	BELIKOV, 1967
	USSR, Moncha-Tundra	3.29	15.5*	–	–	–	–	p = 0.32	BELIKOV, 1967
	USSR, Lake Balkash	3.28	181.8*	–	–	–	–	p = 0.15	BELIKOV, 1967
	Canada, Noranda Mines, Ont., massive	4.59	89.2*	34.1*	0.309*	–	–	–	BIRCH & BANCROFT, 1939
Quartz	USA, Olka, Kans., interlocked with minor calcite	2.72	1.71	–	0.20	328.88	33.78	p = 2.0, secant at 50	BLAIR, 1955
	India, KGF	2.80	58.0*	22.6*	0.29*	–	–	p = 5, vein	RAMANA et al, 1973
	USSR, geosynclines, porphyrites	2.59	53.8	–	0.17	–	–	p = 2.52	BELIKOV, 1967
	USSR, geosynclines, porphyrites, felsitic	2.47	37.9	–	0.21	–	–	p = 6.09	BELIKOV, 1967

Mineral	Location/description								Reference
Quartz	mineral (average)	2.65	96.0	–	–	–	–	–	BELIKOV, 1967
	Canada, Porcupine Mine, porphyry	–	47.0	–	–	167.0	–	–	COCHRANE et al, 1964
	Canada, Porcupine Mine, porphyry	–	44.0	–	–	172.0	–	–	COCHRANE et al, 1964
Quartzite	Mexico, Jaragua Hills	2.67	54.8*	–	0.08*	226.0	–	p = 1.2	RUIZ, 1966
	—fractured	–	64.4–76.4*	–	0.10–0.20*	–	–	–	GEYER & MYUNG, 1970
	S. Africa, fg	–	79.3	0.17	–	–	–	–	LEEMAN, 1964
	USA, Baraboo, massive, fg	2.62	88.4	–	0.11	320.10	11.0	–	MILLER, 1965
	USA, Minn., magnetite bearing taconite	2.75	84.81*	38.61*	0.10	628.80	23.44	p = 0.3, grain = 0.3 mm	MILLER, 1965
	USA, Mich., + slate	–				211.67	8.27	grain = 0.01–0.7 mm	MILLER, 1965
	USA, Mich., + slate	–				297.85	31.03	grain = 0.1–0.7 mm	MILLER, 1965
	USSR, Kursk, magnetic anomaly, ferruginous	–	112.78	–	0.33	350.11	–	p = 0.25	BELIKOV et al, 1967
	USSR, Karelia, Ukraine	–	69.6	–	0.13	373.65	–	p = 0.57	BELIKOV et al, 1967
	Canada, Denison Mine	–	66.19	28.96	0.17	160.99	–	–	MORRISON, 1970
	Canada, Milikan Mine	–	70.33	29.65	0.17	142.72	–	–	MORRISON, 1970
	USA, Flaming George Dam, Utah, fg	2.55	22.06	–	0.75	148.24	3.45	grain = 0.17–0.2 mm	BRANDON, 1974
	India, Kosi Dam, schistose quartzite, mg, slightly fractured	2.60	64.12	–	0.8	216.50	–	grain = 0.16–0.23 mm	BRANDON, 1974

LABORATORY MECHANICAL PROPERTIES OF ROCKS

Rock	Location & Description	Density ϱ (g/cm³)	Modulus of Elasticity E (GPa)	Modulus of Rigidity G (GPa)	Poisson's Ratio υ	Compressive Strength σ_c (MPa)	Tensile Strength σ_t (MPa)	Remarks	Reference
	(1)	(2)	(3)	(4)	(5)	(6)	(7)	(8)	(9)
Quartzite	S. Africa	–	90.0	–	–	263.0	–	–	BIENIAWSKI et al, 1969
	India, River Mahi Dam, Kadana, quartzite, massive	2.75	50.0 / 61.0*	–	0.18	180.0	–	wet	PARASARTHY et al, 1974
	fractured	2.67	–	–	–	138.0	–	wet	PARASARTHY et al, 1974
	altered	2.69	12.5	–	0.33	–	–	–	PARASARTHY et al, 1974
	USA, Northfield Mt. Power Station								
	‖ bedding	–	62.05	–	0.14	–	106.9	–	WILD & McKITTRICK, 1971
	⊥ bedding	–	39.3	–	0.10	–	164.1	–	WILD & McKITTRICK, 1971
	USA, Northfield Mt. Power Station. Mass., micaceous quartzite								
	‖ bedding	–	51.7	–	–	–	146.9	–	WILD & McKITTRICK, 1971
	⊥ bedding	–	39.3	–	–	–	149.6	–	WILD & McKITTRICK, 1971
	USA, Fontana Dam	–	–	–	–	153.0	–	–	BLEE & MEYER, 1955
	Rhodesia, Kariba Dam, left bank fg	2.74	65.2 / 58.2*	–	–	–	–	–	LANE, 1964
	Rhodesia, Kariba Dam, left bank, cg	2.72	83.3 / 65.5*	–	–	–	–	–	LANE, 1964

Quartzite								
Brazil, Janaque Hills	2.63	21.3* 55.9	–	0.08	262.0	–	p = 0.9	RUIZ, 1966
USA, Cliffs shaft, Mich., fg–cg	3.24	–	–	–	147.6	20.0(R)	siderite bonded	BLAIR, 1956
USA, Cliffs shaft, Mich., fg	4.07	97.9*	40.6*	0.2*	293.0	35.2(R)	hematite + chlorite	BLAIR, 1956
USA, Martics Ville, Pa., calca-reous, + biotite schist	3.05	–	–	–	196.5	–	p = 1.7	BLAIR, 1956
Canada, Elliot Lake, Ont., cg	2.66	81.4	–	–	260.0	–	–	COATES & PARSONS, 1966
S. Africa, Naran-dera, argillaceous	–	70.0	–	0.16	190.0	–	layered	GAY, 1973
S. Africa, East Rand Ltd.	–	–	–	–	295.0	–	–	DENKHAUS, 1965
S. Africa, East Rand Ltd.	–	–	–	–	240.0	–	–	DENKHAUS, 1965
S. Africa, City Deep Ltd.	–	–	–	–	220.0	–	–	DENKHAUS, 1965
S. Africa, City Deep Ltd.	–	–	–	–	300.0	–	–	DENKHAUS, 1965
S. Africa, Crown Mines	–	–	–	–	270.0	–	–	DENKHAUS, 1965
S. Africa, Crown Mines	–	–	–	–	220.0	–	–	DENKHAUS, 1965
USA, Cheshire, mg	2.63	53.5 66.0	–	–	–	30.0	p = 0.7	BRACE, 1963
Canada, Rivett, Galena Mines	–	50.33	–	0.29	224.1	17.93(R)	fg, compe-tent	CHAN, 1970
Canada, Rivett, Galena Mines	–	10.69	–	0.27	78.6	10.34(R)	fg, argill., bedded	CHAN, 1970
Canada, Rivett, Galena Mines	–	45.44	–	0.24	77.2	–	mg, minera-lised 15%	CHAN, 1970
Canada, Rivett, Galena Mines	–	11.38	–	0.28	104.1	–	mg–cg, mi-neralised	CHAN, 1970

LABORATORY MECHANICAL PROPERTIES OF ROCKS

Rock	Location & Description	Density ρ (g/cm³)	Modulus of Elasticity E (GPa)	Modulus of Rigidity G (GPa)	Poisson's Ratio ν	Compressive Strength σc (MPa)	Tensile Strength σt (MPa)	Remarks	Reference
	(1)	(2)	(3)	(4)	(5)	(6)	(7)	(8)	(9)
Quartzite	Canada, Rivett, Galena Mines	–	42.47	–	0.25	152.4	–	mg, homogeneous	CHAN, 1970
	USSR, Kursk magnetic anomaly, ferruginous	–	115.0	–	0.33	357.0	–	p = 0.25	BELIKOV et al, 1967
	USSR, Karelia, Ukraine	–	71.0	–	0.13	381.0	–	p = 0.57	BELIKOV et al, 1967
	USSR, Karelia, cg, chlorite + sericite cement	2.68	–	–	–	182.0	–	p = 0.69	TIMCHENKO, 1967
	USSR, Karelia, mg, chlorite + sericite cement	2.67	–	–	–	287.0	–	p = 0.83	TIMCHENKO, 1967
	USSR, Karelia, fg, sericite cement	2.66	–	–	–	409.0	–	p = 0.1	TIMCHENKO, 1967
	USSR, Kursk magnetic anomaly	3.51	119.0	–	0.33	–	–	p = 0.25	BELIKOV, 1967
	USSR, Urals	2.65	70.6	–	0.13	–	–	p = 0.57	BELIKOV, 1967
	USSR, Ukraine	2.65	87.0*	–	–	–	–	p = 0.37	BELIKOV, 1967
	USA, Jasper	2.6	–	–	–	335.77	25.10	p = 0.5	CLARK et al, 1970
	USA, Cliffs shaft, Mich., + slate	–	–	–	–	211.00	8.27(R)	–	BOYUM, 1961
	USA, Cliffs shaft, Mich.	–	–	–	–	297.85	31.03(R)	–	BOYUM, 1961
	Canada, Silver Summit, fg, argillaceous	–	14.75	–	–	107.19	26.14(R)	thick bedded	CHAN & CROCKER, 1971

Quartzite	Canada, Silver Summit, fg–mg, slightly argillaceous	–	20.13	–	–	85.92	26.14(R)	thick beds, microfissures	CHAN & CROCKER, 1971
	Canada, Silver Summit, fg–mg, slightly argillaceous	–	33.16	–	–	150.35	26.14(R)	thin beds, microfissures	CHAN & CROCKER, 1971
	Canada, Silver Summit, vfg	–	89.63	–	–	204.77	26.14(R)	homogeneous	CHAN & CROCKER, 1971
	Canada, Silver Summit, fg	–	43.44	–	–	193.79	26.14(R)	–	CHAN & CROCKER, 1971
	Canada, Silver Summit, fg	–	50.06	–	–	278.71	26.14(R)	–	CHAN & CROCKER, 1971
	Canada, Silver Summit, fg, slightly argillaceous	–	29.16	–	–	171.51	26.14(R)	–	CHAN & CROCKER, 1971
	Canada, Silver Summit, fg	–	61.5	–	–	154.06	26.14	–	CHAN & CROCKER, 1971
	Canada, Silver Summit, + siderite + sulphide	–	–	–	–	–	23.50(R)	45 deg. to bedding	CHAN & CROCKER, 1971
	Canada, Silver Summit, fg, slightly argillaceous	–	–	–	–	–	24.9(R)	80 deg. to bedding	CHAN & CROCKER, 1971
	Canada, Silver Summit, vfg	–	–	–	–	–	35.33(R)	70 deg. to bedding	CHAN & CROCKER, 1971
	Canada, Silver Summit, fg, bedded	–	–	–	–	–	30.05(R)	50 deg. to bedding	CHAN & CROCKER, 1971
	Canada, Silver Summit, slightly, argillaceous, fractures	–	–	–	–	–	23.82(R)	50 deg. to bedding	CHAN & CROCKER, 1971
	Canada, Silver Summit, vfg	–	–	–	–	–	30.63(R)	60 deg. to bedding	CHAN & CROCKER, 1971

410

LABORATORY MECHANICAL PROPERTIES OF ROCKS

Rock	Location & Description	Density ϱ (g/cm³)	Modulus of Elasticity E (GPa)	Modulus of Rigidity G (GPa)	Poisson's Ratio υ	Compressive Strength σc (MPa)	Tensile Strength σt (MPa)	Remarks	Reference
(1)		(2)	(3)	(4)	(5)	(6)	(7)	(8)	(9)
Quartzite	Canada, Silver Summit, fg, slightly argillaceous	–	–	–	–	–	25.06(R)	80 deg. to bedding	CHAN & CROCKER, 1971
	Canada, Silver Summit, fg, slightly argillaceous	–	–	–	–	–	26.51(R)	70 deg. to bedding	CHAN & CROCKER, 1971
	Canada, Silver Summit, fg, fractures	–	–	–	–	–	15.91(R)	50 deg. to bedding	CHAN & CROCKER, 1971
	Canada, Silver Summit, fg	–	–	–	–	–	24.70(R)	45 deg. to bedding	CHAN & CROCKER, 1971
Rhyolite	Japan, Seikan Tunnel	2.42	–	–	–	85.4	5.6	–	MOCHIDA, 1974
	Japan, Taguchi	–	26.0	–	–	–	–	–	IIDA et al, 1960
Rock Salt	Canada, Goderich, Ont.	2.20	4.65	–	–	35.6	–	–	MUIR & COCHRANE, 1966
	Canada, Goderich, Ont.	2.90	51.5	–	–	157.0	–	–	MUIR & COCHRANE, 1966
	Canda, Goderich, Ont.	2.71	44.3	–	–	103.7	–	–	MUIR & COCHRANE, 1966
	Canada, Goderich, Ont.	2.20	4.64	–	–	35.5	2.48	–	MUIR & COCHRANE, 1966
	USA, Diamond crystal, rock salt, massive, cg	2.16	4.89	–	0.73	21.4	0.83	–	MILLER, 1965
	Rumania, rock salt, white	–	–	–	–	25.2	1.14	–	FODOR & TŐKES, 1966

Rock Salt	Rumania, rock salt, white, layered	–	–	–	–	27.6	1.23	–	FODOR & TOKES, 1966
	dark gray	–	25.0	–	–	30.4	1.44	–	FODOR & TOKES, 1966
	Germany	–	–	–	–	50.0	1.5	–	HÖFER & MENZEL, 1964
Salt	USA, Kansas								
	#1	–	28.54	–	0.31	–	–	–	SELLERS, 1970
	#2	–	28.13	–	0.29	–	–	–	SELLERS, 1970
	USA, Ohio								
	#1	–	28.5	–	0.22	–	–	–	SELLERS, 1970
	#2	–	25.3	–	0.21	–	–	–	SELLERS, 1970
	USA, Winnfield, La.	2.16	33.03*	15.58*	0.06	–	–	–	NICHOLLS, 1961
	Germany, Werra, hd	–	–	–	–	36.5	–	–	SPACKELER et al, 1960
	Germany, Werra, hd	–	–	–	–	42.4	–	–	SPACKELER et al, 1960
	Germany, Stassfurt, hd	–	–	–	–	36.2	–	–	SPACKELER et al, 1960
	Germany, Stassfurt, hd	–	–	–	–	47.7	–	–	SPACKELER et al, 1960
	Germany, Sallstedt, hd	–	–	–	–	40.49	1.39	–	HÖFER, 1960
	Germany, Bischofferode, hd	–	–	–	–	38.51	2.66	–	HÖFER, 1960
	Germany, Bleicherode, hd	–	–	–	–	36.82	1.26	–	HÖFER, 1960
	Germany, Sonderhausen, hd	–	–	–	–	40.32	1.78	–	HÖFER, 1960
	Germany, Volkenroda, hd	–	–	–	–	28.65	1.30	–	HÖFER, 1960
	Germany, Rossleben	–	–	–	–	33.43	1.18	–	HÖFER, 1960
Sandstone	USA, Ohio ⊥ bedding	2.06	6.0*	3.17*	–	71.7	3.45(R)	p = 16.0	WUERKER, 1956
	‖ bedding	–	6.69*	3.17*	–	55.16	5.52(R)	–	WUERKER, 1956

LABORATORY MECHANICAL PROPERTIES OF ROCKS

Rock	Location & Description	Density ρ (g/cm³)	Modulus of Elasticity E (GPa)	Modulus of Rigidity G (GPa)	Poisson's Ratio υ	Compressive Strength σc (MPa)	Tensile Strength σt (MPa)	Remarks	Reference
	(1)	(2)	(3)	(4)	(5)	(6)	(7)	(8)	(9)
Sandstone	USA, Ohio, cg ∥ bedding	2.17	10.52*	4.83*	–	38.85	5.17	grain = 0.2–0.5 mm	WUERKER, 1956
	USA, Utah	2.20	21.37*	13.72*	0.22*	106.87	11.03(R)	p = 10.0	WUERKER, 1956
	USA, Ala., ferruginous	3.14	19.10*	10.20*	0.07	166.85	7.58	p = 3.1 grain = 0.1–0.3 mm	WUERKER, 1956
	USA, Ala., fossiliferous red	3.26	30.68*	18.41*	–	154.44	12.41	p = 2.9	WUERKER, 1956
	USA, Wyo.	2.28	22.06	–	0.17	60.74	–	p = 16.4	WUERKER, 1956
	USA, Wyo.	2.37	–	–	–	84.12	–	p = 11.2	WUERKER, 1956
	USA, Ill., ferruginous, weathered	–	5.0	–	–	15.65	–	–	WUERKER, 1956
	USA, Ala., + saltstone + shale	2.76	39.92*	22.68*	–	184.78	15.17	p = 1.7	WUERKER, 1956
	Germany, quartzite, quasi isotropic, agg.	2.65	100.0	47.0	0.07	–	–	–	VOIGT, 1928
	USA, Mont., quartzitic, P = 4000	2.65	97.5	42.4	0.15	–	–	–	BIRCH, 1961
	USA, Pa., quartzitic	2.66	63.6 71.5*	32.4*	11.5	–	–	–	ZISMAN, 1933 IDE, 1936
	Germany, Rügen, flint	–	74.5	34.5	0.08	–	–	–	VOIGT, 1928
	USA, Ohio, feldspathic, P = 70–600	–	15.8	6.1	0.29	–	–	–	ADAMS & COKER, 1906

Sandstone

Locality / description	Density							Reference
USA, Ten Sleep Formation, Wyo., P = 0–35	2.28–2.37	14.0	–	0.06	72.4	–	–	USBR, 1953 also BALMER, 1953
USA, Waterford, Ohio								
⊥ bedding	2.17	7.1	4.00	–	–	–	–	OBERT et al, 1946
∥ bedding	–	10.6–11.2	4.8–4.5	–	–	–	–	OBERT et al, 1946
USA, Amherst, Ohio								
⊥ bedding	2.06	6.0	3.2	–	–	–	–	OBERT et al, 1946
∥ bedding	2.06	6.7–8.8	3.2–4.4	–	–	–	–	
USA, Longmont	2.32	5.0	–	0.14	62.7	1.93	p = 21.4	JUDD, 1969
	2.35	31.2	–	0.08	169.8	–	–	HOSKINS & HORINO, 1969
Brazil, Botucato	2.44	22.8	–	0.11	76.1	–	p = 5.0	RUIZ, 1966
UK, Bunter sandstone	–	–	12.1–23.1	0.15–0.20	12.1	–	p = 16.6	MORGENSTERN & PHUKAN, 1969
Canada, WAC Bennett Dam, hard, cherty								
along strike	–	17.9	–	–	136.5	–	–	IMRIE & JORY, 1968
along dip	–	15.2	–	–	129.6	–	–	IMRIE & JORY, 1968
bedding	–	17.9	–	–	140.0	–	–	IMRIE & JORY, 1968
USA, Berea, massive, fg, slightly porous	2.18	18.3	–	0.38	73.8	1.17	–	MILLER, 1965
USA, Crab Orchard, dense, fg	2.53	39.2	–	0.46	214.0	8.14	–	MILLER, 1965
USA, Navajo, porous, mg–fg	2.02	15.3	–	0.31	43.4	1.24	–	MILLER, 1965
USA, Navajo, Buckhorn Wash Area, Utah, intermediate, case hardened	2.15	11.00* / 17.58	6.14*	0.9* / 0.15	86.87	7.58(R)	p = 12.0	BLAIR, 1955

LABORATORY MECHANICAL PROPERTIES OF ROCKS

Rock	Location & Description	Density ρ (g/cm³)	Modulus of Elasticity E (GPa)	Modulus of Rigidity G (GPa)	Poisson's Ratio υ	Compressive Strength σc (MPa)	Tensile Strength σt (MPa)	Remarks	Reference
	(1)	(2)	(3)	(4)	(5)	(6)	(7)	(8)	(9)
Sandstone	USA, Mich., calcareous, felsitic + quartz	2.60	55.30* 14.00	23.79*	0.16* 0.17	157.89	24.13(R)	p = 5.3	BLAIR, 1955
	USA, Pa., sandstone	2.16	7.31* 14.00	3.93*	−0.09* 0.17	76.53	4.83	p = 11.0	BLAIR, 1955
	USA, sandstone, Pa	2.43	13.79* 3.59	8.69*	0.20* 0.04	102.04	15.86	–	BLAIR, 1955
	Poland, Upper Silesia ⊥ bedding	–	51.0	–	–	80.0	4.0–6.7	–	BORECKI & CHUDEK, 1972
	∥ bedding	–	38.0	–	–	77.0	2.0–5.5	–	BORECKI & CHUDEK, 1972
	Poland, Upper Silesia, with calcium binding	–	32.63	–	0.31	88.0	–	–	BORECKI & CHUDEK, 1972
	Poland, Upper Silesia, clay binder (arkose)	–	22.73	–	0.28	95.0	–	–	BORECKI & CHUDEK, 1972
	USSR, Karelia, Ukraine, Ural, quartzitic	2.65	63.75	–	0.10	254.0	–	p = 1.25	BELIKOV et al, 1967
	USSR, Donbass, quartzose, with hydromica cement	2.63	54.92	–	0.14	254.98	–	p = 2.16	BELIKOV et al, 1967
	USSR, Volga, quartzose, with opal-chalcedory cement	2.48	62.76	–	0.12	299.11	–	p = 1.79	BELIKOV et al, 1967

Sandstone								
USSR, Volga, with opal cement	2.31	49.04	–	0.12	221.64	–	p = 4.75	BELIKOV et al, 1967
USA, Allegheny, Pa., fg	2.7	43.44 / 24.61*	19.79	0.14	158.58	19.31(R)	p = 1.4, secant at 50%	BLAIR, 1956
USA, Woodrow Quarry, Pa., friable, porous ‖ & ⊥ bedding	2.13–2.17	6.96–8.89*	3.52–4.07*	0.01–0.07*	64.81–66.88	–	–	BLAIR, 1956
USA, W.Va., Kanawha Eagle graywacke, mg–fg	2.60	22.34* / 26.20	13.44*	0.17* / 0.18	141.34	11.72(R)	grain = 0.1–0.5 mm secant at 50%	BLAIR, 1955
USA, W.Va., Kanawha Eagle argillaceous, mg–fg	2.80	31.50*	14.82*	0.05*	105.49	4.14(R)	–	BLAIR, 1955
USSR, Kuznetsk Basin high strength	–	35–44	–	0.18–0.22	73–88	9–14	silicitic, 10–20% carbonates	YAGODKIN et al, 1969
medium strength	–	20–29	–	0.18–0.32	44–73	6–11	arkose, graywacke	YAGODKIN et al, 1969
low strength	–	18–20	–	0.36–0.38	29–39	4–8	laminated aleurolitic, coarse	YAGODKIN et al, 1969
USSR, Kuznetsk Basin weak	–	4–9	–	0.4–0.5	10–30	1.5–4	aleurolite, mg–fg	YAGODKIN et al, 1969
UK, Lea Hall	–	10.9	–	–	47.6	–	–	HOBBS, 1970
UK, Pennant, S. Wales, faintly laminated, ⊥ bedding ‖ bedding	2.69	–	–	–	153.3	14.48	–	PRICE, 1958
USA, Valencia County, N.M., Dakota Formation, fg, porous, weathered	2.69	62.05	–	–	138.58	15.17	–	PRICE, 1958
	2.1	3.07	–	–	46.19	–	grain = 0.08–0.33 mm	BRANDON, 1974

LABORATORY MECHANICAL PROPERTIES OF ROCKS

Rock	Location & Description	Density ϱ (g/cm³)	Modulus of Elasticity E (GPa)	Modulus of Rigidity G (GPa)	Poisson's Ratio υ	Compressive Strength σ_c (MPa)	Tensile Strength σ_t (MPa)	Remarks	Reference
	(1)	(2)	(3)	(4)	(5)	(6)	(7)	(8)	(9)
Sandstone	USA, Glen Canyon Dam, Colo., fg-mg, porous, variably cemented	2.02	3.96	–	0.11	24.48	0.72	grain = 0.06–0.36 mm	BRANDON, 1974
	USA, Coconino Dam, Ariz., calcareous, fg, slightly porous	2.45	8.96	–	–	73.77	–	grain = 0.08 mm	BRANDON, 1974
	USA, Bridge Canyon Dam, Ariz., ferruginous, mg, porous, massive	2.39	27.58	–	0.04	90.32	5.17	grain = 0.23–0.46 mm	BRANDON, 1974
	USA, Flaming Gorge Dam, Utah, quartzitic, ferruginous, dense, mg	2.54	14.82	–	0.06	113.07	3.24	grain = 0.06–2.0 mm	BRANDON, 1974
	Australia, Hawkesbury, NSW, mg, high porosity, with clay matrix	–	0.6–5.9	–	–	11.7–21.6 (wet)	–	–	PELLS, 1976
	Australia, Hawkesbury, NSW, mg, high porosity, with clay matrix	–	0.6–5.9	–	–	39.25–48.5 (dry)	–	–	PELLS, 1976
	Japan, Izumi sandstone	2.6	–	–	–	100.0–140.0	8.0–9.0	(dynamic σ_c = 120–200 σ_t = 50–70)	ITO & TERADA, 1970

Sandstone								
Japan, Tako sandstone	2.0	—	—	—	15.0–20.0	2.0–3.0	(dynamic σc = 20–50 σt = 10–20)	ITO & TERADA, 1970
USA, Chambers, Ariz., Arizona sandstone	2.60	11.03	—	0.31	32.82	—	p = 25, tan. 50%	MICHALOPOULOS & TRIANDAFILIDIS, 1976
USA, Chambers, Ariz., Arizona sandstone	2.60	9.45**	—	—	23.86**	—		MICHALOPOULOS & TRIANDAFILIDIS, 1976
USA, Colo., Lyons sandstone	—	4.14	—	0.30	29.4	3.59(R)	grain = 0.1–2.54 mm, p = 11	HEUZE & GOODMAN, 1973
UK, Ormonde, calcareous, massive, ⊥ bedding	2.36	—	—	—	41.44	23.30	tensile strength ‖-lamination, Brazilian	HOBBS, 1964
UK, Cefn Coed, faintly laminated calcareous, ‖ bedding	2.83	—	—	—	257.1	55.85	—	HOBBS, 1964
UK, Darley Dale, Matlock, massive, ⊥ bedding	—	—	—	—	90.32	16.55	tensile strength ‖-lamination, Brazilian	HOBBS, 1964
‖ bedding	—	—	—	—	67.57	4.96	—	PRICE, 1958
‖ bedding	—	—	—	—	37.23	3.31	—	PRICE, 1958
UK, Snowdown, Kent, ‖ bedding	—	72.33	—	—	144.79	11.03	—	PRICE, 1958
⊥ bedding	—	—	—	—	158.58	10.0	—	PRICE, 1958
Canada, Springhill	—	45.51	17.93	0.29	166.58	16.38(t)	t = torsion	MORRISON, 1970
USA, Coconino Dam, Ariz., moderately porous, fg, plagioclase feldspars, moderately altered	2.35	6.55	—	—	78.60	—	grain = 0.15 mm	BRANDON, 1974

LABORATORY MECHANICAL PROPERTIES OF ROCKS

Rock	Location & Description	Density ϱ (g/cm³)	Modulus of Elasticity E (GPa)	Modulus of Rigidity G (GPa)	Poisson's Ratio υ	Compressive Strength σ_c (MPa)	Tensile Strength σ_t (MPa)	Remarks	Reference
	(1)	(2)	(3)	(4)	(5)	(6)	(7)	(8)	(9)
Sandstone	USA, Palisades Dam, Idaho, massive to banded, calcareous, voids up to 3 mm dia. with zeolite and montmorillonite	1.87	35.51	–	0.22	31.72	–	–	BRANDON, 1974
	USA, Colo., Lyons sandstone, cross-bedded with kaolinite 8% + hematite 1%	–	34.48*	–	0.07*		–	–	YOUASH, 1970
	⊥ bedding	–	39.3*	–	0.12*		–	–	YOUASH, 1970
	‖ bedding	–	25.7*	–	0.14*		–	–	YOUASH, 1970
	USA, Golden, Colo., Laramie sandstone, mg–fg, 90% quartz + clay + hematite								
	Germany, Bentheim sandstone	2.63	20.0*	7.5*	0.33*	–	–	p = 23.9	VAN DER VLIS, 1970
	Germany, Gildehaus sandstone	2.64	23.0*	8.6*	0.34*	–	–	p = 20.7	VAN DER VLIS, 1970
	Germany, Mainz sandstone	2.66	24.6*	9.7*	0.28*	–	–	p = 17.7	VAN DER VLIS, 1970
	Germany, Ibbenbüren sandstone	2.67	29.9	11.4	0.31	–	–	p = 15.1	VAN DER VLIS, 1970

Rock	Location / description	density	a	b	c	d	notes	Reference
Sandstone	Germany, Oberkirchen sandstone	2.64	32.5*	12.9*	0.26*	–	$p = 17.4$	VAN DER VLIS, 1970
	USA, Niagara, N.Y., Thorold sandstone	2.49	18.04 / 45.9*	–	0.15 / 0.15*	161.5	–	HOGG, 1964
	USA, Amherst, Ohio, Berea sandstone, sub-granular quartz grains	2.66	9.65–13.8	–	–	–	$p = 20.3$	MESRI et al, 1976
	Switzerland, Rosseus Dam, miocene, mg ⊥ bedding ‖ bedding	–	11.5–2.5	–	–	–	–	GICOT, 1955
	miocene, fg ⊥ bedding	–	2.0–4.0	–	–	–	–	GICOT, 1955
	‖ bedding	–	2.0–4.5	–	–	–	–	GICOT, 1955
	‖ bedding	–	2.5–8.0	–	–	–	–	GICOT, 1955
	Angola, Cambambe Dam, arkose, fg	–	88.6	–	–	63.5	–	SARMENTO & VAZ, 1964
	India, Bhakra Dam, calcareous, subgraywacke	2.67	35.7	–	0.9	84.25	secant modulus, $p = 1.49$–4.23	BHATNAGAR & SHAH, 1964
	India, Bhakra Dam, calcareous, sub-graywacke, weathered	1.78	–	–	–	70.6	secant modulus, $p = 4.04$	BHATNAGAR & SHAH, 1964
	Romania, Carpethian, flysh-sandstone	2.69	14.7	–	–	71.5	–	BANCILA et al, 1961
	Romania, Carpethian, flysh-sandstone	2.71	67.0	–	–	134.0	–	BANCILA et al, 1961
	Yugoslavia, with shale	–	0.058	0.031	–	–	$\tau = 0$–0.3 MPa	DVORAK, 1967

LABORATORY MECHANICAL PROPERTIES OF ROCKS

Rock	Location & Description	Density ϱ (g/cm³)	Modulus of Elasticity E (GPa)	Modulus of Rigidity G (GPa)	Poisson's Ratio ν	Compressive Strength σ_c (MPa)	Tensile Strength σ_t (MPa)	Remarks	Reference
	(1)	(2)	(3)	(4)	(5)	(6)	(7)	(8)	(9)
Sandstone	UK, Scammonden Dam	–	9.4	–	–	56.6	–	21.6 m depth	PENMAN & MITCHELL, 1970
	UK, Scammonden Dam	–	19.2	–	–	80.6	–	53.0 m depth	PENMAN & MITCHELL, 1970
	USA	–	21.37 / 40.95*	–	–	82.2	–	–	AVEDISSIAN & WOOD, 1968
	USA	–	17.9 / 19.9*	–	–	55.2	–	–	AVEDISSIAN & WOOD, 1968
	USA	–	7.7 / 13.2*	–	–	53.8	–	–	AVEDISSIAN & WOOD, 1968
	USA, Estancia Valley	–	29.3* / 30.3	12.9* / 10.0	0.14* / 0.13	–	–	–	BRATTON & PRATT, 1968
	USA, Estancia Valley	–	50.3	–	0.27	96.5	–	secant at 21 MPa	BRATTON & PRATT, 1968
	USA, Estancia Valley	–	11.7	–	0.12	84.3	–	secant at 21 MPa	BRATTON & PRATT, 1968
	Japan, Kawazu	2.08	–	–	–	25.1	3.7	–	MITANI & KAWAI, 1974
	Japan, Seikan Tunnel	1.44	–	–	–	4.1	0.5	–	MOCHIDA, 1974
	Brazil, Botucatu Cuesta	2.32	43.7* / 23.2	–	0.11	83.0	–	p = 2.8	RUIZ, 1966
	Poland, Podhale, thin bedded, fg	2.73	–	–	–	153.4	17.8(R)	p = 3.81	BORETTI-ONYSZKIEWICZ, 1966
	Poland, Podhale, thick bedded, fg	2.77	–	–	–	128.3	16.7(R)	p = 5.91	BORETTI-ONYSZKIEWICZ, 1966
	Poland, Podhale, thin bedded, fg, laminated	2.76	–	–	–	95.9	10.1(R)	p = 6.42	BORETTI-ONYSZKIEWICZ, 1966
	Poland, Podhale, thick bedded, stratified	270	–	–	–	95.5	9.6(R)	p = 10.42	BORETTI-ONYSZKIEWICZ, 1966

Sandstone	Poland, Podhale					70.7	7.5(R)	p = 12.14	BORETTI-ONYSZKIEWICZ, 1966
	2.74	–	–	–	–				
USA, Morrison, Colo., dry, mg, hard	2.51	17.4* 12.7	–		0.19	–	–	–	BLAIR, 1956
USA, Morrison, Colo., wet, mg, hard	2.53	13.9* 22.4	8.27		0.16	–	–	–	BLAIR, 1956
USA, Morrison, Colo., dry, mg, hard	2.49	27.3* 22.5	–		0.06	–	–	well interlocked	BLAIR, 1956
USA, Morrison, Colo., wet, mg, hard	2.68	27.8*	8.14*		0.36*	–	–	well interlocked	BLAIR, 1956
USA, Morrison, Colo., dry, fg	2.52	12.6* 9.8	–		0.06	–	–	friable, clay like	BLAIR, 1956
USA, Morrison, Colo., wet, fg	2.54	17.6*	9.1*		0.04	–	–	friable, clay like	BLAIR, 1956
USA, Morrison, Colo., dry, fg, hard	2.56	9.01* 30.1	–		0.07	–	–	dense	BLAIR, 1956
USA, Morrison, Colo., wet, fg, hard	2.56	7.31*	–		–	–	–	dense	BLAIR, 1956
USA, Morrison, Colo., wet, fg	2.26	11.79* 8.0	8.62*		0.31*	–	–	friable	BLAIR, 1956
USA, Morrison, Colo., dry, hard	2.17	12.6* 10.1	–		0.16	–	–	friable, dense	BLAIR, 1956
USA, Morrison, Colo., wet, fg	2.20	11.7*	9.1*		0.36*	–	–	friable	BLAIR, 1956
USA, Morrison, Colo., wet, fg	2.29	6.62*	6.27*		0.47*	–	–	friable	BLAIR, 1956
USA, Morrison, Colo., wet, fg	2.25	7.58*	6.89*		0.45*	–	–	oily, porous	BLAIR, 1956
USA, Morrison, Colo., wet, fg	2.28	8.14*	8.27*		0.51	–	–	brown	BLAIR, 1956
USA, Tulsa County, Okla., fg	2.17	–	–		–	43.4	–	p = 16.9	BLAIR, 1956

LABORATORY MECHANICAL PROPERTIES OF ROCKS

Rock	Location & Description	Density ρ (g/cm³)	Modulus of Elasticity E (GPa)	Modulus of Rigidity G (GPa)	Poisson's Ratio ν	Compressive Strength σc (MPa)	Tensile Strength σt (MPa)	Remarks	Reference
	(1)	(2)	(3)	(4)	(5)	(6)	(7)	(8)	(9)
Sandstone	USA, Turley, Okla., mg	2.14	–	–	–	75.15	–	p = 16.8	BLAIR, 1956
	USA, Carteroil, Tulsa, Okla., mg	2.26	–	–	–	52.4	–	friable p = 20.5	BLAIR, 1956
	USA, Carteroil, Tulsa, Okla., fg, dense	–	–	–	–	59.3	–	silty	BLAIR, 1956
	USA, Carteroil, Tulsa, Okla., fg, dense, silty	2.5	21.9*	10.82*	–	74.5	–	p = 17	BLAIR, 1956
	USA, Pa., mg	2.45	21.0* 19.2	11.2*	0.06* 0.18	107.6	10.34(R)	p = 7.8	BLAIR, 1956
	USA, Pa., porous, mg	2.20	8.27* 16.4	4.69*	0.11* 0.27	86.9	4.1(R)	p = 12.0	BLAIR, 1956
	USA, Pa., porous, mg	2.21	5.58* 11.86	3.17*	0.12* 0.30	66.9	3.45(R)	p = 11.1	BLAIR, 1956
	Australia, Hawkesbury, NSW	–	34.5	–	0.11	57.23	–	–	O'BRIEN, 1969
	Australia	–	–	–	–	39.14	4.65	–	SINGH, 1969
	Australia	–	–	–	–	29.1	2.0	–	SINGH, 1969
	Australia, Stawell, Vic.	–	–	–	–	121.8	6.47	–	SINGH, 1969
	Canada, McGillivray Mine	–	11.7	–	–	81.4	–	–	HARDY, 1959
	Canada, McGillivray MIne	–	20.7	–	–	142.0	0.08	–	HARDY, 1959
	Canada, McGillivray Mine	–	13.1	–	–	84.8	–	–	HARDY, 1959

Sandstone								Reference
UK, Gt. Mt. Colliery, mg, dry	–	–	–	–	203.4	–	calcareous, p = 2.5	PRICE, 1960
UK, Park Colliery, fg, dry	–	–	–	–	296.5	–	massive calcareous p = 2.5	PRICE, 1960
UK, Fforchaman Colliery, fg, dry	–	–	–	–	317.2	–	p = 2.5	PRICE, 1960
UK, Markham Colliery, fg, dry	–	–	–	–	108.9	–	p = 6.0 bedded	PRICE, 1960
UK, Park gate rock, dry, fg, massive	–	–	–	–	111.7	–	p = 10.0	PRICE, 1960
UK, Darlydale, mg	–	–	–	–	64.8	–	p = 19.5 friable	PRICE, 1960
UK, Pennant, fg, dry	–	–	–	–	137.9	–	p = 2.5	PRICE, 1960
UK, Graig Fawr Colliery, mg, dry	–	–	–	–	139.3	–	massive p = 3.5	PRICE, 1960
UK, Snowdown Colliery, fg, dry	–	–	–	–	144.8	–	p = 6.5 massive	PRICE, 1960
Germany, red	–	13.0	–	–	10.0	1.2	–	HÖFER & MENZEL, 1964
UK, Ormond, massive, # 3	–	–	–	–	80.9	23.1	–	HOBBS, 1964
UK, Ormond, massive, mg, # 4	–	–	–	–	90.5	24.2	–	HOBBS, 1964
UK, Darley Dale, massive, mg	–	–	–	–	50.1	35.7	–	HOBBS, 1964
Canada, Ottawa, fg	2.28	36.0	–	–	92.0	–	–	COATES & PARSONS, 1966
Japan, Osaka, mg	–	10.0	–	–	85.0	–	–	NISHIHARA, 1957
USA, Tenn.	–	–	–	–	122.38	–	–	HARTMAN, 1966
USA, N.M., Abo-, Julian, Wash.	2.11	8.69*	–	–	–	–	banded, fg–mg	RICKETTS & GOLDSMITH, 1970

LABORATORY MECHANICAL PROPERTIES OF ROCKS

Rock	Location & Description	Density ϱ (g/cm³)	Modulus of Elasticity E (GPa)	Modulus of Rigidity G (GPa)	Poisson's Ratio υ	Compressive Strength σ_c (MPa)	Tensile Strength σ_t (MPa)	Remarks	Reference
	(1)	(2)	(3)	(4)	(5)	(6)	(7)	(8)	(9)
Sandstone	S. Africa, mg	–	35.0	–	0.19	100.0	–	ferruginous	GAY, 1973
	USA, Berea, ⊥ bedding	2.12	15.3	–	–	57.9	1.17	p = 19.8	HAWKES et al, 1973
	Canada, Springhill	–	33.82	–	0.17	–	–	–	EMERY, 1963
	Canada	2.28	26.9	–	–	51.4	–	–	CLARK & CAUDLE, 1963
	Canada	2.14	7.58	–	–	97.9	–	–	CLARK & CAUDLE, 1963
	USA, mg #1	2.06	6.0*	3.2*	0.06*	71.71	3.12(R)	p = 16 weakly cemented	OBERT et al, 1946
	#2	2.06	6.7*	3.2*	0.05*	55.15	5.19(R)	p = 16 weakly cemented	OBERT et al, 1946
	#3	2.06	8.8*	4.4*	0.03*	53.09	5.59(R)	p = 16 weakly cemented	OBERT et al, 1946
	USA, mg #1	2.17	7.1*	4.0*	0.14*	42.06	2.96(R)	p = 16 weakley cemented	OBERT et al, 1946
	#2	2.17	10.6*	4.8*	0.11*	35.85	4.99(R)	p = 16 weakly cemented	OBERT et al, 1946
	#3	2.17	11.2*	4.48*	0.17*	365.16	5.52(R)	p = 16 weakly cemented	OBERT et al, 1946
	USSR, Prokopevsk, Kieselevsk coal, cg, siliceous	2.54	22.4	–	0.22	113.7	6.4		YAGODKIN et al, 1969

Sandstone	USSR, Prokopevsk, Kieselevsk coal, mg, siliceous-argillaceous	2.57	27.0	–	0.24	136.0	8.1		YAGODKIN et al, 1969
	USSR, Prokopevsk, Kieselevsk coal, fg, argillaceous, siliceous	2.66	32.0	–	0.28	149.5	9.9		YAGODKIN et al, 1969
	USSR, Prokopevsk, Kieselevsk coal, mg, siliceous-carbonaceous	2.80	44.0	–	0.30	183.0	11.6		YAGODKIN et al, 1969
	USSR, Prokopevsk, Kieselevsk coal, fg, siliceous-carbonaceous	2.68	22.0	–	0.22	152.0	9.1		YAGODKIN et al, 1969
	USSR, Prokopevsk, Kieselevsk coal, mg, siliceous, argillaceous	2.45	26.0	–	0.27	129.0	7.1		YAGODKIN et al, 1969
	USSR, Prokopevsk, Kieselevsk coal, fg, carbonaceous, argillaceous	2.55	36.0	–	0.30	152.0	6.7		YAGODKIN et al, 1969
	USSR, Prokopevsk, Kieselevsk coal, fg, siliceous, argillaceous	2.7	35.0	–	0.33	139.0	9.3		YAGODKIN et al, 1969
	USSR, Prokopvsk, Kieselevsk coal, cg, aleurolites, argillaceous	2.80	30.0	–	0.34	99.8	9.0		YAGODKIN et al, 1969
	USSR, Leninsk, Kuznetsk Basin	–	7.0	–	0.22	50.0	5.2	–	YAGODKIN et al, 1969

LABORATORY MECHANICAL PROPERTIES OF ROCKS

Rock	Location & Description	Density ϱ (g/cm³)	Modulus of Elasticity E (GPa)	Modulus of Rigidity G (GPa)	Poisson's Ratio υ	Compressive Strength σ$_c$ (MPa)	Tensile Strength σ$_t$ (MPa)	Remarks	Reference
	(1)	(2)	(3)	(4)	(5)	(6)	(7)	(8)	(9)
Sandstone	USSR, Leninsk, Kuznetsk Basin	–	6.0	–	0.30	60.0	8.0	–	Yagodkin et al, 1969
	USSR, Lysaya gora	–	–	–	–	40.0–60.0	6.4–11.5	–	Koifman, 1969
	USSR, Kara-ganda	–	–	–	–	63.0	4.0	–	Koifman & Chirkov, 1969
	USSR, Kuzbass, fg, stable	–	–	–	–	60.0	5.1	–	Koifman & Chirkov, 1969
	fg, stable	–	–	–	–	93.0	6.0	–	Koifman & Chirkov, 1969
	fg, stable	–	–	–	–	50.0	3.3	–	Koifman & Chirkov, 1969
	fg, stable	–	–	–	–	39.0	2.4	–	Koifman & Chirkov, 1969
	USSR, Karelia, Ukraine, quartzitic	–	65.0	–	0.10	259.0	–	p = 1.25	Belikov et al, 1967
	USSR, Donbass, quartzitic + mica cement	–	56.0	–	0.14	260.0	–	p = 2.16	Belikov et al, 1967
	USSR, Volga, with opal-chalcedony cement	–	64.0	–	0.12	305.0	–	p = 1.79	Belikov et al, 1967
	USSR, Volga, with opal cement	–	50.0	–	0.12	226.0	–	p = 4.75	Belikov et al, 1967
	USSR, Jotnian, quartzitic, + prophylite-glauconite cement	2.63	–	–	–	250.0	–	–	Timchenko, 1967

Rock	Description							Reference
Sandstone	USSR, Jotnian, fg, quartzitic, iron oxide + pyrophyllite	2.66	–	–	328.0	–	p = 0.36	TIMCHENKO, 1967
	USSR, Donbass, cg, quartzose + illite	2.56	–	–	261.0	–	p = 3.39	TIMCHENKO, 1967
	USSR, Donbass, mg, quartzose, illite + CO_3	2.68	–	–	270.0	–	p = 2.48	TIMCHENKO, 1967
	USSR, Donbass, fg, quartzose + illite	2.54	–	–	152.0	–	p = 4.72	TIMCHENKO, 1967
	USSR, Donbass, fg, quartzose + illite + CO_3	2.69	–	–	292.0	–	p = 0.43	TIMCHENKO, 1967
	USSR, Donbass, fg, quartzose + illite	2.67	–	–	317.0	–	p = 0.23	TIMCHENKO, 1967
	USSR, Donbass, fg, quartzose + illite	2.67	–	–	284.0	–	p = 0.69	TIMCHENKO, 1967
	USSR Donbass, mg, quartzose, illite + CO_3	2.68	–	–	266.0	–	p = 0.2	TIMCHENKO, 1967
	USSR, Donbass, cg, quartzose + illite	2.67	–	–	352.0	–	p = 0.23	TIMCHENKO, 1967
	USSR, Donbass, fg, quartzose + illite	2.63	–	–	281.0	–	p = 1.44	TIMCHENKO, 1967
	USSR, Donbass, cg, quartzose, + monthermite	2.50	–	–	131.00	–	p = 6.7	TIMCHENKO, 1967

LABORATORY MECHANICAL PROPERTIES OF ROCKS

Rock	Location & Description	Density ϱ (g/cm³)	Modulus of Elasticity E (GPa)	Modulus of Rigidity G (GPa)	Poisson's Ratio υ	Compressive Strength σ_c (MPa)	Tensile Strength σ_t (MPa)	Remarks	Reference
	(1)	(2)	(3)	(4)	(5)	(6)	(7)	(8)	(9)
Sandstone	USSR, Volga, mg,siliceous, quartzitic, opal + quartzine	2.40	–	–	–	271.0	–	p = 3.61	TIMCHENKO, 1967
	USSR, Volga, mg, siliceous, quartzitic, opal + quartzine	2.41	–	–	–	277.0	–	p = 1.66	TIMCHENKO, 1967
	USSR, Volga, cg, siliceous, quartzitic	2.42	–	–	–	237.0	–	p = 3.44	TIMCHENKO, 1967
	USSR, Volga, mg, siliceous, quartzitic	2.43	–	–	–	216.0	–	p = 3.46	TIMCHENKO, 1967
	USSR, Volga, fg, siliceous, quartzitic, opal + quartzine	2.45	–	–	–	272.0	–	p = 2.72	TIMCHENKO, 1967
	USSR, Volga, mg, siliceous, quartzitic	2.44	–	–	–	265.0	–	p = 3.22	TIMCHENKO, 1967
	USSR, Volga, mg, siliceous, quartzitic	2.45	–	–	–	290.0	–	p = 3.18	TIMCHENKO, 1967
	USSR, Volga, fg, siliceous, quartzitic, opal + quartzine	2.46	–	–	–	284.0	–	p = 3.74	TIMCHENKO, 1967

Sandstone

USSR, Volga, mg, siliceous, quartzitic	2.39	–	–	250.0	–	p = 4.15	Timchenko, 1967
USSR, Volga, mg, siliceous, quartzitic	2.34	–	–	221.0	–	p = 5.74, veined	Timchenko, 1967
USSR, Sangileev, cg, siliceous, quartzitic, lenticular, banded	2.35	–	–	261.0	–	p = 5.56	Timchenko, 1967
USSR, Sangileev, mg, siliceous, quartzitic, lenticular, banded	2.37	–	–	256.0	–	p = 3.70	Timchenko, 1967
USSR, Sangileev, siliceous, quartzitic, lenticular, banded	2.37	–	–	233.0	–	p = 3.68	Timchenko, 1967
USSR, Sangileev, mg, siliceous, quartzitic	2.30	–	–	251.0	–	p = 7.87, lenticular-banded	Timchenko, 1967
USSR, Sangileev, fg, quartzitic	2.25	–	–	147.0	–	p = 8.74, lenticular-banded	Timchenko, 1967
USSR, Sangileev, mg, siliceous, quartzitic	2.35	–	–	280.0	–	p = 5.47, lenticular-banded	Timchenko, 1967
USSR, Sangileev, mg, siliceous, quartzitic	2.34	–	–	209.0	–	p = 9.99, opal + chalcedony	Timchenko, 1967
USSR, Sangileev, cg, siliceous, quartzitic	2.41	–	–	272.0	–	p = 6.17	Timchenko, 1967
USSR, Sangileev, mg, siliceous, quartzitic	2.31	–	–	212.0	–	p = 9.96	Timchenko, 1967

LABORATORY MECHANICAL PROPERTIES OF ROCKS

Rock	Location & Description	Density ϱ (g/cm³)	Modulus of Elasticity E (GPa)	Modulus of Rigidity G (GPa)	Poisson's Ratio υ	Compressive Strength σc (MPa)	Tensile Strength σt (MPa)	Remarks	Reference
	(1)	(2)	(3)	(4)	(5)	(6)	(7)	(8)	(9)
Sandstone	USSR, Sangileev, mg, siliceous, quartzitic	2.25	–	–	–	200.0	–	p = 11.44	Timchenko, 1967
	USSR, Sangileev, mg, siliceous, quartzitic	2.33	–	–	–	190.0	–	p = 9.36	Timchenko, 1967
	USSR, Sangileev, cg, siliceous, quartzitic	2.36	–	–	–	360.0	–	p = 4.05, chalcedony	Timchenko, 1967
	USSR, Donbass, carboniferous	2.63	55.2	–	0.14	–	–	p = 3.16	Belikov, 1967
	USSR, Urals, quartzitic	2.65	62.3	–	0.10	–	–	p = 1.24	Belikov, 1967
	USSR, Turkmenia, calcitic cement	2.60	53.5	–	–	–	–	p = 2.97	Belikov, 1967
	USSR, Volga, quartzitic opal cement	2.31	49.4	–	0.12	–	–	p = 4.75	Belikov, 1967
	USSR, Volga, quartzitic, chalcedony cement	2.48	64.1	–	0.12	–	–	p = 1.79	Belikov, 1967
	USA, Berea	2.2	–	–	–	59.57	4.64	p = 11.4	Clark et al, 1970
	USA, St. Peter	2.3	–	–	–	38.75	2.08	p = 12.3	Clark et al, 1970
	USA, Massillon	2.03	9.72	–	0.24	30.9	1.27	–	Singh & Huck, 1970
	UK, Darley Dale, mg, hd	–	11.03	–	0.12	–	–	–	Wilson, 1961
	USA, Berea	–	13.10	–	0.20	71.71	5.17	p = 18.8	Haimson & Fairhurst, 1969

Sandstone	USA, Pottsville, mg	2.27	–	52.05	–	3.03	p = 16.3	SCHWARTZ, 1964
	USA, Gnome nuclear explosion site	–	–	71.4	–	–	p = 15.4	KNUTSON & BOHOR, 1964
	USA, Gnome nuclear explosion site	–	–	45.3	–	–	p = 18.3	KNUTSON & BOHOR, 1964
	USA, Gnome nuclear explosion site	–	–	33.9	–	–	p = 19.4	KNUTSON & BOHOR, 1964
	USA, Gnome nuclear explosion site	–	–	7.3	–	–	p = 24.4	KNUTSON & BOHOR, 1964
	USA, Gnome nuclear explosion site	–	–	67.0	–	–	p = 14.1	KNUTSON & BOHOR, 1964
	USA, Gnome nuclear explosion site	–	–	2.1	–	–	p = 24.2	KNUTSON & BOHOR, 1964
	USA, White Pine, S-W Mine, dry	2.73	–	148.9	–	–	–	PARKER & SCOTT, 1964
	USA, White Pine, S-W Mine, wet	2.80	49.99	122.73	0.11	13.1	–	PARKER & SCOTT, 1964
	USA, White Pine, S-W Mine, + siltstone, wet	2.87	–	144.10	–	–	–	PARKER & SCOTT, 1964
	Canada, Bennet Dam, along strike	–	17.93	136.52	–	–	–	IMRIE & JORY, 1968
	Canada, Bennet Dam, along strike	–	17.24	113.07	–	–	–	IMRIE & JORY, 1968
	Canada, Bennet Dam, along dip	–	16.55	133.07	–	–	–	IMRIE & JORY, 1968
	Canada, Bennet Dam, along dip	–	15.17	129.62	–	–	–	IMRIE & JORY, 1968

LABORATORY MECHANICAL PROPERTIES OF ROCKS

Rock	Location & Description	Density ϱ (g/cm³)	Modulus of Elasticity E (GPa)	Modulus of Rigidity G (GPa)	Poisson's Ratio υ	Compressive Strength σ_c (MPa)	Tensile Strength σ_t (MPa)	Remarks	Reference
	(1)	(2)	(3)	(4)	(5)	(6)	(7)	(8)	(9)
Sandstone	Canada, Bennet Dam								
	⊥ bedding	–	15.86	–	–	122.04	–	–	IMRIE & JORY, 1968
	⊥ bedding	–	17.93	–	–	139.95	–	–	IMRIE & JORY, 1968
	Poland, Upper Silesia, calcitic	–	32.66	–	–	88.0	–	–	ZNAŃSKI, 1960
	Poland, Upper Silesia, arkose	–	22.73	–	–	95.0	–	–	ZNAŃSKI, 1960
	Czechoslovakia	–	15.0	–	0.17	120.0	–	–	BORDIA, 1972
	Holland, Wilhelmina Colliery	–	35.0–45.0	–	0.12	100.0–140.0	–	carboniferous	CREUELS & HERMES, 1956
	Germany, Concordia Mine	–	55.0	–	–	–	–	carboniferous	EVERLING, 1956
	Germany, Girondelle seam	–	38.0–53.0*	–	–	–	–	carboniferous	EVERLING, 1956
	USSR, Donbass	–	35.5	–	0.17	106.0	–	–	TSCHERNJAK et al, 1967
	USSR, Donbass	–	30.7	–	0.20	97.5	–	–	TSCHERNJAK et al, 1967
	USSR, Donbass	–	36.0	–	0.18	139.5	–	–	TSCHERNJAK et al, 1967
	USSR, Donbass	–	26.9	–	0.27	91.0	–	–	TSCHERNJAK et al, 1967
Schist	USA, Montezuma Tunnel, Colo., biotite + pegmatite	2.70	39.8	–	0.05	68.1	–	p = 1.44	BALMER, 1953
	USA, Montezuma Tunnel, Colo., biotite + pegmatite	2.70	59.29*	25.51*	0.16	–	–	p = 1.44	BALMER, 1953
	USA, Montezuma Tunnel, Colo., biotite + pegmatite, high strength	2.71	–	–	–	83.43	–	p = 0.59	BALMER, 1953

Schist								
USA, Montezuma Tunnel, Colo., biotite + chlorite, low medium strength, P = 140	2.69–2.71	68.0	–	0.20	36.47–81.29	–	p = 0.81–0.86	BALMER, 1953
USA, Montezuma Tunnel, Colo., biotite + chlorite, high strength	2.75	68.26	–	0.20	117.21	–	p = 0.52	BALMER, 1953
USA, Montezuma Tunnel, Colo., biotite + sillimanite	2.71–2.72	21.2	–	–	8.0–33.99	–	p = 1.0	BALMER, 1953
USA, Montezuma Tunnel, Colo., with quartz injectives	2.70–2.71	23.3	–	0.09	8.62–31.16	–	p = 2.8	BALMER, 1953
USA, Sly Park Dam, Colo., seritic quartzose	2.47	8.96 / 17.93*	6.21*	0.12 / 0.44*	15.03	–	–	WUERKER, 1956
USA, Ariz., seritic	2.70	59.98*	–	26.20*	162.03	–	–	WUERKER, 1956
USA, Framingham, Mass., chlorite-epidote	2.95	70.5*	31.5*	0.12*	–	–	–	BIRCH & BANCROFT, 1940
USA, Framingham, Mass., quartz-muscovite	2.70	54.4*	23.0*	0.18*	–	–	–	BIRCH & BANCROFT, 1940
USA, Idaho Spring Formation, Colo., biotite, P = 0–35	2.68–2.71	40.0	–	0.01	–	–	–	USBR, 1953
—	2.80	34.9	–	0.11	50.3	–	p = 3.74	JUDD, 1969

LABORATORY MECHANICAL PROPERTIES OF ROCKS

Rock	Location & Description	Density ϱ (g/cm³)	Modulus of Elasticity E (GPa)	Modulus of Rigidity G (GPa)	Poisson's Ratio υ	Compressive Strength σ_c (MPa)	Tensile Strength σ_t (MPa)	Remarks	Reference
	(1)	(2)	(3)	(4)	(5)	(6)	(7)	(8)	(9)
Schist	USA, Fremont Canyon Dam, vfg	–	69.0	–	0.19	–	17.4	–	NESBITT, 1960
	USA, Luther Falls, micaceous, ⊥ foliation	2.81	20.7	–	0.31	55.2	0.55	at 50% fracture	MILLER, 1965
	‖ foliation	2.82	58.1	–	0.81	82.7	5.24	at 50% fracture	MILLER, 1965
	USA, Morrow Point Dam, quartz-mica augen-schist, hard, mg	2.72	28.5	–	0.06	107.5	–	p = 0.7	USBR, 1965
	USA, Morrow Point, quartz-mica, mg-cg	2.74	8.20	–	0.04	44.0	–	p = 2	USBR, 1965
	USA, Morrow Point, cg, muscovite-biotite	2.83	5.93	–	0.02	24.9	–	p = 1.7	USBR, 1965
	USA, Morrow Point, hornblende schist, mg	2.74	61.0	–	0.14	198.7	–	p = 0.4	USBR, 1965
	India, Kolar Gold Field, hornblende schist	–	87.56	17.93	0.29	244.07	10.69	–	MORRISON, 1970
	USA, Swan Lake Dam, Alaska, hornblende, garnet, biotite, mg, fractured	2.89	39.30	–	0.11	129.62	5.52	–	BRANDON, 1974

Schist									
	USA, Swan Lake Dam, Alaska, hornblende biotite, mg, slightly fractured	2.73	16.55	–	0.13	157.20	–	–	BRANDON, 1974
	India, Mahi River, Kadana, quartz-mica, fresh	2.78	49.5 / 73.0*	–	0.18	60.0	–	–	PARASARTHY et al, 1974
	altered	2.73	10.2 / 21.0*	–	0.22	12.0	–	–	PARASARTHY et al, 1974
	USA, Northfield Mt. Power Station, Mass., biotite-, ‖ bedding	–	51.02	–	0.22	–	150.3	–	WILD & MCKITTRICK, 1971
	⊥ bedding	–	39.9	–	0.12	–	153.7	–	WILD & MCKITTRICK, 1971
	USA, Lake Mead Chamber, Nev., amphibolitic schist	–	91.7	–	0.18	78.6–251.7	–	–	RISING & ERICKSON, 1971
	USA, Lake Mead Chamber, Nev., biotite schist	–	–	–	–	158.6	–	–	RISING & ERICKSON, 1971
	USA, Lake Mead Chamber, Nev., amphibole-chlorite	–	80.67	–	0.23	–	48.61–105.8	–	RISING & ERICKSON, 1971
	UK, Monar Dam	–	56.0	–	–	50.5	–	p = 0.1	HENKEL et al, 1964
	USA, biotitic + pegmatite	–	39.99* / 59.29	–	0.01	–	–	highly porous	RINEHART et al, 1961
	USA, mg, hd, slightly weathered, sericitic	–	7.58 / 17.93*	–	0.02	–	–	porous	RINEHART et al, 1961

LABORATORY MECHANICAL PROPERTIES OF ROCKS

Rock	Location & Description	Density ρ (g/cm³)	Modulus of Elasticity E (GPa)	Modulus of Rigidity G (GPa)	Poisson's Ratio ν	Compressive Strength σc (MPa)	Tensile Strength σt (MPa)	Remarks	Reference
	(1)	(2)	(3)	(4)	(5)	(6)	(7)	(8)	(9)
Schist	USA, Peters Creek, Pa., chlorite-, foliated	2.72	–	–	–	137.90	–	–	WEST & PERRY, 1969
Schistose Gneiss	–	2.78	54.6	–	0.37	92.7	1.12	–	ROSENBLAD, 1968
Scoria	USA, Mohave Desert	2.22	7.58 / 16.55*	–	–	28.96	4.31	–	RICKETTS & GOLDSMITH, 1970
Sediments	USA, Auborn Dam, meta-sediments	2.77	63.43	–	–	113.76	–	p = 0.8	WALLACE et al, 1969
Shale	–	2.21	21.9	–	0.18	67.6	2.41	p = 16.2	JUDD, 1969
	USA, Marble Canyon Dam, Ariz., calcareous	2.67	12.41 / 24.48*	15.86*	0.04	35.93	–	p = 1.8 / 0–70 bars	BALMER, 1953
	USA, Marble Canyon Dam, Ariz., quartzose	2.69	13.79 / 22.06*	11.72*	0.07	122.52	–	p = 6.6	BALMER, 1953
	USA, Murdock, Ill.	–	7.52	–	0.103	34.27	10.64	grain size = 0.01 mm	WUERKER, 1956
	USA, Utah, shale / USA, Utah, silicified	2.81 / 2.80	58.19* / 68.05*	26.61* / 30.47*	0.09* / 0.12*	215.81 / 230.97	17.24(R) / 14.48(R)	p = 0.9 / p = 0.57	WUERKER, 1956 / WUERKER, 1956
	USA, Utah, mineralised + limestone	2.92	65.71*	28.06	0.17*	239.94	16.55(R)	p = 0.62	WUERKER, 1956
	USA, Monticello Dam, Colo., Chico Siltstone Formation, P = 0–140	2.50	13.0 / 27.0*	12.0*	0.12	–	–	–	USBR, 1953

Shale								
Canada, W.A.C. Bennett Dam, NS, silty, soft, along strike	—	38.6	—	—	94.4	—	—	IMRIE & JORY, 1968
along dip	—	30.4	—	—	81.3	—	—	IMRIE & JORY, 1968
bedding	—	36.5	—	—	112.3	—	—	IMRIE & JORY, 1968
Poland, Upper Silesia, sandy shale ⊥ bedding	—	54.7	—	—	71.0–94.0	—	—	BORECKI & CHUDEK, 1972
‖ bedding	—	63.7	—	—	46.0–67.0	—	—	BORECKI & CHUDEK, 1972
Poland, Upper Silesia, clayey shale	—	11.32	—	0.23	70.0	—	—	BORECKI & CHUDEK, 1972
USSR, Vegorukskii, Karelia, tuffaceous	—	88.26	—	—	274.6		p = 0.52	BELIKOV et al, 1967
Poland, Upper Silesia, Kenel shale	—	21.87	—	0.38	503.08	—	—	ZNAŃSKI, 1964
USA, Allegheny, Pa., slate mudstone, fg	2.72	31.16 / 15.31*	15.38*	0.14*	101.35	1.38	p = 1.7 secant at 50%	BLAIR, 1956
USA, Kanawha, W.Va., Eagle coal seam, carbonaceous	2.74	10.48* / 21.30	10.69*	0.15	99.97	0.69	secant at 50%	BLAIR, 1955
USA, Kanawha, W.Va., above Eagle coal seam, siderite banded	2.76	13.31* / 22.68	11.72*	0.16	112.38	2.76	secant at 50%	BLAIR, 1955
UK, Hucknall	—	10.2	—	—	58.7	—	—	HOBBS, 1970
USA, Marble Canyon Dam, Ariz., fg, medium porosity	2.52	22.06	—	0.09	88.94	2.62	—	BRANDON, 1974

LABORATORY MECHANICAL PROPERTIES OF ROCKS

Rock	Location & Description	Density ϱ (g/cm³)	Modulus of Elasticity E (GPa)	Modulus of Rigidity G (GPa)	Poisson's Ratio ʋ	Compressive Strength σc (MPa)	Tensile Strength σt (MPa)	Remarks	Reference
	(1)	(2)	(3)	(4)	(5)	(6)	(7)	(8)	(9)
Shale	USA, Flaming Gorge Dam, Utah, sandy-silty shale, dense, moderate porosity, flakes in contact with water	2.46	5.52	–	0.25	35.16	0.21	–	BRANDON, 1974
	USA, Green River, Wyo., lean								
	⊥ bedding	–	39.9*	0.22*	–	–	–	–	YOUASH, 1970
	‖ bedding rich	–	42.7*	0.27*	–	–	–	–	YOUASH, 1970
	⊥ bedding	–	24.1*	0.26*	–	–	–	–	YOUASH, 1970
	‖ bedding	–	47.4*	0.30*	–	–	–	–	YOUASH, 1970
	USA, Niagara, N.Y., Rochester shale		9.83*	–	0.32	–	–	–	HOGG, 1964
			45.19*						
	USA, Niagara, N.Y., Grinsby sand & shale	–	15.10	–	0.08	110.32	–	–	HOGG, 1964
			39.58*		0.08*				
	USA, Rifle, Colo., Oil shale, vfg = 4 μ	2.35	–	–	–	–	17.0*	–	SCHULER et al, 1976
	USA, Rock Springs, Wyo., USBM site 9, vfg = 4 μ	2.22	–	–	–	–	23.0*	–	SCHULER et al, 1976
	Canada, Bearpaw, Sask.-River Dam, soft	2.74–2.76	–	–	–	0.06–0.63	–	grain = 0.005 mm w ≈ 32	RINGHEIM, 1964

Shale		hard	2.76	0.13	–	–	–	1.2–3.2	–	grain = 0.05 mm w = 23	RINGHEIM, 1964
	USA, Oahe Dam	1.60		0.14–0.97	–	–	–	–	–	w = 18.27	UNDERWOOD, 1961
	USA	–		5.17 9.93*	–	–	43.1	–	–	AVEDISSIAN & WOOD, 1968
	USA	–		6.76 10.3*	–	–	48.3	–	–	AVEDISSIAN & WOOD, 1968
	USA, Estancia Valley	–		11.1 6.21	4.21*	0.32 0.15	59.3	–	–	BRATTON & PRATT, 1968
	USA, Estancia Valley, + sandstone	–		5.31	–	0.13	72.4	–	–	BRATTON & PRATT, 1968
	Japan, Seikan Tunnel	1.82		–	–	–	62.9	10.7	–	MOCHIDA, 1974
	Japan, Shimane Nuclear Plant	2.6		1.8	–	0.3	–	–	c = 0.2, Ø = 30°	KITAHARA et al, 1974
	Japan, Shimane Nuclear Plant	2.6		1.0	–	0.3	–	–	c = 0.15, Ø = 28°	KITAHARA et al, 1974
	Japan, Shimane Nuclear Plant	2.6		0.5	–	0.3	–	–	c = 0.1, Ø = 25°	KITAHARA et al, 1974
	Japan, Shimane Nuclear Plant	2.4		1.50	–	0.2	–	–	c = 0.2, Ø = 24°	KITAHARA et al, 1974
	USA, Brunswick, N.J.	1.8–2.5		0.2–3.3	–	–	1.63–7.52	–	–	JUMIKIS, 1966
	USA, Ebensburg, Ohio, Mine 1, silty, fg	2.67		–	–	–	101.3	–	–	BLAIR, 1956
	USA, Jonathan mine, Ohio, micaceous	2.56		11.1*	7.9*	0.29	75.2	2.07	kaolinite + sericite	BLAIR, 1956

LABORATORY MECHANICAL PROPERTIES OF ROCKS

Rock	Location & Description	Density ϱ (g/cm^3)	Modulus of Elasticity E (GPa)	Modulus of Rigidity G (GPa)	Poisson's Ratio υ	Compressive Strength σ_c (MPa)	Tensile Strength σ_t (MPa)	Remarks	Reference
	(1)	(2)	(3)	(4)	(5)	(6)	(7)	(8)	(9)
Shale	USA, Pine Creek Mine, Tenn., fg	2.30	13.86* 15.72	6.55*	0.001* 0.27	112.4	2.07(R)	kaolin + quartz p = 1.6	BLAIR, 1956
	USA, Saligo Mine, Tenn.	2.30	13.44* 11.93	7.10*	0.02* 0.22	110.3	2.76(R)	kaolin + quartz p = 1.7	BLAIR, 1956
	USA, Pine Creek Mine, Tenn., fg, silty	2.53	17.31* 12.41	5.17*	0.14	83.4	0.69(R)	carbonaceous	BLAIR, 1956
	USA, Saligo Mine, Tenn., fg, silty	–	–	–	–	84.81	–	carbonaceous	BLAIR, 1956
	Canada, McGillivray Mine, Alta,	–	30.34	–	–	155.1	0.07	–	HARDY, 1959
	Canada, McGillivray Mine, Alta.,	–	25.5	–	–	140.7	0.08	–	HARDY, 1959
	sandy		23.4	–	–	170.3	0.05	–	HARDY, 1959
	Canada, Brown, vfg	2.90	35.2	–	–	150.0	–	carbonaceous	COATES & PARSONS, 1966
	Japan, Kyushu	–	18.0	–	–	110.0	–	–	NISHIHARA, 1952
	Japan, Kyushu	–	13.0	–	–	60.0	–	–	NISHIHARA, 1952
	Japan, Kyushu, ⊥ bedding	–	13.0	–	–	60.0	–	–	NISHIHARA, 1952
	Japan, Kyushu, ‖ bedding	–	13.0	–	–	60.0	–	–	NISHIHARA, 1952
	Canada, Bearpaw Formation, wet	2.07	–	–	–	70.2	–	+ montmorillonite	COATES & PARSONS, 1966

Shale

S. Africa, Daggafontein Gold Ltd.	–	–	–	–	–	160.0	–	–	DENKHAUS, 1965
S. Africa, Kimberley, Western Deep Ltd.	–	–	–	–	–	170.0	–	–	DENKHAUS, 1965
USSR, Donbass, argillaceous	–	–	–	–	–	39.0	6.0	–	KOIFMAN & CHIRKOV, 1969
USSR, Donbass, argillaceous	–	–	–	–	–	54.0	5.8	compact	KOIFMAN & CHIRKOV, 1969
USSR, Baltic, combustible	–	–	–	–	–	27.5	2.0	–	KOIFMAN & CHIRKOV, 1969
USSR, Baltic, combustible	–	–	–	–	–	23.5	2.2	limestone	KOIFMAN & CHIRKOV, 1969
USSR, Karelia, tuffaceous	–	90.0	–	–	–	280.0	–	p = 0.52	BELIKOV et al, 1967
USA, quartzose, med. hard	–	16.55 22.06*	–	0.08	–	–	–	med. laminated	RINEHART et al, 1961
USA, calcareous, fg, med. hd.	–	15.86 24.82	–	0.02	–	–	–	laminated	RINEHART et al, 1961
Canada, Bennett Dam, along strike	–	38.61	–	–	–	94.45	–	–	IMRIE & JORY, 1968
Canada, Bennett Dam, along dip	–	30.34	–	–	–	81.36	–	–	IMRIE & JORY, 1968
Canada, Bennett Dam ⊥ bedding	–	35.54	–	–	–	112.38	–	–	IMRIE & JORY, 1968
Poland, Upper Silesia, clay	–	113.2	–	–	–	70.0	–	–	ZNAŃSKI, 1960
Czechoslovakia, Ziska Colliery	–	–	–	–	–	4.8	0.24	carbonaceous	SIBEK, 1960
Czechoslovakia, Ziska Colliery, clay	–	–	–	–	–	3.7	0.19	–	SIBEK, 1960
Holland, Wilhelmina Colliery.	–	12.0–15.0	–	0.12	–	55.0–80.0	–	carbonaceous	CREUELS & HERMES, 1956

LABORATORY MECHANICAL PROPERTIES OF ROCKS

Rock	Location & Description	Density ϱ (g/cm^3)	Modulus of Elasticity E (GPa)	Modulus of Rigidity G (GPa)	Poisson's Ratio υ	Compressive Strength σ_c (MPa)	Tensile Strength σ_t (MPa)	Remarks	Reference
	(1)	(2)	(3)	(4)	(5)	(6)	(7)	(8)	(9)
Shale + Clay	Angola, Cambambe Dam	–	10.0	–	–	20.2	–	–	SARMENTO & VAZ, 1964
	Yugoslavia, Algonkian	–	4.8	2.6	–	–	–	$\tau = 0$–1.5	DVORAK, 1967
Shale + Sandstone	USA, Estancia Valley	–	5.31	–	0.13	72.4	–	secant at 21 MPa	BRATTON & PRATT, 1968
	USA, Estancia Valley	–	4.21	–	0.39	69.3	–	secant at 21 MPa	BRATTON & PRATT, 1968
Siderite	Canada, Wawa, Ont., fg	3.59	90.5	–	–	279.0	–	–	COATES & PARSONS, 1966
	Canada, Macleod Mine	–	114.5	–	0.32	–	–	–	HERGET, 1973
Siltstone	–	2.67	32.6	–	0.23	108.2	–	p = 16.6	JUDD, 1969
	USA, Hackensack, N.Y., siltstone, dense, massive, clayey, cemented	2.59	26.2	–	0.22	122.7	2.96	–	MILLER, 1965
	USA, Monticello Dam, Colo., porous, fg	2.50	13.1 26.89*	12.41*	0.12	24.1	–	p = 10.3	BALMER, 1953
	USA, Ala., + shale	2.76	53.23*	25.30*	–	256.48	18.62(R)	p = 0.8	WUERKER, 1956
	USA, Ala., + shale	2.77	64.33*	28.61*	–	315.78	34.47(R)	p = 1.0	WUERKER, 1956
	USA, Ala., + sandstone + shale	2.76	39.92	22.68	–	184.78	15.17(R)	p = 1.7	WUERKER, 1956

Rock	Location / description								Reference
Siltstone	USA, Allegheny, Pa., fg	2.76	30.60 7.10*	14.62	0.13	113.07	2.76(R)	p = 1.8, secant at 50%	BLAIR, 1956
	UK, Chislet, Ormonde siltstone	–	20.3	–	–	68.9	–	–	HOBBS, 1970
	UK, Chislet, Kent, siltstone ‖ bedding	–	34.47	0.21	–	90.32	7.93	–	PRICE, 1958
	UK, Chislet, siltstone	–	19.7	–	–	61.7	–	–	HOBBS, 1970
	UK, Donisthorpe siltstone, laminated, ⊥ lamination	2.70	–	–	–	73.15	29.85	tensile strength ‖ laminations (Brazilian)	HOBBS, 1970
	UK, Donisthorpe siltstone II, laminated ⊥ lamination	2.65	–	–	–	96.66	47.57	tensile strength ‖ laminations (Brazilian)	HOBBS, 1970
	UK, Snowdown, Kent, ⊥ bedding	–	–	–	–	111.65	6.55	–	PRICE, 1958
	‖ bedding	–	57.9	–	0.22	93.08	5.38	–	PRICE, 1958
	USA, Sanford Dam, Tex., highly porous, highly friable, slakes on wetting	2.17	0.72	–	0.27	4.67	–	grain = 0.05 mm	BRANDON, 1974
	India, Bhakra Dam	2.63	40.4	–	0.10	95.77	–	secant modulus, p = 1.1	BHATNAGAR & SHAH, 1964
	UK, Scammondan Dam	–	16.1	–	–	77.1	–	49.1 m depth	PENMAN & MITCHELL, 1970
	USA	–	7.45 13.65*	–	–	48.5	–	–	AVEDISSIAN & WOOD, 1968

LABORATORY MECHANICAL PROPERTIES OF ROCKS

Rock	Location & Description	Density ϱ (g/cm³)	Modulus of Elasticity E (GPa)	Modulus of Rigidity G (GPa)	Poisson's Ratio υ	Compressive Strength σ_c (MPa)	Tensile Strength σ_t (MPa)	Remarks	Reference
	(1)	(2)	(3)	(4)	(5)	(6)	(7)	(8)	(9)
Siltstone	USA	–	25.4 40.7*	–	–	106.9	–	–	AVEDISSIAN & WOOD, 1968
	Japan, Seikan Tunnel	2.18	–	–	–	47.4	3.3	–	MOCHIDA, 1974
	Japan, Seikan Tunnel	1.86	–	–	–	58.3	5.4	–	MOCHIDA, 1974
	Japan, Seikan Tunnel	1.11	–	–	–	4.9	0.9	–	MOCHIDA, 1974
	USA, Ebensburg Mine-1, fg	2.66	74.7*	–	–	118.5	34.4(R)	dynamic tensile, E = flexure	BLAIR, 1956
	USA, Jonathan Mine, Ohio, fg, wet, ∥bedding	2.66	31.0	–	0.42	36.5	7.58(R)	kerogenous– micaceous	BLAIR, 1956
	⊥ bedding	2.68	38.7	–	0.21	34.5	10.3(R)	kerogenous– micaceous	BLAIR, 1956
	USA, Carteroil Tulsa, Okla., fg, argillaceous	–	–	–	–	55.9	–	laminated	BLAIR, 1956
	UK, Ffaldan Colliery, cg, dry	–	–	–	–	203.4	–	calcareous, dry, p = 2.5%	PRICE, 1960
	UK, Cwmtillery Colliery, fg, dry	–	–	–	–	344.7	–	mass., calcareous, p = 2.5	PRICE, 1960
	UK, Deep Navigation Colliery, mg, dry	–	–	–	–	164.1	–	p = 3.5 massive	PRICE, 1960

Siltstone	UK, Snowdown Colliery, fg, dry	–	–	–	93.1	–	p = 3.5 laminated	PRICE, 1960
	UK, Chislet Colliery, fg, dry	–	–	–	90.3	–	p = 3.5 laminated	PRICE, 1960
	UK, Markham Colliery, cg, dry	–	–	–	93.8	–	p = 3.5 laminated	PRICE, 1960
	UK, Chatterley White Field, cg, dry	–	–	–	111.7	–	p = 3.5 laminated	PRICE, 1960
	UK, Cannock Chase	–	–	–	61.4	–	p = 3.5 argil., laminated	PRICE, 1960
	USA, fg, hd	–	13.1 26.89*	0.05 0.08*	–	–	porous	RINEHART et al, 1961
	UK, Chislet Colliery, fg, hd	–	34.47	0.08	–	–	–	WILSON, 1961
	USA, White Pine, S-W Mine, + shale	2.83	–	–	195.8	–	–	PARKER & SCOTT, 1964
	USA, White Pine, S-W Mine, + shale	2.87	53.02	0.09	142.03	4.14(R)	–	PARKER & SCOTT, 1964
	Hungary, 4 seam, sandy, ⊥ bedding	2.02	–	–	8.20	–	p = 34.05 w = 18.0	MARTOS, 1965
	‖ bedding	2.02	–	–	6.90	–	p = 34.5 w = 18.0	MARTOS, 1965
	⊥ bedding	2.00	–	–	8.40	–	p = 37.0 w = 20.8	MARTOS, 1965
	‖ bedding	2.00	–	–	8.30	–	p = 37.0 w = 20.8	MARTOS, 1965
	Hungary, Lyuko, ⊥ bedding	2.00	–	–	6.3	–	p = 31.5 w = 17.5	MARTOS, 1965
	‖ bedding	2.00	–	–	6.2	–	p = 31.5 w = 17.5	MARTOS, 1965

LABORATORY MECHANICAL PROPERTIES OF ROCKS

Rock	Location & Description	Density ϱ (g/cm³)	Modulus of Elasticity E (GPa)	Modulus of Rigidity G (GPa)	Poisson's Ratio υ	Compressive Strength σ_c (MPa)	Tensile Strength σ_t (MPa)	Remarks	Reference
	(1)	(2)	(3)	(4)	(5)	(6)	(7)	(8)	(9)
Siltstone	$\perp$ bedding	2.40	–	–	–	7.6	–	p = 32.5, w = 16.6	Martos, 1965
	‖ bedding	2.40	–	–	–	7.3	–	p = 32.5, w = 16.6	Martos, 1965
	Hungary, Szuha-kallo,								
	$\perp$ bedding	2.17	–	–	–	1.8	–	w = 29.3	Martos, 1965
	‖ bedding	2.17	–	–	–	3.0	–	w = 17.5	Martos, 1965
	$\perp$ bedding	2.15	–	–	–	7.65	–	p = 29.0, w = 16.0	Martos, 1965
	‖ bedding	2.15	–	–	–	8.10	–	p = 29.0, w = 16.0	Martos, 1965
	Hungary,								
	$\perp$ bedding	2.15	–	–	–	3.70	–	p = 29.0, w = 16.0	Martos, 1965
	‖ bedding	2.15	–	–	–	0.80	–	p = 29.0, w = 16.0	Martos, 1965
	$\perp$ bedding	2.20	–	–	–	5.25	–	p = 27.5, w = 12.5	Martos, 1965
	‖ bedding	2.20	–	–	–	5.50	–	p = 27.5, w = 12.5	Martos, 1965
	$\perp$ bedding	1.96	–	–	–	0.20	–	p = 39.6, w = 20.5	Martos, 1965
	‖ bedding	1.96	–	–	–	0.15	–	p = 39.6, w = 20.5	Martos, 1965
	$\perp$ bedding	2.10	–	–	–	0.25	–	p = 37.8, w = 21.5	Martos, 1965
	‖ bedding	2.10	–	–	–	0.25	–	p = 37.8, w = 21.5	Martos, 1965
	$\perp$ bedding	1.97	–	–	–	0.25	–	p = 40.0, w = 21.5	Martos, 1965

Siltstone	‖ bedding	1.97	–	–	–	0.30	–	p = 40.0 w = 21.5	MARTOS, 1965
	Hungary, ⊥ bedding	2.20	–	–	–	3.85	–	p = 31.5 w = 15.0	MARTOS, 1965
	‖ bedding	2.20	–	–	–	3.45	–	p = 31.5 w = 15.0	MARTOS, 1965
	⊥ bedding	2.00	–	–	–	0.45	–	p = 41.0 w = 23.5	MARTOS, 1965
	‖ bedding	2.00	–	–	–	0.30	–	p = 41.0 w = 23.5	MARTOS, 1965
	⊥ bedding	2.15	–	–	–	0.20	–	p = 35.0 w = 19.0	MARTOS, 1965
	‖ bedding	2.15	–	–	–	0.95	–	p = 35.0 w = 19.0	MARTOS, 1965
	⊥ bedding	2.14	–	–	–	16.10	–	p = 39.5 w = 14.0	MARTOS, 1965
	‖ bedding	2.14	–	–	–	12.65	–	p = 39.5 w = 14.0	MARTOS, 1965
	⊥ bedding	1.97	–	–	–	0.50	–	p = 48.5 w = 18.2	MARTOS, 1965
	‖ bedding	1.97	–	–	–	0.60	–	p = 48.5 w = 18.2	MARTOS, 1965
Slate	USA, Bangor, Pa. ⊥ bedding	2.74	–	–	–	209.60	–	grain = 0.3 mm	WUERKER, 1956
	‖ bedding	–	94.18	27.51	–	155.13	–	–	WUERKER, 1956
	USA, Slate, Mich.	2.93	75.84*	33.51*	–	180.64	25.51(R)	p = 0.6 grain = 0.05–2.0 mm	WUERKER, 1956
	USA, Everett, Mass.	2.67	48.7*	21.8*	0.12	–	–	–	BIRCH & BANCROFT, 1940
	USA, Brookline, Mass., argillite	2.76	76.9*	34.8*	0.11	–	–	–	BIRCH & BANCROFT, 1940
	USA, Ishpeming, Mich., white	2.93	76.0	33.0	–	–	–	–	WINDES, 1949

LABORATORY MECHANICAL PROPERTIES OF ROCKS

Rock	Location & Description	Density ρ (g/cm³)	Modulus of Elasticity E (GPa)	Modulus of Rigidity G (GPa)	Poisson's Ratio ν	Compressive Strength σc (MPa)	Tensile Strength σt (MPa)	Remarks	Reference
	(1)	(2)	(3)	(4)	(5)	(6)	(7)	(8)	(9)
Slate	Japan, Inai-machi, Sendai,								
	#1	–	71.5	26.9	0.22	324.5	37.7	grain = 0.1–0.2 mm	NISHIMATSU, 1970
	#2	–	63.0	27.9	0.38	293.4	13.9		NISHIMATSU, 1970
	#3	–	82.1	24.9	0.28	221.4	20.4		NISHIMATSU, 1970
	Austria, Wallsee Dam,								
	0°	2.69	–	–	–	1.9	–	p = 0.3	FENZ et al, 1970
	45°	2.69	–	–	–	3.0	–	p = 0.3	
	90°	2.69	–	–	–	3.4	–	p = 0.3	
	Austria, Ottensheim Dam	2.71	–	–	–	2.45	–	p = 0.3	FENZ et al 1970
	Austria, Ottensheim Dam	2.71	–	–	–	2.30	–	p = 0.3	FENZ et al 1970
	Australia, Mintaro, SA,								
	#1	–	19.72	–	–	195.0	6.34(R)	–	BAMFORD, 1969
	#2	–	32.9	–	–	148.0	6.34(R)	–	BAMFORD, 1969
	#3	–	28.8	–	–	123.5	6.34(R)	–	BAMFORD, 1969
	UK, North Wales, hardblue								
	#1	–	46.3*	16.0*	0.44*	–	–	–	ATTEWELL, 1970
	#2	–	44.1*	15.4*	0.43*	–	–	–	ATTEWELL, 1970
	#3	–	30.1*	10.6*	0.43*	–	–	–	ATTEWELL, 1970
	UK, North Wales, green								
	∥ c-axis	–	40.5*	13.9*	0.46*	–	–	–	ATTEWELL, 1970
	⊥ c-axis	–	23.9*	8.1*	0.47*	–	–	–	ATTEWELL, 1970

Rock	Locality								Reference
Slate	USA, fg, #1	2.74	–	–	–	209.60	–	p = 1, calcareous	OBERT et al, 1946
	#2	2.74	94.2*	27.5*	0.71*	155.13	–	p = 1, calcareous	OBERT et al, 1946
	#3	2.74	83.5*	21.86*	0.63*	210.29	–	p = 1, calcareous	OBERT et al, 1946
	USA, Cliffs shaft, Mich.	2.93	75.84	33.51	0.14	180.64	25.5(R)	p = 0.6	BOYUM, 1961
Spessartite	USA	2.89	61.36 110.32*	–	–	372.32	17.24	–	RICKETTS & GOLDSMITH, 1970
Syenite	Canada, Montreal, Que., nepheline, P = 70–600	–	62.9	25.1	0.26	–	–	–	ADAMS & COKER, 1906
	Germany, Dresden, P = 100–900	–	67.1	–	–	–	–	–	BREYER, 1930
	Germany, Freital, P = 100–900	–	67.8	–	–	–	–	–	BREYER, 1930
	Germany, Riesa-Groba	–	86.3	–	–	–	–	–	BREYER, 1930
	Canada, Ont. #1, P = 1	2.79	17.1	–	–	–	–	–	BIRCH & BANCROFT, 1938
	#1, P = 1	–	15.8	–	–	–	–	–	BIRCH & BANCROFT, 1938
	Canada, Noranda Mines, Ont., porphyry	2.65	75.5*	32.9*	0.180*	–	–	–	BIRCH & BANCROFT, 1939
	Canada, Kirkland Lake, Ont.	2.70	71.9*	30.34*	–	434.37	–	–	WINDES, 1949
	Canada, Kirkland Lake, Ont., mafic	2.82	73.7*	28.27*	–	303.37	–	–	WINDES, 1949
	USA, N.Y., gneiss, granite	2.75	42.2*	20.68*	–	215.12	15.86(R)	p = 0.9	WUERKER, 1956
	USA, N.Y., gneiss, quartz	2.65	33.92*	16.34*	–	293.72	19.99(R)	p = 0.8	WUERKER, 1956

LABORATORY MECHANICAL PROPERTIES OF ROCKS

Rock	Location & Description	Density ρ (g/cm³)	Modulus of Elasticity E (GPa)	Modulus of Rigidity G (GPa)	Poisson's Ratio ν	Compressive Strength σc (MPa)	Tensile Strength σt (MPa)	Remarks	Reference
	(1)	(2)	(3)	(4)	(5)	(6)	(7)	(8)	(9)
Syenite	USA, N.Y., gneissic, dark, to shonkinite	2.81	55.23*	24.13*	–	257.17	18.62(R)	p = 0.7	WUERKER, 1956
	USA, N.Y., granite + shonkinite	2.70	68.46*	28.48*	–	237.87	22.06(R)	p = 0.6	WUERKER, 1956
	Italy, Servo River (Piedmont), P = 140	–	–	–	–	15.7	7.0	–	FUMAGALLI, 1966
	Sweden, Kiruna, porphyritic	–	–	–	–	254.0	12.8	–	HANSAGI, 1966
	Sweden, Kiruna, porphyritic	–	–	–	–	177.7	11.0	–	HANSAGI, 1966
		–	–	–	–	210.6	13.9	–	HANSAGI, 1966
Taconite Ore	–	–	–	–	–	197.88	–	–	BAUER & CALDER, 1967
Tactite	USA, Utah, epidote	2.87	61.36*	27.72*	0.11*	266.14	18.62(R)	p = 1.5	WUERKER, 1956
	USA, Anaconda Mines, Calif., fg-cg	2.92	–	–	–	397.8	–	–	BLAIR, 1956
Tonalite	USA, Calif.	2.76	93.0*	37.0*	0.27	–	–	–	BIRCH, 1961
	USA, Cedar City, Utah	2.63	15.86 / 3.72*	–	0.22	60.74 / 20.06**	–	p = 13.2	MICHALOPOULOS & TRIANDAFILIDIS, 1976
	USA, Cedar City	2.48	–	–	–	5.52	–	insitu test	COOPER & BLOUIN, 1970
	USA, Cedar City	2.48	8.27	–	–	55.16	–	lab. test	COOPER & BLOUIN, 1970
Trap	USA, Mich., amygdaloidal	2.81	64.6*	270.17	–	183.4	17.24	p = 1.5, grain = 0.5–3.0 mm	WUERKER, 1956
Tremolite Schist	India, KGF	3.01	92.7* / 89.6	36.0*	0.29* / 0.11	235.4	–	p = 6	RAMANA et al, 1973

	Location								Reference
Tuff	Canada, Kirkland, Ont.	2.78	86.87*	32.41*	–	289.58	–	p = 1.5	WINDES, 1949
	USA, Howard Prairie Dam, Ore., lithic	1.45	1.38	–	0.11	3.65	–	p = 42.9	USBR, 1953
	USA, Oak Spring Formation, Nev., bedded	1.6	4.2*	2.1*	–	–	–	p = 37.0	ROBERTSON (Unpublished)
	USA, Oak Spring Formation, Nev., bedded	–	4.9	0.17	0.08	–	–	tension	ROBERTSON (Unpublished)
	USA, Oak Spring Formation, Nev., bedded	–	7.6	3.4	0.11	–	–	compression	ROBERTSON (Unpublished)
	USA, Nev., welded	2.2	10.2*	4.1*	–	–	–	p = 14.0	ROBERTSON (Unpublished)
	USA, Nev., welded	2.39	3.65	–	0.19	11.3	1.17	p = 19.8	STOWE, 1969
	Canada, Noranda Mines, Ont., silicified	2.74	81.5*	37.7*	0.07*	–	–	–	BIRCH & BANCROFT, 1939
	USA, AEC Nevada test site, red to red yellow	1.92	3.45	–	0.24	9.65	–	w = 19.3	CORDING, 1967
	—, yellow	2.0	15.6	–	0.09	35.3	–	w = 17.5	CORDING, 1967
	—, red & yellow	1.60	6.34	–	0.15	22.3	–	w = 4.6	CORDING, 1967
	USA, NTS-E Tunnel, porous, cemented	1.61	5.03	–	0.21	24.1	1.45	–	MILLER, 1965
	Canada, Lakeshore	–	76.53	31.03	0.23	262.92	105.49	–	MORRISON, 1970
	Canada, Helen Mine, vertical	–	82.74	–	0.27	155.13	17.31	–	MORRISON, 1970

LABORATORY MECHANICAL PROPERTIES OF ROCKS

Rock	Location & Description	Density ρ (g/cm³)	Modulus of Elasticity E (GPa)	Modulus of Rigidity G (GPa)	Poisson's Ratio υ	Compressive Strength σc (MPa)	Tensile Strength σt (MPa)	Remarks	Reference
	(1)	(2)	(3)	(4)	(5)	(6)	(7)	(8)	(9)
Tuff	USA, McDowell Dam, Ariz., lithic tuff, fg, altered, salt matrix	–	1.38	–	–	15.86	–	grain = 1–2 mm	BRANDON, 1974
	USA, AEC Nevada test site, lapilli tuff, glass with volcanic constituents	–	9.71*	–	0.10*	–	–	variable pore size	YOUASH, 1970
	Mexico, Soledad Dam	1.9 / 2.2	1.7* / 2.7*	–	–	–	–	seismic field	ULLAO, 1964 / ULLAO, 1964
	USA, Green Peters Dam	–	7.4	0.19	–	23.0	–	c = 7.0, Ø = 24°	CORNS & NESBITT, 1967
	USA, Green Peters Dam	–	7.0	0.20	–	22.0	–	c = 7.7, Ø = 19°	CORNS & NESBITT, 1967
	USA, AEC Nevada test site	–	3.72	–	0.19	11.31	1.17	–	STOWE & AINSWORTH, 1968
	Japan, Seikan Tunnel	2.19	–	–	–	21.9	2.2	–	MOCHIDA, 1974
	Japan, Seikan Tunnel, tuff-breccia	2.36	–	–	–	28.6	3.9	–	MOCHIDA, 1974
	Japan, Shimane Nuclear Plant	2.4	3.0	–	0.2	–	–	c = 0.4, Ø = 30°	KITAHARA et al, 1974
	Japan, Shimane Nuclear Plant	2.4	2.0	–	0.2	–	–	c = 0.3, Ø = 35°	KITAHARA et al, 1974
	Japan	–	7.14	–	–	34.7	–	–	YAMAGUCHI, 1968
	Japan, Aoishi	–	68.0	–	–	33.8	4.31	sandy	YAMAGUCHI, 1968

	Locality								Reference
Tuff	Japan, Aoishi	1.91	76.0	–	–	36.0	4.31	sandy	YAMAGUCHI, 1968
	Japan, Aoishi	–	70.0	–	–	34.2	4.45	sandy	YAMAGUCHI, 1968
	Canada, Macleod Mine	–	63.43	–	0.24	–	–	–	HERGET, 1973
	Canada, Macleod Mine	–	74.5	–	0.25	–	–	–	HERGET, 1973
	Canada, Macleod Mine	–	61.4	–	0.26	–	–	–	HERGET, 1973
	Canada, Macleod Mine	–	71.7	–	0.28	–	–	–	HERGET, 1973
	USSR, geosynclines + breccias	2.49	22.7	–	0.13	–	–	p = 8.31	BELIKOV, 1967
	Hungary, rhyolite, #1	2.08	–	–	–	9.4	–	p = 30.0 w = 16.5	MARTOS, 1965
	#2	2.08	–	–	–	9.8	–	p = 30.0 w = 16.5	MARTOS, 1965
Uralite Basalt	India, KGF	3.06	104.7* 78.5	41.0*	0.28* 0.15	215.5	–	p = 9	RAMANA et al, 1973
	India, KGF	2.67	62.4*	26.7*	0.17*	–	–	p = 7	RAMANA et al, 1973
	India, KGF	3.16	82.0* 91.0	31.0*	0.32* 0.25	198.8	–	p = 3	RAMANA et al, 1973
Uralite Diabase	India, KGF	3.00	105.2*	41.3*	0.27*	–	–	–	RAMANA et al, 1973
	India, KGF	3.05	101.5*	45.1*	0.13*	–	–	–	RAMANA et al, 1973
	India, KGF	3.03	59.1*	21.4*	0.38*	–	–	p = 3	RAMANA et al, 1973

References to Appendix II

1. ADAMS, F. D. and COKER, E. G.: An investigation into the elastic constants of rocks, especially with reference to cubic compressibility. Carnegie Inst., Washington, Pub. No. 46, 1906, 69 p.

2. ATCHISON, T. C., DUVALL, W. I. and PETKOF, B.: Rock breakage in quarry blasting. Proc. 4th Symp. Rock Mech., Univ. Park, Penn., 1961, pp. 163–169.

3. ATTEWELL, P. B.: Triaxial anisotropy of wave velocity and elastic moduli in slate and their axial concordance with fabric and tectonic symmetry. Int. J. Rock Mech. Min. Sci., Vol. 7, 1970, pp. 193–207.

4. AVEDISSIAN, Y. M. and WOOD, L. E.: Prediction of compressive strength of rock from its sonic properties. Proc. 10th Symp. Rock Mech., Univ. Texas, Austin, Texas, 1968, pp. 55–71.

5. AVILA, F. P.: Some applications of seismic field tests in rock media. Proc. 1st Cong. Int. Soc. Rock Mech., Lisbon, 1966, Vol. I, pp. 3–6.

6. AVRAMOVA-TACHEVA, E. and CHESHANKOVA, K.: A case of especially deformation behaviour of a brittle rock after it cracking. Proc. Symp. Rock Fracture, Nancy, 1971, Paper II–17.

7. BALMER, G. G.: Physical properties of some typical foundation rocks. U.S. Bureau of Reclamation, Concrete Lab. Rep. SP-59, 1953, 150 p.

8. BAMFORD, W. E.: Anisotropy, and natural variability of rock properties. Proc. Symp. Rock Mech., Univ. Sydney, 1969, pp. 1–10.

9. BANCILA, I., DIACON, A., GEORGESCU, M. and RADULESCU, D.: Les conditions de fondation d'un barrage dans Le flysch des Carpathes Orientales. Trans. 7th Cong. Large Dams, Rome, 1961, Vol. II, pp. 331–350.

10. BAUER, A. and CALDER, P. N.: Open pit drilling factors influencing drilling rates. Proc. 4th Can. Rock Mech. Symp., Ottawa, 1967, pp. 1–33.

11. BELIKOV, B. P.: Plastic constants of rock-forming minerals and their effect on the elasticity of rocks. In Physical and Mechanical Properties of Rocks (Ed. ZALESSKII, B. V.), Jerusalem, Israel Program for Scientific Translations, 1967, pp. 124–140.

12. BELIKOV, B. P., ZALESSKII, B. V., ROZANOV, Yu. A., SANINA, E. A. and TIMCHENKO, I. P.: Methods of studying the physico-mechanical properties of rocks. In Physical and Mechanical Properties of Rocks (Ed. ZALESSKII, B. V.), Jerusalem, Israel Program for Scientific Translations, 1967, pp. 1–58.

13. BHATNAGAR, P. S. and SHAH, S. R.: Experiments of Bhakra dam foundations. Trans. 8th Cong. Large Dams, Edinburgh, 1964, Vol. 1, pp. 1081–1108.

14. BIENIAWSKI, Z. T.: Deformation behaviour of fractured rock under multiaxial compression. Proc. Int. Conf. Struct., Solid Mech. and Eng. Design in Civ. Eng. Mater., Southampton, 1969, Paper 55.

15. BIENIAWSKI, Z. T., DENKHAUS, H. G. and VOGLER, U. W.: Failure of fractured rock. Int. J. Rock Mech. Min. Sci., Vol. 6, 1969, pp. 323–341.

16. BIENIAWSKI, Z. T. and VOGLER, U. W.: Load deformation behaviour of coal after failure. Proc. 2nd Cong. Int. Soc. Rock Mech., Belegrade, 1970, Vol. 1, pp. 345–351.

17. BIRCH, F.: 1940, Unpublished.

18. BIRCH, F.: The velocity of compressional waves in rocks to 10 kilobars. J. Geophys. Res., Vol. 66, 1961, pp. 2199–2224.
19. BIRCH, F. and BANCROFT, D.: The effect of pressure on the rigidity of rocks. J. Geophys. Res., Vol. 46, 1938, pp. 59–87, 113–141.
20. BIRCH, F. and BANCROFT, D.: Elasticity and internal friction in a long column of granite. Bull. Seism. Soc. Am., Vol. 28, 1938, pp. 243–254.
21. BIRCH, F. and BANCROFT, D.: The elasticity of certain rocks and minerals. Am. J. Sci., Vol. 237, 1939, pp. 2–6.
22. BIRCH, F. and BANCROFT, D.: New measurements of the rigidity of rocks at high pressure. J. Geol., Vol. 48, 1940, pp. 752–766.
23. BLAIR, B. E.: Physical properties of mine rock, Part III. U.S.B.M. R.I. 5130, 1955, 69 p.
24. BLAIR, B. E.: Physical properties of mine rock, Part IV. U.S.B.M. R.I. 5244, 1956, 69 p.
25. BLEE, C. E. and MEYER, A. A.: Measurement of settlements at certain dams of the TVA system and assumptions for earth loading for dams in the TVA area. Trans. 5th Cong. Large Dams, Paris, 1955, Vol. III.
26. BODONYI, J.: Deformation at failure as determined by laboratory tests on brittle rocks. Proc. Symp. Rock Fracture, Nancy, 1971, Paper II–19.
27. BORDIA, S. K.: Complete stress – volumetric strain equation for brittle rock up to strength failure. Int. J. Rock Mech. Min. Sci., Vol. 9, 1972, pp. 17–24.
28. BORECKI, M. and CHUDEK, M.: Mechanika Gorotworu. „Slack", Katowice, 1972.
29. BORETTI-ONYSZKIEWICZ, W.: Joints in the flysch sandstone on the ground of strength examinations. Proc. 1st Cong. Int. Soc. Rock Mech., Lisbon, 1966, Vol. I, pp. 153–157.
30. BOYUM, B. H.: Subsidence case studies in Michigan mines. Proc. 4th Symp. Rock Mech., Univ. Park, Penn., 1961, pp. 19–57.
31. BRACE, W. F.: Brittle fracture of rocks. Proc. Int. Conf. State of Stress in the Earth's Crust, Santa Monica, California, 1963, pp. 110–174.
32. BRANDON, T. R.: Rock mechanics properties of typical foundation rock types. U.S. Bureau of Reclamation, Denver, Colorado, Rep. REC–ERC. 74–10, 1974.
33. BRATTON, J. L. and PRATT, H. R.: Two dimensional finite difference calculations of dynamic insitu response of layered geologic media to a large explosive load. Proc. 10th Symp. Rock Mech., Univ. Texas, Austin, Texas, 1968, pp. 149–198.
34. BREYER-KASSEL, H.: Über die Elastizität von Gesteinen. Zeitschr. Geophysik, Vol. 6, 1930, pp. 98–111.
35. BROWN, E. L., CHARLAMBAKIS, S., CREPEAU, P. M. and LE FRANCOIS, P.: Les fondations du barrage Daniel-Johnson (Manicougan 5). Trans. 10th Cong. Large Dams, Montreal, 1970, Vol. II, pp. 631–650.
36. CHAN, S. S. M.: Deformation behaviour of Revett quartzite under uniaxial and triaxial loading. Proc. 6th Can. Rock Mech. Symp., Montreal, 1970, pp. 9–31.
37. CHAN, S. S. M. and CROCKER, T. J.: A case study of insitu rock deformation behaviour in the Silver Summit mine, Coeur D'Alene mining district. Proc. 7th Can. Rock Mech. Symp., Edmonton, 1971, pp. 135–160.
38. CHENEVERT, M. E.: The deformation failure characteristics of laminated sedimentary rocks. Ph. D. Thesis, Univ. Texas, Austin, Texas, 1964, 203 p.
39. CLARK, S. P. (Editor): Handbook of Physical Constants. Geol. Soc. Am. Mem. 97, 1966.
40. CLARK, G. B. and CAUDLE, R. D.: Failure of homogeneous rock under dynamic compressive loading. Proc. Int. Conf. State of Stress in the Earth's Crust, Santa Monica, California, 1963, pp. 300–320.
41. CLARK, G. B., ROLLINS, R. R., BROWN, J. W. and KALIA, H. N.: Rock penetration by jets from lined cavity explosive charges. Proc. 12th Symp. Rock Mech., Rolla, Missouri, 1970, pp. 621–651.

457

42. Coates, D. F. and Parsons, R. C.: Experimental criteria for classification of rock substances. Int. J. Rock Mech. Min. Sci., Vol. 3, 1966, pp. 181–189.
43. Cochrane, T. S., Carter, O. F. and Barron, K.: Studies of ground behaviour in a metal mine. Proc. 4th Int. Conf. Strata Control and Rock Mech., New York, 1964, pp. 123–139.
44. Cooper, H. F. and Blouin, S. E.: Dynamic insitu rock properties from buried high explosive arrays. Proc. 12th Symp. Rock Mech., Rolla, Missouri, 1970, pp. 45–70.
45. Cording, E. J.: The stability during construction of three large u. g. openings in rock. Ph. D. Thesis, Univ. Ill., Urbana, Ill., 1967.
46. Corns, C. F. and Nesbitt, R. H.: Sliding stability of three dams on weak rock foundations. Trans. 9th Cong. Large Dams, Istanbul, Vol. I, pp. 463–486.
47. Creuels, F. H. and Hermes, J. M.: Measurement of the changes in rock pressure in the vicinity of a working face. Proc. Int. Strata Control Cong., Essen, 1956.

48. D'Andrea, D. V. and Condon, J. L.: Dye penetrant studies of fractures produced in laboratory cratering. Proc. 12th Symp. Rock Mech., Rolla, Missouri, 1970, pp. 479–495.
49. Deklotz, E. J., Brown, J. W. and Stemler, O. A.: Anisotropy of a schistose gneiss. Proc. 1st Cong. Int. Soc. Rock Mech., Lisbon, 1966, Vol. I, pp. 465–470.
50. Denkhaus, H. G.: Strength of rock material and rock systems. Int. J. Rock Mech. Min. Sci., Vol. 2, 1965, pp. 111–126.
51. Drude, P. and Voight, W.: Bestimmung der Elasticitätsconstanten einiger dichter Mineralien. Ann. Phys. Chem., Wied, Vol. 42, 1891, pp. 537–548.
52. Dvorak, A.: Tangential deformations of rocks at the foundations of hydrotechnic structures. Trans. 9th Cong. Large Dams, Istanbul, 1967, Vol. 1, pp. 343–351.

53. Emery, C. L.: Strain energy in rocks. Proc. Int. Conf. State of Stress in the Earth's Crust, Santa Monica, California, 1963, pp. 234–260.
54. Evans, I. and Pomeroy, C. D.: The strength of cubes of coal in uniaxial compression. Proc. Conf. Mech. Prop. of Non-metallic Brittle Materials, London, 1958, pp. 5–28.
55. Everell, M. D., Gill, D. E. and Sirois, L. L.: Relation of grinding selection functions to physicomechanical properties of rocks. Proc. 6th Can. Rock Mech. Symp., Montreal, 1970, pp. 177–193.
56. Everling, G.: Calculation of stress from measurements made in boreholes in the seam and in the rock. Proc. Int. Strata Control Cong., Essen, 1956.

57. Fai, J.: Bull Azrestiya Acad. Sci., U.S.S.R., Geophy., Ser. 10, 1961.
58. Fayed, L. A. and Khattab, A. F.: Investigation and construction of Wadi El Rayan tunnel, Egypt. Int. J. Rock Mech. Min. Sci. & Geomech. Abstr., Vol. 10, 1973, pp. 97–103.
59. Fenz, R., Kobilka, J. G. and Makovec, F. F.: Problems encountered in the slate foundations of the Wallsee and Ottensheim power plants on the Danube in Austria. Trans. 10th Cong. Large Dams, Montreal, 1970, Vol. II, pp. 551–569.
60. Fodor, I. and Tokes, T.: Neue Forschungen über die physikalischen und mechanischen Eigenschaften des rumänischen Steinsalzes. Proc. 1st Cong. Int. Soc. Rock Mech., Lisbon, 1966, Vol. I, pp. 705–709.
61. Fox, P. P., Meyer, A. A. and Talobre, J. A.: Foundations of the Pahlavi dam on the Dez river. Trans. 8th Cong. Large Dams, Edinburgh, 1964, Vol. I, pp. 1–22.
62. Fumaglli, E.: Stability of arch dam rock abutments. Proc. 1st Cong. Int. Soc. Rock Mech., Lisbon, 1966, Vol. II, pp. 503–508.

63. Gassmann, Weber and Vogtli: Verh. Schweiz Naturforsch. Ges., Bern, 1952.
64. Gay, N. C.: Fracture growth around openings in thick-walled cylinders of rock subjected to hydrostatic compression. Int. J. Rock Mech. Min. Sci. & Geomech. Abstr., Vol. 10, 1973, pp. 209–233.

458

65. GEYER, R. L. and MYUNG, J. I.: The 3-D velocity log; A tool for insitu determination of the elastic moduli of rocks. Proc. 12th Symp. Rock Mech., Rolla, Missouri, 1970, pp. 71–107.
66. GICOT, H.: Influence du sol de fondation sur les deformations du barrage de Rossens. Trans. 5th Cong. Large Dams, Paris, 1955, Vol. III, pp. 475–493.

67. HAIMSON, B. C. and FAIRHURST, C.: Insitu stress determination at great depth by means of hydraulic fracturing. Proc. 11th Symp. Rock Mech., Berkeley, California, 1969, pp. 559–584.
68. HAIMSON, B. C. and FAIRHURST, C.: Some bit penetration characteristics in Pink Tennessee marble. Proc. 12th Symp. Rock Mech., Rolla, Missouri, 1970, pp. 547–559.
69. HANSAGI, I.: Eine praktische Methode der Gebirgsklassifizierung. Proc. 7th Cong. Int. Bur. Rock Mech., Leipzig, 1965, pp. 209–213.
70. HANSAGI, I.: Bergmännisches Verfahren der Gebirgsfestigkeitsbestimmung und der Gebirgsklassifizierung. Proc. 1st Cong. Int. Soc. Rock Mech., Lisbon, 1966, Vol. I, pp. 179–183.
71. HARDY, H. R.: Standardised procedures for the determination of the physical properties of mine rock under short-period uniaxial compression. Dept. Mines & Tech. Surveys, Mines Branch, Canada, TB-8, Dec. 1959.
72. HARTMAN, H. L.: The effectiveness of indexing in percussion and rotary drilling. Int. J. Rock Mech. Min. Sci., Vol. 3, 1966, pp. 265–278.
73. HAWKES, I., MELLOR, M. and GARIEPY, S.: Deformation of rocks under uniaxial tension. Int. J. Rock Mech. Min. Sci. & Geomech. Abstr., Vol. 10, 1973, pp. 493–507.
74. HENKEL, D. J., KNILL, J. L., LLOYD, D. G. and SKEMPTON, A. W.: Stability of the foundations of Monar dam. Trans. 8th Cong. Large Dams, Edinburgh, 1964, Vol. I, pp. 425–441.
75. HEREL, J.: Festigkeits- und Formänderungsverhalten charakterisierender Eigenschaften der Magnesite, die an den Probekörpern von verschiedenem Schlankheitsgrad festgestellt wurden. Proc. 1st Cong. Int. Soc. Rock Mech., Lisbon, 1966, Vol. I, pp. 503–507.
76. HERGET, G.: Variation of rock stresses with depth at a Canadian iron mine. Int. J. Rock Mech. Min. Sci., Vol. 10, 1973, pp. 37–51.
77. HEUZE, F. E. and GOODMAN, R. E.: Finite element and physical model studies of tunnel reinforcement in rock. Proc. 15th Symp. Rock Mech., Rapid City, South Dakota, 1973, pp. 37–67.
78. HOBBS, D. W.: A simple method for assessing the uniaxial compressive strength of rock. Int. J. Rock Mech. Min. Sci., Vol. 1, 1964, pp. 5–15.
79. HOBBS, D. W.: The tensile strength of rocks. Int. J. Rock Mech. Min. Sci., Vol. 1, 1964, pp. 385–396.
80. HOBBS, D. W.: Stress-strain-time behaviour of a number of coal measure rocks. Int. J. Rock Mech. Min. Sci., Vol. 7, 1970, pp. 149–170.
81. HÖFER, K. H.: Methoden gebirgsmechanischer Festigkeitsuntersuchungen für die Deutsche Demokratische Republik. Proc. 2nd Cong. Int. Bur. Rock Mech., Leipzig, 1960, pp. 30–37.
82. HÖFER, K. H. and MENZEL, W.: Comparative study of pillar loads in potash mines established by calculation and by measurements below ground. Int. J. Rock Mech. Min. Sci., Vol. 1, 1964, pp. 181–198.
83. HOGG, A. D.: Some engineering studies of rock movement in the Niagara area. In Engineering Geology Case Histories, Numbers 1–5, Geol. Soc. Am. (Ed. TRASK, P. and KIERSCH, G. A.), 1964.
84. HOLLAND, C. T.: The strength of coal in mine pillars. Proc. 6th Symp. Rock Mech., Rolla, Missouri, 1964, pp. 450–466.

85. Hoskins, J. R. and Horino, F. G.: Influence of spherical head size and specimen diameters on the uniaxial compressive strength of rocks. U.S.B.M. R.I. 7234, 1969, 16 p.

86. Houska, J.: Beitrag zu den Spaltungsversuchen an Gesteinen. Proc. 10th Cong. Int. Bur. Rock Mech., Leipzig, 1968, pp. 137–148.

87. Huck, P. J. and Singh, M. M.: Effect of specimen size on confined compression testing of rock cores. IIT Research Centre, Project No. D 6059, 1972.

88. Ide, J. M.: Comparison of statically and dynamically determined Young's modulus of rocks. Proc. Nat. Acad. Sci. U.S.A., Vol. 22, 1936, pp. 81–92.

89. Ide, J. M.: The elastic properties of rocks. A correlation of theory and experiment. Data. Proc. Nat. Acad. Sci. U.S.A., Vol. 22, 1936, pp. 482–596.

90. Iida, K., Wada, T., Iida, Y. and Shichi, R.: Measurements of creep in igneous rocks. J. Earth Sciences, Nagoya Univ., Vol. 8, No. 1, 1960, pp. 1–16.

91. Iliev, I. G.: An attempt to estimate the degree of weathering of intrusive rocks from their physico-mechanical properties. Proc. 1st Cong. Int. Soc. Rock Mech., Lisbon, 1966, Vol. I, pp. 109–114.

92. Imrie, A. S. and Jory, L. T.: Behaviour of the underground power house arch at the WAC Bennett dam during excavation. Proc. 5th Can. Rock Mech. Symp., Toronto, 1968, pp. 29.–37.

93. Ito, I. and Terada, M.: Experimental study of the dynamic behaviour of rocks. Rock Mech. in Japan, Vol. I, 1970, pp. 47–49.

94. Jahns, H.: Messung der Gebirgsfestigkeit in situ bei wachsendem Maßstabsverhältnis. Proc. 1st Cong. Int. Soc. Rock Mech., Lisbon, 1966, Vol. I, pp. 477–482.

95. James, L. S.: Stresses and deflections: Maretai dam. Trans. 5th Cong. Large Dams, Paris, 1955, Vol. III, pp. 273–291.

96. Judd, W. R.: Statistical methods to compile and correlate rock properties. Sch. Civ. Eng., Purdue Univ., Lafayette, Indiana, 1969.

97. Jumikis, A. R.: Some engineering aspects of Brunswick shale. Proc. 1st Cong. Int. Soc. Rock Mech., Lisbon, 1966, Vol. I, pp. 99–102.

98. Kendorski, F. S. and Mehtab, M. A.: Fracture patterns and anisotropy of San Manual Quartz monozite. Bull. Assoc. Eng. Geol., Vol. XIII, No. 1, 1976, pp. 23–52.

99. Kidybinski, A.: Mechaniczne wlasnosci skal Karbonskich Zaglebia Gornoslaskiego. Przeglad Gorniczy, Vol. 25, No. 11, 1969, pp. 517–523.

100. Kitahara, Y., Fujiwara, Y. and Kawamura, M.: The stability of slope during excavation – the method of analysis and observation. Rock Mech. in Japan, Vol. II, 1974, pp. 187–189.

101. Kluth, D. J.: Rock stress measurements in the Jor underground power station of the Cameroun Highlands hydro-electric scheme, Malaya. Trans. 8th Cong. Large Dams, Edinburgh, 1964, Vol. I, pp. 103–119.

102. Knopoff, L.: Seismic wave velocities in Westerly granite. Trans. Am. Geophys. Union, Vol. 35, 1954, pp. 969–973.

103. Knutson, C. F. and Bohor, B. F.: Micro deformation of sedimentary rocks subjected to Gnome nuclear explosion. Proc. 6th Symp. Rock Mech., Rolla, Missouri, 1964, pp. 152–184.

104. Ko, H. Y. and Gerstle, K. H.: Elastic properties of two coals. Int. J. Rock Mech. Min. Sci. & Geomech. Abstr., Vol. 13, 1976, pp. 81–90.

105. Koifman, M. I.: Quick combined method of determining mechanical properties of rocks. In Mechanical Properties of Rocks by Protodyakonov, M. M., Koifman, M. I. and others. Jerusalem, Israel Program for Scientific Translations, 1969, pp. 74–86.

106. KOIFMAN, M. I. and CHIRKOV, S. E.: Prophilographic investigations and correct data on mechanical properties of rocks. In Mechanical Properties of Rocks by PROTODYAKONOV, M. M., KOIFMAN, M. I. and others. Jerusalem, Israel Program for Scientific Translations, 1969, pp. 9–14.

107. KRUSE, G. H., ZERNEKE, K. L., SCOTT, J. B., JOHNSON, W. S. and NELSON, J. S.: Approach to classifying rock for tunnel liner design. Proc. 11th Symp. Rock Mech., Berkeley, California, 1969, pp. 169–192.

108. LADANYI, B. and DON, N.: Study of strains in rock associated with brittle failure. Proc. 6th Can. Rock Mech. Symp., Montreal, 1970, pp. 49–64.

109. LADANYI, B. and ROY, A.: Some aspects of bearing capacity of rock mass. Proc. 7th Can. Rock Mech. Symp., Edmonton, 1971, pp. 161–190.

110. LAMA, R. D.: Elasticity and strength of coal seams in situ and an attempt to determine the energy in pressure bursting of roadsides. D. Sc. Tech. Thesis, Faculty of Mining, Academy of Min. & Metall., Cracow, Poland, 1966.

111. LANE, R. G.: Rock foundations – Diagnosis of mechanical properties and treatment. Trans. 8th Cong. Large Dams, Edinburgh, 1964, Vol. I, pp. 141–165.

112. LAROCQUE, G. E. and FAVREAU, R. F.: Blasting research at the Mines Branch. Proc. 12th Symp. Rock Mech., Rolla, Missouri, 1970, pp. 341–358.

113. LEARMONTH, A. P. and GARRETT, B. K.: Strength and deformation characteristics of Victorian Silurian mudstones. Proc. Symp. Rock Mech., Univ. Sydney, 1969, pp. 14–17.

114. LECOMTE, P.: Methods for measuring the dynamic properties of rocks. Proc. 2nd Can. Rock Mech. Symp., Kingston, Ontario, 1963, pp. 15–24.

115. LEEMAN, E. R.: Absolute rock stress measurements using a borehole trepanning stress-relieving technique. Proc. 6th Symp. Rock Mech., Rolla, Missouri, 1964, pp. 407–426.

116. LEHNHOFF, T. F., PATEL, M. R. and CLARK, G. B.: A thermal rock fragmentation model. Proc. 15th Symp. Rock Mech., Rapid City, South Dakota, 1973, pp. 523–555.

117. LIVINGSTON, C. W.: The natural arch, the fracture pattern, and the sequence of failure in massive rocks surrounding an underground opening. Proc. 4th Symp. Rock Mech., Univ. Park, Penn., 1961, pp. 197–204.

118. LOZINSKA-STEPIEN, H.: Deformations during compression of the Cretaceous clay limestones and marls in an interval of loads from 0 to the boundary of proportionality in the light of laboratory examinations. Proc. 1st Cong. Int. Soc. Rock Mech., Lisbon, 1966, Vol. I, pp. 381–384.

119. LUNDBORG, N.: The strength-size relation of granite. Int. J. Rock Mech. Min. Sci., Vol. 4, 1967, pp. 269–272.

120. MARTOS, F.: Wassergehalt und Gesteinsfestigkeit. Proc. 7th Cong. Int. Bur. Rock Mech., Leipzig, 1965, pp. 222–243.

121. MATSUSHIMA, S.: On the flow and fracture of igneous rocks. Kyoto Univ. Disaster Prevent. Res. Inst. Bull., Vol. 10, No. 36, 1960.

122. MESRI, G., ADACHI, K. and ULLRICH, C. R.: Pore pressure response in rock to undrained change in all round stress. Geotechnique, Vol. 26, No. 2, June, 1976, pp. 317–330.

123. MICHALOPOULOS, A. P. and TRIANDAFILIDIS, G. E.: Influence of water on hardness, strength and compressibility of rock. Bull. Assoc. Eng. Geol., Vol. XIII, No. 1, 1976, pp. 1–22.

124. MILLER, R. P.: Engineering classification and index properties for intact rock. Ph. D. Thesis, Univ. Ill., Urbana, Ill., 1965.

125. MITANI, T. and KAWAI, T.: Experimental study on relationship between engineering properties of rock and mechanical drilling characteristics. Rock Mech. in Japan, Vol. II, 1974, pp. 124–126.

126. MOCHIDA, Y.: Rock properties in the Seikan tunnel. Rock Mech. in Japan, Vol. II, 1974, pp. 161–163.

127. MORGENSTERN, N. R. and PHUKAN, A. L. T.: Non-linear stress-strain relations for a homogeneous sandstone. Int. J. Rock Mech. Min. Sci., Vol. 6, 1969, pp. 127–142.

128. MORRISON, R. G. K.: A philosophy of ground control. Ontario Dept. Mines, Toronto, Canada, 1970, 170 p.

129. MORUZI, G. A. and PASIEKA, A. R.: Evaluation of a blasting technique for destressing ground subject to rockbursting. Proc. 6th Symp. Rock Mech., Rolla, Missouri, 1964, pp. 185–204.

130. MOYE, D. G.: Rock mechanics in the investigation and construction of T. 1 underground power station, Snowy Mountains, Australia. In Engineering Geology Case Histories, Numbers 1–5, Geol. Soc. Am., (Ed. TRASK, P. and KIERSCH, G. A.), 1964.

131. MUIR, W. G. and COCHRANE, T. S.: Rock mechanics investigations in a Canadian salt mine. Proc. 1st Cong. Int. Soc. Rock Mech., Lisbon, 1966, Vol. II, pp. 411–416.

132. MURRELL, S. A. F.: The strength of coal under triaxial compression. Proc. Conf. Mech. Prop. Non-metallic Brittle Materials, London, 1958.

133. NESBITT, R. H.: Laboratory tests of rock cores from Fremont Canyon tunnel area – Glendo Unit – Missouri River Basin Project, Wyoming. U.S.B.R., Denver, Colo., Lab. Rep. C945, 1960.

134. NICHOLLS, H. R.: In situ determination of the dynamic elastic constants of rock. Proc. Int. Symp. Min. Res., Rolla, Missouri, 1961, Vol. 2, pp. 727–738.

135. NISHIHARA, M.: Creep of shale and sandy shale. J. Geol. Soc. Japan, Vol. 58, No. 683, 1952, pp. 373–377.

136. NISHIHARA, M.: Stress-strain relation of rocks (Rheological properties of rocks I). Doshisha Eng. Rev., Vol. 8, No. 2, 1957, pp. 32–55.

137. NISHIMATSU, Y.: The torsion test and elastic constants of the orthotropic rock substance. Proc. 2nd Cong. Int. Soc. Rock Mech., Belgrade, 1970, Vol. 1, pp. 479–484.

138. OBERT, L., WINDES, S. L. and DUVALL, W. I.: Standardised tests for determining the physical properties of mine rock U S.B.M. R.I. 3891, 1946, 67 p.

139. OBERTI, G. and REBAUDI, A.: Bedrock stability behaviour with time at the Place Moulin arch gravity dam. Trans. 9th Cong. Large Dams, Istanbul, 1967, Vol. I, pp. 849–872.

140. O'BRIEN, T.: Physical and geological characteristics of Hawkesbury sandstone. Proc. Symp. Rock Mech., Univ. Sydney, 1969, pp. 17–19.

141. PANOV, S. I., SAPEGIN, D. D. and KHRAPKOV, A. A.: Some specific features of deformability of rock masses adjoining a gallery. Proc. 2nd Cong. Int. Soc. Rock Mech., Belgrade, 1970, Vol. 1, pp. 485–490.

142. PARASARTHY, A., SAHAH, S. D. and LIMAYE, R. G.: Certain engineering geological studies pertaining to Kadana region, Gujrat, India. Proc. 2nd Int. Cong. Int. Assoc. Eng. Geol., Sao Paulo, 1974, Vol. II, Paper V1–31.

143. PARASKEVOV, R. D., KOLEV, K. L. and KAZAKOV, B. N.: Untersuchung der physikalisch-mechanischen Eigenschaften des Gesteins der Braunkohlenlagerstätte „Bistrica" – V.R. Bulgarien, bei Kurz- und Dauerbelastung, Schlußfolgerungen für die Gewährleistung der Standsicherheit von Grubenbauen. Proc. 10th Cong. Int. Bur. Rock Mech., Leipzig, 1968, pp. 126–136.

144. PARKER, J. and SCOTT, J. J.: Instrumentation for room and pillar workings in a copper mine of the Copper Range Company, White Pine, Michigan. Proc. 6th Symp. Rock Mech., Rolla, Missouri, 1964, pp. 669–720.

145. PELLS, P. J. N.: Strength and deformation characteristics of Hawkesbury sandstone. Univ. Sydney, Sch. Civ. Eng., Res. Rep. No. R 287, 1976.

146. PENG, S. and JOHNSON, A. M.: Crack growth and faulting in cylindrical specimens of Chelmsford granite. Int. J. Rock Mech. Min. Sci., Vol. 9, 1971, pp. 37–86.

147. PENMAN, A. D. M. and MITCHELL, P. B.: Initial behaviour of Scammonden dam. Trans. 10th Cong. Large Dams, Montreal, 1970, Vol. 1, pp. 723–747.

148. PESELNICK, L.: Elastic constants of Solenhofen limestone and their dependence upon density and saturation. J. Geophys. Res., Vol. 67, 1962, pp. 4441–4448.

149. PHELINES, R. F.: Measures and procedures adopted to ensure the stability and safety of foundations for the Pongolapoort dam. Trans. 9th Cong. Large Dams, Istanbul, 1967, Vol. 1, pp. 619–646.

150. POMEROY, C. D.: Laboratory experiments to investigate the effect of overburden pressure on coal ploughing. Proc. 3rd Int. Conf. Strata Control, Paris, 1960, pp. 197–209.

151. PRATT, H. R., BLACK, A. D., BROWN, W. S. and BRACE, W. F.: The effect of specimen size on the mechanical properties of unjointed diorite. Int. J. Rock Mech. Min. Sci., Vol. 9, 1972, pp. 513–529.

152. PRICE, N. J.: A study of rock properties in conditions of triaxial stress. Proc. Conf. Mech. Prop. Non-metallic Brittle Materials, London, 1958, pp. 106–122.

153. PRICE, N. J.: The compressive strength of coal measure rocks. Coll. Eng., Vol. 37, 1960, pp. 283–292.

154. RAD, P. F. and McGARRY, F. J.: Thermally assisted cutting of granite. Proc. 12th Symp. Rock Mech., Rolla, Missouri, 1970, pp. 721–757.

155. RAMANA, Y. V. and VENKATANARAYANA, B.: Laboratory studies on Kolar rocks. Int. J. Rock Mech. Min. Sci. & Geomech. Abstr., Vol. 10, 1973, pp. 465–489.

156. REGULA, W.: Seismische Untersuchungen des Geophysikalischen Instituts in Göttingen. Zeitschr. Geophysik, Vol. 16, 1940, pp. 40–56.

157. REICHMUTH, D. R.: Correlation of force-displacement data with physical properties of rock for percussive drilling systems. Proc. 5th Symp. Rock Mech., Minneapolis, Minn., 1962, pp. 33–59.

158. RICHARDS, T. C.: On elastic constants of rocks with seismic application. Proc. Phys. Soc., Vol. 45, No. 246, 1933, pp. 70–81.

159. RICKETTS, T. E. and GOLDSMITH, W.: Dynamic properties of rocks and composite structural materials. Int. J. Rock Mech. Min. Sci., Vol. 7, 1970, pp. 315–335.

160. RINEHART, J. S., FORTIN, J. P. and BURGIN, L.: Propagation velocity of longitudinal waves in rocks. Effect of state of stress, stress level of the wave, water content, porosity, temperature, stratification and texture. Proc. 4th Symp. Rock Mech., Univ. Park, Penn., 1961, pp. 119–135.

161. RINGHEIM, A. S.: Experiences with the Bearpaw shale at the south Saskatchewan River dam. Trans. 8th Cong. Large Dams, Edinburgh, 1964, Vol. I, pp. 529–550.

162. RISING, R. R. and ERICKSON, G. A.: Design of underground pumping chamber at Lake Mead. Proc. Symp. Underground Rock Chambers, Phoenix, Arizona, 1971, pp. 381–405.

163. ROBERTSON, W. E.: (Unpublished).

164. ROSENBLAD, J. L.: Development of a rocklike model material. Proc. 10th Symp. Rock Mech., Univ. Texas, Austin, Texas, 1968, pp. 331–361.

165. RUIZ, M. D.: Some technological characteristics of 26 Brazilian rock types. Proc. 1st Cong. Int. Soc. Rock Mech., Lisbon, 1966, Vol. I, pp. 115–119.

166. SALAS, J. A. J.: Mechanical resistance. Proc. Int. Symp. Rock Mech., Madrid, 1968, pp. 115–129.

167. SAMALIKOVA, M.: Constructive weathering of granite on the dam profile near Liberec, Czechoslovakia. Proc. 2nd Int. Cong. Int. Assoc. Eng. Geol., Sao Paulo, 1974, Vol. I, Paper IV–8.

168. SARMENTO, G. and VAZ, L.: Cambambe dam. Problems posed by the foundation ground and their solution. Trans. 8th Cong. Large Dams, Edinburgh, 1964, Vol. I, pp. 443–464.

169. SCHNITTER, N. J. and SCHNEIDER, T. R.: Abutment stability investigations for Emosson arch dam. Trans. 10th Cong. Large Dams, Montreal, 1970, Vol. II, pp. 69–87.

170. SCHULER, K. W., LYSNE, P. C. and STEVENS, A. L.: Dynamic mechanical properties of two grades of oil shales. Int. J. Rock Mech. Min. Sci. & Geomech. Abstr., Vol. 13, 1976, pp. 91–95.

171. SCHWARTZ, A. E.: Failure of rock in triaxial shear test. Proc. 6th Symp. Rock Mech., Rolla, Missouri, 1964, pp. 109–152.

172. SELDENRATH, IR. TH. R., and GRAMBERG, J. IR.: Stress – strain relations and breakage of rocks. Proc. Conf. Mech. Prop. Non-metallic Brittle Materials, London, 1958.

173. SELLERS, J. B.: The measurement of rock stress changes using hydraulic borehole gauges. Int. J. Rock Mech. Min. Sci., Vol. 7, 1970, pp. 423–435.

174. SHIH, TSO-MIN.: Investigation of the physical properties of a Gaspé skarn. Proc. 2nd Can. Rock Mech. Symp., Kingston, Ontario, 1963, pp. 75–79.

175. SIBEK, V.: Rock pressure and deformation of tertiary rock during excavation of thick brown coal seams by longwall method in strips with caving. Proc. 3rd Int. Conf. Strata Control, Paris, 1960, pp. 422–441.

176. SIBEK, V., SIMANE, J. and BUBEN, J.: Methods of research into rock bursts in the Czechoslovak Socialistic Republic. Proc. 4th Int. Conf. Strata Control and Rock Mech., New York, 1964, pp. 420–433.

177. SIMMONS, G.: Velocity of shear waves in rocks to 10 kilobars, 1. J. Geophys. Res., Vol. 69, 1964, pp. 1123–1130.

178. SIMMONS, G.: Velocity of compressional waves in various minerals at pressures to 10 kilobars. J. Geophys. Res., Vol. 69, 1964, pp. 1117–1121.

179. SINGH, D. P.: Drillability and physical properties of rocks. Proc. Symp. Rock Mech., Univ. Sydney, 1969, pp. 29–34.

180. SINGH, R. D.: Compressive strength of some Indian coals. Trans. Min. Geol. Met. Inst. India, Vol. 62, No. 1, 1965, pp. 43–66.

181. SINGH, M. M. and HUCK, P. J.: Correlation of rock properties to damage effected by water jet. Proc. 12th Symp. Rock Mech., Rolla, Missouri, 1970, pp. 681–695.

182. SOMERTON, W. H., WARD, S. H. and KING, M. S.: Physical properties of Mohole test site basalt. J. Geophys. Res., Vol. 68, 1963, pp. 849–856.

183. SPACKELER, G., GIMM, W., HÖFER, K. H. and DUCHROW, G.: New information on rockbursts in potash mines. Proc. 3rd Int. Conf. Strata Control, Paris, 1960, pp. 551–563.

184. STASSEN, P. and HAUSMAN, A.: The rock conditions in a deposit and their influence on the operations of development, planning of winning, and roof control. Proc. Int. Strata Control Cong., Essen, 1956.

185. STEWART, J. W.: Compression of solidified gases to 20,000 kg/cm² at low temperature. Phys. and Chem. Solids, Vol. 1, 1956, pp. 146–158.

186. STOWE, R. L.: Strength and deformation properties of granite, basalt, limestone and tuff at various loading rates. U.S. Army Corps Eng., Waterways Exp. Station, Vicksburg, Miss., Misc. Paper C-69-1, 1969.

187. STOWE, R. L. and AINSWORTH, D. L.: Effect of rate of loading on strength and YOUNG's modulus of elasticity of rock. Proc. 10th Symp. Rock Mech., Univ. Texas, Austin, Texas, 1968, pp. 3–34.

188. THIRUMALAI, K.: Potential of internal heating method for rock fragmentation. Proc. 12th Symp. Rock Mech., Rolla, Missouri, 1970, pp. 697–719.

189. TIMCHENKO, I. P.: Physicomechanical properties of certain types of massive, essentially quartzose rocks. In Physical and Mechanical Properties of Rocks (Ed. ZALESSKII, B. V.), Jerusalem, Israel Program for Scientific Translations, 1967, pp. 92–116.

190. TINCELIN, E.: Research on rock pressure in the iron mines of Lorraine (France). Proc. Int. Conf. Rock Pressure and Support in the Workings, Liege, 1951, pp. 158–175.
191. TINCELIN, E. and SINOU, P.: Control of weak roof strata in the iron ore mines of Lorraine. Int. J. Rock Mech. Min. Sci., Vol. 1, 1964, pp. 341–383.
192. TSCHERNJAK, I. L. and PARASCHKEVOV, R. D.: Verformungsmechanismus und Gebirgszerstörung um bergmännische Hohlräume. Proc. 9th Cong. Int. Bur. Rock Mech., Leipzig, 1967, pp. 188–200.
193. ULLAO, A.: Field observations at La Soledad dam. Trans. 8th Cong. Large Dams, Edinburgh, 1964, Vol. II, pp. 821–839.
194. UNDERWOOD, L. B.: Tunnelling by mechanical miners in faulted shale at Oahe dam. Trans. 7th Cong. Large Dams, Rome, 1961, Vol. II, pp. 375–404.
195. UNDERWOOD, L. B.: Chalk foundations at four major dams in the Missouri River basin. Trans. 8th Cong. Large Dams, Edinburgh, 1964, Vol. I, pp. 23–47.
196. U.S. Bureau of Reclamation: Concrete Lab. Rep. No. SP-39, 1953.
197. U.S. Bureau of Reclamation: 1960.
198. U.S. Bureau of Reclamation: Morrow point dam and power plant foundation investigations, Colorado River Storage Project. U.S.B.R., Denver, Colo., 1965, 190 p.

199. VAN DER VLIS, A. C.: Rock classification by a simple hardness test. Proc. 2nd Cong. Int. Soc. Rock Mech., Belgrade, 1970, Vol. 2, pp. 23–30.
200. VOIGT, W.: Lehrbuch der Kristallphysik. Leipzig, 1928.

201. WALLACE, G. B., SLEBIR, E. J. and ANDERSON, F. A.: Foundation testing for Auburn dam. Proc. 11th Symp. Rock Mech., Berkeley, California, 1969, pp. 461–498.
202. WEATHERBY, R. B. and FAUST, L. Y.: Influence of geologic factors on longitudinal seismic velocities. Bull. Am. Assoc. Pet. Geol., Vol. 19, No. 1, 1935, pp. 1–8.
203. WEIR-JONES, I.: The modification of operating dimensions within an existing gypsum-anhydrite mine. Proc. 7th Can. Rock Mech. Symp., Edmonton, 1971, pp. 119–134.
204. WEST, L. J. and PERRY, R. M.: Rock mechanics studies for a high cut at a nuclear power station in Pennsylvania. Proc. 11th Symp. Rock Mech., Berkeley, California, 1969, pp. 445–459.
205. WILD, P. A. and McKITTRICK, D. P.: Northfield mountain underground power station. Proc. Symp. Underground Rock Chambers, Phoenix, Arizona, 1971, pp. 287–331.
206. WILLIAM and NELSON: 1970. (Unpublished).
207. WILSON, A. H.: A laboratory investigation of a high modulus borehole plug gage for the measurement of rock stress. Proc. 4th Symp. Rock Mech., Univ. Park, Penn., 1961, pp. 185–195.
208. WINDES, S. L.: Physical properties of mine rock, Part 1. U.S.B.M. R.I. 4459, 1949, 79 p.
209. WINDES, S. L.: Physical properties of mine rock, Part 2. U.S.B.M. R.I. 4727, 1950, 37 p.
210. WISEMAN, G., LEVY, J. and AISENSTEIN, B.: Strength studies of chalk specimens under moderate confining pressures. Proc. 1st Cong. Int. Soc. Rock Mech., Lisbon, 1966, Vol. I, pp. 483–487.
211. WUERKER, R. G.: Annotated tables of strength of rock. Trans. A.I.M.E., 1956, Pet. Paper N-663-G, 12 p.

212. YAGODKIN, G. I., TEDER, R. I. and CHIRKOV, S. E.: Determination of strength and elastic properties of country rock and coal of the Leninsk-Kuznetsk bituminous coal deposits of the Kuznetsk basin (Kuzbass). In Mechanical Properties of Rocks by PROTODYAKONOV, M. M., KOIFMAN, M. I. and others. Jerusalem, Israel Program for Scientific Translations, 1969, pp. 64–73.
213. YAMAGUCHI, U.: On the relationship between uniaxial compressive strength and YOUNG's modulus of rock. J. Soc. Mat. Sci., Japan, Vol. 17, No. 181, 1968, pp. 56–61.

214. YAMAGUCHI, U. and MORITA, M.: Simple measurement technique of dynamic YOUNG's modulus of rock by means of sound propagation method. In Japanese. J. Min. Met. Inst. Japan, Vol. 85, 1969, pp. 319–323.
215. YAMANO, T.: Fill-type dam and soft rocks. Rock Mech. in Japan, Vol. II, 1974, pp. 64–66.
216. YOUASH, Y. Y.: Dynamic physical properties of rocks: Part II, Experimental results. Proc. 2nd Cong. Int. Soc. Rock Mech., Belgrade, 1970, Vol. 1, pp. 185–195.

217. ZISMAN, W. A.: YOUNG's modulus and POISSON's ratio with reference to geophysical application. Proc. Nat. Acad. Sci. U.S.A., Vol. 19, 1933, pp. 653–665.
218. ZISMAN, W. A.: Compressibility and anisotropy of rocks at and near the earth's surface. Proc. Nat. Acad. Sci. U.S.A., Vol. 19, 1933, pp. 666–679.
219. ZNANSKI, J.: Criteria of the gradual and sudden collapses of surrounding rock into a mining excavation. Proc. 3rd Int. Conf. Strata Control, Paris, 1960, pp. 211–225.
220. ZNANSKI, J.: Sklonnosciskal do Tapania. Archiwum Gornictwa, Vol. 2, No. 2, 1964, pp. 225–240.

Series on Rock & Soil Mechanics
Vol. 2 (1974/77) No. 4

THE PRESSUREMETER AND FOUNDATION ENGINEERING

by **F. BAGUELIN, J. F. JÉZÉQUEL, D. H. SHIELDS,** France and Canada

January 1978, 624 pages, 314 figs. US Dollar 52.00 (or sFr. 130.00) cloth

PREFACE

The design and construction of foundations require a thorough knowledge of the behaviour of soils and rocks in the field. Since even elaborate laboratory tests on large subsurface samples can at best only approximate the field conditions, in-situ tests are often preferable. The pressuremeter is probably the most versatile in-situ testing device available at present for investigating static and cyclic strength and deformation properties of soils and rocks.

Based on the authors' comparisons between the results of standardized pressuremeter tests and both static and standard penetration tests under different site conditions, the merits and limitations of the various methods of field investigations can readily be assessed. At the same time the extensive experience gained by these reliable, practical and semi-empirical methods of using pressuremeter data becomes available to other types of field investigations to their mutual benefit. These approaches require mature engineering judgment and sound experience based on performance observations on structures during and after construction. In this way pressuremeter tests can lead to safe and economical solutions to many geotechnical problems, as shown in this warmly recommended book.

G. G. MEYERHOF

TRANS TECH PUBLICATIONS

Trans Tech House CH-4711 Aedermannsdorf Switzerland